THE INDUSTRIAL REVOLUTIONS

General Editors R. A. Church and E. A. Wrigley

Volume 1
Pre-industrial Britain
Edited by J. A. Chartres

Volumes 2 and 3
The Industrial Revolution in Britain
Edited by Julian Hoppit and E. A. Wrigley

Volumes 4 and 5
The Industrial Revolution in Europe
Edited by P. K. O'Brien

Volume 6
Industrialization in North America
Edited by Peter Temin

Volume 7
The Industrialization of Japan
Edited by W. J. Macpherson

Volume 8
The Textile Industries
Edited by D. T. Jenkins

Volume 9
The Metal Fabrication and Engineering Industries
Edited by Sidney Pollard

Volume 10
The Coal and Iron Industries
Edited by R. A. Church

Volume 11
Commercial and Financial Services
Edited by R. C. Michie

The Industrial Revolution in Europe I

Edited with an introduction by

P. K. O'Brien

The Economic History Society

BLACKWELL
Oxford UK & Cambridge USA

First published 1994

Blackwell Publishers
108 Cowley Road, Oxford OX4 1JF
UK

238 Main Street
Cambridge, Massachusetts 02142
USA

British Library Cataloguing in Publication Data

A CIP catalogue record for this book is available from the British Library.

Library of Congress Cataloging-in-Publication Data

The Industrial Revoluton in Europe/edited, with an introduction, by P. K. O'Brien.
p. cm. — (The Industrial revolutions : v. 4-5)
"The Economic History Society."
Includes bibliographical references.
ISBN 0-631-18073-7 (v. 1). —— ISBN 0-631-18145-8 (v. 2)
ISBN 0-631 17486 9 (11-volume set).
1. Europe—Industries—History. 2. Technological innovations—Europe—History. 3. Europe—Economic conditions. I. O'Brien Patrick Karl. II. Economic History Society. III. Series.
HD2321.I533 1994 vol. 4-5
[HC240]
338.09 s—dc20
[338.094]

93-40550
CIP

Introduction typeset in 11 on 12 point Plantin by Typesetting Solutions, Slough, Berks., UK.
Printed in Great Britain by T.J. Press Ltd. Padstow, Cornwall.

Table of Contents

VOLUME 4

VOLUME 5

General Editors' Introduction

R. A. CHURCH and E. A. WRIGLEY

The aim of this series of volumes is to bring together journal articles and chapters from collections of essays that deal with some aspect of the industrial revolution and may be regarded as 'classics'. Some of the articles were published recently, but writings from each decade since the Second World War are well represented and there is a scattering of articles that appeared in the interwar period. Very few libraries contain all the journal series, collections of essays, and research monographs from which the material reproduced in these volumes was drawn.

As well as indicating the latest trends in research, these volumes provide an insight into the development of thinking about the issues that have informed work on the industrial revolution over the last half-century. Each piece is reproduced with its original pagination, but with continuous pagination added for ease of reference. In every case but one the original was reproduced photographically. In the one remaining case (volume 4, article 7), the text was reset. Each volume contains an introduction which explains the choice of material reproduced, describes the main debates to which they relate, and provides a guide to further reading that should make it possible to pursue particular questions effectively.

The series consists both of topical volumes dealing with an individual industry or area of economic activity (for example, textiles or metal fabrication and engineering industries), and of volumes that cover the occurrence of the industrial revolution in a particular country or continent (Britain or North America).

By 'industrial revolution', a loose and imprecise term, we mean principally what is often referred to as the first industrial revolution, the period of the cotton mill and the steam engine, rather than that of electrical engineering, fine chemicals, and the internal combustion engine (though the first volume deals with the period immediately before the conventional dating of the industrial revolution in Britain). The industrial revolution was an event of such profound and continuing importance for all societies, both those initially affected and those which were remote from the main centres of early growth, that achieving a better understanding of its nature should be a prominent element in the reflections of all those concerned about society today as well as in the past. We hope that this series may prove of use to those who are persuaded of the truth of this assertion.

Introduction

P. K. O'BRIEN

Alexander Gerschenkron, the greatest historian of modern European industrialization, maintained that students could not really comprehend the economic history of one country unless they understood the economic history of at least two others.[1] In 1993 there is no longer any need to make the case for studying any number of economies about to ratify a Maastricht treaty for the integration of European states into a single economic community. Nevertheless Gerschenkron advocated much more than a programme to escape from the cultural bias of national history. In effect his suggestion implies that the process and causes of European industrialization could only be fully explained through systematic and rigorous comparisons across several countries.[2] That now familiar and acceptable injunction, which informed his own writing and teaching at Harvard, had, moreover, been developed in theory and practice a generation earlier by Marc Bloch and Lucien Febvre, founders of the *Annales* school of French historians.[3] The writings of that now influential school demonstrated that any style of history (economic, social or cultural) that wishes to accomplish more than a reconstruction of the past for its own sake but seeks instead to account for developments over time must adopt a comparative method.[4] Chronological histories explicitly use a nation's past as their point of reference. To escape from the 'conceits of time and place' comparative histories juxtapose the experience of two or more countries in order to provide perspective, illumination, and debatable explanations for long-term economic change.

For these two volumes, articles have been selected which have charted and analysed the pace and pattern of national transitions to industrial economies as they proceeded within western Europe from around the middle of the eighteenth century down to and beyond the Great War. Those decades (*c.* 1750 to *c.* 1914) are conventionally demarcated as the relevant era for the intensive study of industrialization because during that time all the countries considered here experienced an accelerated rate of structural change often loosely labelled as industrial revolutions or as the beginnings of industrialization. Structural change refers to the fact that their accumulating stocks of capital, workforces, and technologies became discernibly less and less engaged with producing food and agrarian raw materials, or with servicing agriculture, and more involved

[1] Gerschenkron, *Continuity in history*, pp. 1–8.
[2] Skocpol and Somers, 'The uses of comparative history'.
[3] Sewell, 'Marc Bloch'.
[4] On the *Annales* school, see Burke, *New perspectives*, and Stoianovich, *French historical methods.*

with industry, especially manufacturing, and with the trade, transport, finance, and construction related to industry. Structural changes which appeared in the composition of European national outputs, in the allocation of labour across sectors of economies, and in relation to the modernity of the machines, tools, and forms of organization utilized to produce goods and services were invariably accompanied by population growth and urbanization, the spread of literacy, the integration of markets, closer involvement in international commerce, the diffusion of advanced technology, institutional and political reform, and several other familiar features of modern economic growth.[5] By 1993 the bibliography of research published in English since 1950 concerned with the industrialization of Europe from 1750 onwards runs into hundreds of items and the principles of selection required to compress a sample even of the best articles and essays into just 30 entries had to be elastic, eclectic, and personal.[6]

I started with the assumption that readers of these volumes would be engaged in studying several national histories of economic growth (which are for the most part written within the geopolitical framework of European nation states) but that they read in order to understand the pace and pattern of industrialization as that process unfurled across Europe from around the mid eighteenth century onwards.[7] Thus they would expect comparisons to be illuminating and to contribute to an explanation of why some countries (e.g. Belgium) industrialized before others (e.g. Italy). Bilateral or multilateral comparisons would not, however, be sufficient and the search for explanations and understanding would depend rather heavily upon theories borrowed largely from economics but also from other social sciences, including geography, politics, sociology, and anthropology.[8] Students will also expect to be introduced to conceptual and theoretical essays that have initiated and informed recent debates among economic historians of Europe's industrialization.[9] Only major issues could be covered, but the articles in volume 4 exemplify three general and overlapping approaches to the study of European industrialization before 1914. First are articles concerned to make the historians' central argument that the pace and pattern of industrialization after 1750 rested upon historical foundations or preconditions, namely: (a) the prior development of a framework of property rights, institutions, and security within which effective markets for commodities and factors of production could operate (articles 4, 8, 9, and 10); (b) the concentration

[5] On the theory and measurement of structural change, see Kuznets, *Modern economic growth*, and *idem*, *Economic growth of nations*. For exemplary cases of the Kuznetsian approach see **Heikkinan and Hjerppe,** reproduced as article 25 and **Jorberg,** reproduced as article 27. But see Kaelble, 'Was Prometheus most unbound in Europe?' for the argument that the early industrialization of the workforce was peculiarly European.

[6] For bibliography, see Aldcroft and Rodger, *Bibliography of European economic history;* Hacken, *Central European economic history;* Adelman and Morris, *Comparative patterns of economic development.*

[7] As exemplified in standard modern textbooks by Goodman and Honeyman, *Gainful pursuits;* Milward and Saul, *Development of the economies of continental Europe;* Trebilcock, *Industrialization of the continental powers.*

[8] Skocpol, *Vision and method in historical sociology;* Pounds, *Historical geography.*

[9] Pollard, *Typology of industrialization processes,* is the most recent survey of these debates.

advanced at varying speeds in several countries down to 1914.[12] As usual, both the context and the chronology within which European industrialization has been treated continue to promote debate. Braudel, Cameron, Jones, Wallerstein, and others argue explicitly (and implicitly in the way that they deal with the economic success of western Europe as a whole) that the geographical context for analysis should be widened to include contrasts and, in their view, critical interactions with other continents—particularly with Asia and the Americas, but also with Africa and eastern Europe.[13] Historical comparisons on an inter-continental scale certainly expose very clearly western Europe's natural and political advantages compared with other parts of the globe. Furthermore, Wallerstein's books and other world systems' approaches provide a forceful and eloquent reminder that the long-run economic development of western Europe for more than four centuries after the voyages of discovery owed something—Wallerstein, Solow, Woolf, Frank, Amin, and others would argue that it owed its main impetus—to the successful deployment of military force and colonization which turned the conditions and terms for trade with the rest of the world heavily in Europe's favour.[14]

Nevertheless, the profits from territorial conquest in the Americas and Asia, from slavery and plunder, and above all from regularized transcontinental trade and investment were never equally distributed among the nations of western Europe.[15] At various times Iberia, Holland, France, and Britain obtained large shares of the spoils and gains from participation in the global economy as it evolved after 1492.[16] Other European countries which confined most of their international transactions to commerce within their own continent managed to industrialize without any significant and direct involvement with the rest of the world economy.[17] Furthermore, Portugal and Spain, which were dominant in the initial stages of the expansion of trade with Africa, Asia, and the Americas, failed to carry their industrialization very far before the twentieth century; while Holland and France, active in the seventeenth and eighteenth centuries, lagged behind Britain down to 1914.[18] It is also not clear just how important commerce with other continents and with eastern Europe was for the British industrial revolution.[19]

[12] Knowles, *Economic development;* Bowden, Karpovitch, and Usher, *Economic history of Europe;* Clapham, *Economic development of France and Germany;* Clough and Cole, *Economic history of Europe;* Heaton, *Economic history of Europe;* Henderson, *Industrial revolution on the continent;* and see above, n. 7.

[13] Braudel, *Civilization and capitalism,* esp. vols. II and III; Jones, *European miracle; idem, Growth recurring;* Wallerstein, *Modern world system.* Cameron's recent *tour de force* suggests that economic historians should take the world and its chronology back to paleolithic times as their oyster: Cameron, *Concise economic history of the world.*

[14] Wallerstein, *Modern world system;* Woolf, *Europe and the people without a history;* Frank, *World accumulation;* Amin, *Accumulation on a world scale;* Solow, 'Slavery and colonization'.

[15] Tracy, ed., *Rise of merchant empires;* Fischer *et al., Emergence of a world economy.*

[16] Tracy, ed., *Political economy of merchant empires.* This recent collection and the earlier one cited in n. 15 contain excellent surveys and bibliographies for the study of European expansion overseas.

[17] O'Brien, 'European economic development' argues that for Europe as a whole these connections were not significant for long-run industrialization.

[18] See **Crouzet,** reproduced as article 4; Israel, *Dutch primacy in world trade;* Liss, *Atlantic empires.*

[19] O'Brien and Engerman, 'Exports and the growth of the British economy', but see their critics referenced in this essay.

Scholars who insist on the wider geographical context and longer time span are correct in that western Europe's industrialization after 1750 should not be represented as a self-contained endogenous process. The evolving global economy from 1492 to 1914 certainly impinged more and more, as time went on and especially for the first industrial nation. Intercontinental exchanges mattered far less for the long-term growth of the rest of western Europe, and on balance may have retarded the transformation of the Portuguese, Spanish, and perhaps even the Dutch economy.[20] Furthermore, and with hindsight, at this end of the twentieth century it is becoming accepted that Britain's sustained strategic and economic involvement with its empire and with oceanic trade acted after, say, 1875 to restrain the growth of the domestic economy. That persistent involvement is now put forward as an important component of that country's relative economic decline in the later nineteenth and twentieth centuries.[21] Thus interconnections between western Europe and other parts of the global economy through conquest and colonization, as well as through peaceable investment and commerce, may turn out to be neither sufficient, necessary, nor even significant elements of historical explanations for its earlier transformation into a group of conspicuously affluent industrial societies.[22]

Real insights can be derived from placing the European industrialization in a global context and within time spans that predate the post-1750 and modern phase of accelerated structural change by some three centuries. In that context, well-specified contrasts in ecological and demographic endowments, in political and legal institutions conducive to the spread of competitive markets and an appreciation of differences accorded by other cultures and religions to science, technology, innovation, and risk may continue (as Max Weber suggested) to be a heuristically more powerful approach to understanding the endogenous and specific forces that promoted industrialization in one part of one continent.[23] Western Europe's political and economic interactions with other areas of the globe seem to have been both negative and positive. Over the long run the gains from commerce rarely flowed in one direction, and the development of underdevelopment in Asia, Africa, and southern America does not logically entail that anything like the *primus mobile* for European industrialization can be found in its relations with other and less fortunate continents.[24]

20 Phillips, 'Trade in the Iberian empires'. On Dutch decline see **de Vries,** reproduced as article 20, and Riley, 'Dutch economy after 1650'.

21 Davis and Huttenback, *Mammon and the pursuit of empire;* O'Brien, 'Costs and benefits of British imperialism'.

22 A recent collection of essays with good bibliographies that address this theme is Baechler *et al.*, eds., *Europe and the rise of capitalism.*

23 Jones, *European miracle; idem, Growth recurring;* Mann, *Sources of social power* are recent and excellent introductions to this grand theme.

24 Pohl, *European discovery of the world,* addresses the interactions.

II Measurement within the national accounts framework

Meanwhile the modern economic history of western Europe remains dominated by comparisons confined to a group of countries whose transition to industrial market economies began early, and which are now perceived in international terms as affluent and successful.

Within this group, and well before 1914, contemporaries appreciated that economic transformation had proceeded much further in some European countries (e.g. Holland) than in others (e.g. Spain). But that perception was investigated in quantitative terms after Keynesian economics and the data required for macroeconomic management began to be utilized by governments during and after the Second World War. From around mid century onwards economic historians began to import macro social accounting into their studies of the past in order to assess and to analyse the progress of national economies over long spans of time.[25] Three indicators quickly became central for the observation and evaluation of relative economic achievements at nationwide levels: per caput income which, subject to familiar caveats, served to measure average standards of living across countries; output per worker employed, which provided the basic index of a nation's productive efficiency; and the composition of national outputs, which showed how far a given economy had been transformed from an agrarian to an industrial base.[26] Once they had located or constructed accounts for national incomes, population, and workforces, economic historians measured rates of growth in incomes per caput, output per worker, and long-run structural changes within countries; and compared relative levels of achievement across countries for benchmark periods of time.[27]

Serious research and imagination has gone into the making of basic indicators which are a precondition for the kind of macroeconomic analysis conducted by economists who work upon the comparative growth and development of European, North American, and Japanese economies for the postwar years.[28] Tabulations claiming to represent levels of national and per caput incomes, outputs per worker employed in industry, agriculture, and services, the composition of national outputs, and the allocation of workforces across sectors and industries of European economies for the period before 1914 have confronted historians with unexpected contrasts and comparisons that called for explanation.[29] Certainly the value of this major research programme (inspired by Simon Kuznets) which aims to build up datasets for outputs and inputs country by

[25] Compare standard textbooks in European economic history (cited above, nn. 7, 12) written before and after mid century.

[26] Kuznets, *Modern economic growth.*

[27] Maddison, *Economic growth,* is a pioneering study written in 1964. Following an innovative essay by Marczewski, 'Le produit physique', O'Brien and Keyder, *Economic growth,* attempted systematic comparisons using a national accounts framework for Britain and France, 1780–1914.

[28] e.g., Boltho, *European economy;* Kendrick and Vacara, *New developments;* Van Ark, 'Comparative levels'.

[29] The current text with bibliography and major statistical series for European economies is Maddison, *Dynamic forces.*

country in standardized form, resides in focusing attention upon differences in those key economic variables. Such facts now set much of the agenda for major questions in European economic history; such as why and to what extent was British per caput income higher than per caput incomes elsewhere in Europe down to the Great War?[30] When did other national economies (Belgium, France, Switzerland, Germany, Holland, and Scandinavia) begin to converge towards that lead country's standards of living and economic efficiency?[31] Why did convergence not occur to anything like the same degree for Italy, Spain, Portugal, and parts of the Austro-Hungarian Empire until well into the twentieth century?[32] Thus progress in the construction of comparable sets of basic data has been effective in drawing national economic histories into a European frame of reference, and national income accounting also provides taxonomies for the location of observed variations across countries, and for the speed of change and of convergence through time.[33]

For example, within the social economic framework differences in measured per caput incomes across countries depend upon contrasts in participation ratios, gaps in labour productivities, and the allocation of available labour time among sectors and industries of national economies. In symbols $(Y/P)^n$ is a function of W/P: $x(O/L)^a$; $y(O/L)^i$, and $z(O/L)^s$ — where $(Y/P)^n$ is per caput income in country n; W/P is the share of its population in the workforce; O/L is output produced per unit of labour input; a, i, and s represent agriculture, industry, and services; and x, y, and z are the shares of the workforce employed in these three broad sectors of national economies.[34] Thus for any decade before 1913 per caput income could be higher in, say, Britain than Germany because Britain's participation ratio (i.e. the share of its workforce in fulltime employment) was higher than Germany's; because the productivities of labour inputs employed in similar sectors were superior; or because large shares of the British workforce happened to be concentrated in sectors and industries of superior productivity to similar industries located in Germany.[35]

If and when labour productivities can be measured sector by sector, industry by industry, then the sources of any European economy's superior performance over any other can be located and its importance for overall differences in output per man hour of labour input and, by extension, income per caput estimated.[36]

[30] Feinstein, 'Economic growth since 1870'.

[31] Abramovitz, 'Catching up', represents a useful introduction to the convergence thesis.

[32] Tortella and Prados de la Escosura, 'Patterns of economic modernization'; Komlos, *Economic development*.

[33] Maddison, *Phases of capitalist development* and *idem*, *Dynamic forces* have been particularly influential.

[34] Fremdling and O'Brien, eds., *Productivity in the economies of Europe* is a collection of essays built around this taxonomical framework.

[35] Fremdling, 'Productivity comparisons'.

[36] An exemplary set of articles for the twentieth century built around productivity comparisons across countries is Crafts and Broadberry, eds., *European productivity*. For agriculture see O'Brien and Prados de la Escosura, 'Agricultural productivity'.

To set out the objectives of the national accounts programme serves to highlight how useful that framework might eventually become in providing some sense of significance for observed variations across countries. For example, the importance of the pronounced superiority in the productivity of labour employed in British cotton textiles compared with workers employed in the same industry in France will be proportionate to the measured gap in output per worker across the two economies weighted by cotton's share in national income. If, for example, in 1830–5 the observed differential amounted to 50 per cent and cotton's share in French national output was 1 per cent, the 'failure' of this particular industry to achieve British standards of labour productivity reduced French national income by up to 0.5 (1 per cent) or 0.5 per cent below a potentially attainable level. Alternatively, and given that cotton's share of British national output comes to around 6 per cent, of which about two-thirds was exported, the question might be posed as to why France and not Britain obtained a large share of the world market in cotton textiles. Counterfactually, *if* French businessmen and investors had managed to create a cotton textile industry as large (in absolute terms) and as productive as the British industry, then by 1830–5 the country's national output might conceivably have been 0.5 (3 per cent) or 1.5 per cent higher. Simple arithmetical exercises of this kind expose the significance of gaps in productivity, industry by industry, and raise interesting questions of scale and of comparative and natural advantages behind the long-term growth of particular industries.[37] Thus British superiority seems to reside more in the overall size of its cotton textile industry than in higher productivity as such. Could France conceivably have developed a far larger cotton industry by 1830–5? What set of historical circumstances restrained that development over time? Were opportunities missed along the way that represent examples of national and entrepreneurial failure?[38]

Once acceptable figures are to hand, the national accounts framework facilitates chains of focused historical investigations; and it also positions the great volume of micro research concerned with the history of firms, industries, and localities in particular countries within a larger and potentially more interesting question about European industrialization as a whole.[39]

Although theoretical and heuristic gains from a research programme in European economic history conducted within the parameters of the national accounts framework are potentially very illuminating, the empirical difficulties of producing estimates for relative living standards, for structural differences, and above all for levels of productivity for years before the last years of the nineteenth century are daunting. There is still a great deal of archival work to be

[37] McCloskey, *Economic history,* is particularly instructive on how to use and to infer significance from numbers of this kind.

[38] These kind of questions focused on particular industries are central to the debate about the relative backwardness of France compared to Britain. For recent surveys see Cameron and Freedman, 'French economic growth'; Heywood, *Development of the French economy;* Aldrich, 'Latecomer or early starter?'.

[39] Ultimately questions concerned with productivity differentials can only be resolved at the level of particular industries and firms; Prados de la Escosura, 'Spain's gross domestic product'.

completed before numerous and essential national series are available in any form, especially for decades before 1900. Bold and imaginative attempts to manufacture figures for purposes of international comparisons are, however, already in print.[40] Nevertheless, the debate that surrounds many of the initial attempts to produce usable statistics warns students not to assume that the published data are now accepted as accurate within the margins of error required to draw firm conclusions about salient, let alone less marked, differentials across the economies of western Europe.[41]

Alas, the database has not yet been constructed securely enough for economic historians to accept even the available facts, as they find them.[42] Obviously the range and quality of basic but potentially quantifiable macroeconomic information varies by country and period.[43] League tables, even for such standard indicators as incomes per caput, real wages, distributions of income and wealth, levels of industrialization, investment, shares, the productivities of labour employed in major industries and in agriculture, capital output and capital labour ratios are proving difficult to construct.[44] For several countries nationwide data are not available and, even when flawed but usable statistics exist, the transformation of quantities expressed or valued in national currencies into a common currency (e.g. sterling or dollars) to facilitate comparisons continues to pose formidable conceptual and statistical problems. Those issues have been extensively discussed in recently published literature. Several plausible solutions have been devised, including the construction of purchasing power/parity rates of exchange, backward extrapolation from cross-country comparisons, made for recent decades in international dollars, and the more common use of published rates of exchange for purposes of conversion. All three solutions introduce immeasurable degrees of bias into conversions into any common unit of value, and that bias may vitiate the bounds of potential error required for meaningful comparisons.[45] Furthermore, other less common but analytically relevant indicators for degrees of commodity and factor market integration, shares of monetarized transactions, labour participation rates, educational, occupational, and skill distributions among national workforces, *et sim.*, are unlikely to emerge except at micro or local levels that may or may not

[40] Bairoch, 'Europe's gross national product'; *idem*, 'International industrialization levels'.

[41] Maddison, 'Comparison of levels of G.D.P.'; Crafts, 'Gross national product'.

[42] Critical commentaries on existing estimates need to be treated more seriously and sensitivity tests conducted before any national data are included in European wide league tables and comparisons, but see Maddison, *Dynamic forces.*

[43] This becomes evident from recent and sometimes quite drastic revisions to previously accepted estimates for national income. See Krantz, 'New estimates'; Gregory, *Russian national income;* Holtfrerich, 'Growth of net domestic product'.

[44] But research continues to produce and to revise national accounts. See Nunes, Mata, and Valerio, 'Portuguese economic growth'; Prados de la Escosura, 'Spain's gross domestic product'; Toutain, 'Le produit intérieur'; Scholliers, *Real wages.*

[45] Van Zanden, 'First green revolution' and O'Brien and Prados de la Escosura, 'Agricultural productivity' contain references to the theoretical issues involved.

be representative of particular national economies as a whole.[46] Meanwhile the figures at present available certainly represent hard-won and useful responses to challenges to produce some kind of statistically based conclusions quickly. In making these estimates or conjectures their authors have exposed both the variety and the fragility of sources available for macroeconomic measurement before 1914. With all their imperfections, the figures are certainly useful because a proper consideration of what is required to produce properly quantified answers serves to educate students in statistically based modes of macroeconomic thought about long-run and comparative economic development. That rigorous and demanding conceptual and statistical framework is, moreover, a salutary reminder that other mainstream approaches to the economic history of Europe before 1914 remain essentially impressionistic, can offer only partial and tentative conclusions, and fall short of the standards of empirical validation which are mandatory for the comparative analysis of European growth in the postwar or even the interwar periods.[47]

III Regions, towns, and proto-industrialization

In turning to that currently dominant approach to the subject, students can call upon a formidable bibliography of secondary sources in English that contrast in style with the writings by learned historians of the previous generation, which seem designed to introduce them to the enormous complexity and variety of continental experiences with industrialization. Modern texts, monographs, and articles published since mid century are more self-consciously analytical.[48] Most of the theory and much of the vocabulary deployed is borrowed from economics and the geographical frame of references is explicitly or seems implicitly European, even when a national economy or a sector, region, and town within a given country comes under review.

Although they accept national economies and the political boundaries of European states as appropriate geographical units for study, many, if not most, economic historians hark back to earlier modes of scholarly investigation and insist that the foundations of modern (nineteenth-century) European industry will be found in regional, local, and urban economies which had developed proto forms of manufacturing activity centuries before steam-powered machinery and factories appeared on the continent.[49]

When it emerged after 1750, mechanized industry did not spread randomly over the map of Europe but 'clustered' within defined geographical boundaries.[50]

[46] For a bold but unsatisfactory attempt to manufacture quantitative indicators for these and other important economic variables, see Adelman and Morris, *Comparative patterns*. For comparative analysis of the role of literacy and education in European industrialization, see Tortella, *Education and economic development;* Graff, *The legacies of literacy*.

[47] Matthews, *Slower growth;* Dowrick and Gemmell, 'Industrialization'.

[48] There is a useful bibliography in Goodman and Honeyman, *Gainful pursuits*.

[49] The *locus classicus* is Pollard, *Peaceful conquest*.

[50] Pounds, *Historical geography of Europe*.

Insights are gained by observing the factors and circumstances which led industry to locate, grow, and decay in some regions rather than others; and in some towns and villages before others.[51] Clearly the geographical matrix for the analysis of economic change should be relevant to the problem the historian is trying to address.[52] The rise and decline of regions has long formed an important component of national economic history. Where and when markets for commodities and factors of production were bounded in space, those particular areas, rather than national or multinational economies, constitute obvious units for study.[53] There should be no question that a regional approach is complementary to and not competitive with national or even global perspectives. Yet there are serious problems in defining regions and in deciding what are endogenous or exogenous components of ostensibly local developments.

Research continues into regions and localities of proto-industrialization premised on the implicit assumption that it will create the factual basis for a typology which will help historians to explain the long-term evolution of major branches of manufacturing (particularly textiles, but also leather goods, finished metal wares, clocks and watches, and furniture) into mechanized factory industries over the nineteenth century.[54] Nevertheless, there now appears to be no single model which could conceivably account for even a majority of the examples that historians have uncovered of tendencies for industries to concentrate geographically, to specialize, and to engage in long-distance trade.[55]

This realization has developed because the set of initial conditions which contributed to the capacity and willingness of local workers (males, females, and children) to turn from dependence upon agriculture to become involved (either part- or full-time) in manufacturing for national or international markets differed from place to place.[56] At present the lengthy list of local preconditions specified by historians of proto-industrialization includes infertile land, inheritance systems, size of farms, communal rights to land, the cost of alternative supplies of food, the productivity of local agriculture, and competition within agriculture. Populations living close to subsistence levels, on relatively small holdings and afflicted by high densities of labour to cultivable land were predictably among the first to embrace opportunities to supplement family incomes provided by industrial work, but the same path was followed by landless and seasonally underemployed labourers from richer and more commercialized agricultural areas.[57]

Elastic supplies of labour within agrarian economies are rarely sufficient to account for the emergence and subsequent growth of regions of industrial concentration. Mercantile organization and capital were also required, and both

[51] Leon *et al.*, *L'industrialisation en Europe.*

[52] There seems no need to present them as alternative and more realistic approaches to industrial history: Pollard, 'Industrialization and the European economy'.

[53] An excellent integration of the two approaches is Soderberg, 'Regional economic disparity'.

[54] **Mendels,** reproduced as article 7 and Kriedte *et al.*, *Industrialization before industrialization.*

[55] Coleman, 'Proto-industrialization'.

[56] Berg *et al.*, *Manufacture in town and country.*

[57] Parker and Jones, eds., *European peasants and their markets;* Kisch, *From domestic manufactures.*

equipment, bricks, glass, and minerals) as well as some important consumers' goods (such as flour, beer, and soap), were not manufactured under conditions that approximate to a putting-out system. If the programme is to fulfil its promise and provide explanations for transitions to modern industry, we need to know more about the varieties and adaptabilities of organizational forms that are encompassed under the heading of proto-industrialization and far more about the causes of success and failure both before and, *above all,* during the transition to the factory system.[72]

Apart from the rediscovery of industry before industrialization, local, urban, and regional history may be leading European economic history away from any foreseeable prospects for generalization. An entirely scholarly body of publications has not apparently brought to light new elements in the growth process, or persuaded economic historians to revise their perceptions of the relative importance of familiar inputs and institutions behind the development of modern industry. Nevertheless it will please geographers to observe the emphasis accorded by the new regional history to cheap energy and raw materials as the primary force behind the growth and concentration of modern industrial activity before 1914.[73]

By changing spatial parameters, village, urban, and regional history invariably enlarges the evidence for economic history but that may turn out to be a mixed blessing. Alas, it is often quite impossible to quantify inputs and outputs on a local basis and such studies inevitably become rather specific and descriptive in character. Alternatively, the partial indicators available fail to meet the needs of production function analysis, social accounting, and macro growth models.[74] For example, regional or urban income is usually impossible to measure, and regional balance of payments accounts (which could render Europe's regions such a fertile testing ground for trade theory) cannot be constructed. Perhaps the central weakness of this particular research agenda is exemplified whenever groups of historians cooperate to compile national histories on a regional basis. As otherwise excellent collections of scholarship on Belgium and Germany illustrate, reaggregation becomes acutely difficult and the approach simply multiplies descriptions of relatively advanced and backward economies.[75] Although the integration of regions and towns into both national and international economies has long been recognized as an integral component of nineteenth-century industrialization, its significance for any European country has too rarely been taken up by historians researching within a pre-selected regional or urban framework.[76] All too often their work suffers from self-imposed geographical boundaries that comparative economic history

[72] Mokyr, 'Growing up'; Schremmer, 'Proto industrialization'; Van der Wee, *Rise and decline.*

[73] Pollard, *Peaceful conquest;* Wrigley, *Industrial growth.*

[74] These difficulties are discussed by Lee, *Regional economic growth.*

[75] Tipton, *Regional variations;* Lebrun *et al., Histoire quantitative;* Mokyr, 'Industrialization in two languages'.

[76] Notable exceptions are Soderberg, 'Long-term perspective'; *idem,* 'Regional economic disparity'; De Rosa, 'Urbanization and industrialization'.

seeks to avoid. This style of history can also become over-committed to the derogation of central governments and is given to exaggerating local losses from the formation of wider markets[77]. How enclaves extended themselves outwards, how advanced and backward regions interacted, how, why, and when their paths to higher incomes converged would appear to be interesting questions to pursue for historians interested in placing regional history into an international frame of reference.[78] Unfortunately, the concentration on explaining observed locational patterns and the rise and decline of particular regions often leaves these questions somewhat suspended and unanswered.[79]

IV Agrarian foundations

Investigation into the proto, regional, and urban origins of modern industry also leads us back to its agrarian foundations. Agriculture's 'parental' role is well understood. The 'primary' sector which dominated most European economies until well into the nineteenth century could stifle, restrain, or nurture industrialization in several ways: (a) by supplying food and raw materials to industries and towns at competitive and preferably falling prices; (b) by releasing labour and capital for employment in industry and services; (c) by paying 'disproportionate' shares of the taxes required to fund the central and local government; (d) by acting as a market for the increasing volume of commodities produced by manufacturing industry.[80]

How and why agricultural sectors located within some national or regional boundaries managed to perform these functions more effectively than others represents a key question for the economic history of Europe. Some agricultures (e.g. those in England and Flanders) arrived at plateaux of productivity from where they released factors of production and supplied industrial and urban populations with food and raw materials long before others (such as Italy and Hungary).[81]

Three general answers to this key question are now offered in the voluminous literature on European agrarian history, which is normally not particularly concerned with connections between agriculture and industry but far more with the sources of agrarian progress.[82] First, there is the traditional discussion of the evolution of property rights and tenurial institutions within agriculture. That literature deals, for example, with land ownership, inheritance systems, estate management, farm size, tenurial contracts, common rights, rents, local taxes,

[77] The affliction besets historians of multinational empires. See Ashworth, 'Typologies and evidence'; Arnez, *Slovenian lands;* **Good,** reproduced as article 22. Komlos, *Habsburg monarchy,* avoids nationalist history. For Europe see Ashworth, 'Industrialization and the economic integration of Europe'.

[78] Good models are Dumke, 'Political economy of German unification'; Zamagni, *Industrializione esquilibria regionali in Italia;* Sanchez-Albornoz, 'Convergence among Spanish economic regions'.

[79] Williamson, 'Regional inequality' provides the theory, and see **Komlos,** reproduced as article 23.

[80] O'Brien, 'Agriculture and the industrial revolution'.

[81] **Vandenbroeke,** reproduced as article 18; O'Brien and Toniolo, 'Poverty of Italy'.

[82] Campbell and Overton, eds., *Land, labour and livestock* includes most of the modern bibliography.

and other facets of the legal and institutional arrangements within which cultivation of land and the sale of agricultural produce were conducted.[83] Following some version of the Marxian theory of stages of development, many social historians deal with property and tenurial rights under the heading of transitions from feudalism to capitalism, and suggest that the earlier developments of well-functioning markets (for land, farms, agricultural labour, tenants, and agricultural inputs) promoted higher rates of productivity growth required first for the development of a rural workforce, then for its shift into manufactures, and later for the progression from proto to factory forms of manufacturing.[84] Although 'feudalism'—i.e. legal and institutional impediments to agricultural efficiency and to structural change—certainly operated throughout Europe, debate continues about the precise significance of these impediments for changes in crop yields and labour productivities within national agricultures.[85] Even the conspicuous and very early success of British farming (which is too often explained in terms of a supposedly unique institutional development) is recognized to be strongly related to that kingdom's endowments of soils and climate which made so much of its farmland hospitable to the spread of nitrogen fixing fodder crops and the accumulation of stocks of animals.[86] Geographers and soil scientists have made historians aware of the real ecological constraints upon the opportunities open to farmers for raising productivity across different regions of Europe.[87] Regardless of particular tenurial institutions, patterns of landownership (and even access to urban markets for farm produce), soils, climates, elevations, and other geographical conditions went a long way towards determining the relevance and the scale of productivity gains that could be obtained from a slowly evolving body of agrarian knowledge.[88]

Agrarian reductionism, into which this particular line of argument may well degenerate, has, however, been challenged by research into the provisioning of early modern towns and cities.[89] This work suggests that several of Europe's agricultural regions, even in medieval times, had achieved grain yields per hectare and outputs per worker that were sufficient to sustain far higher levels of urbanization than actually occurred at that time.[90] Furthermore, agriculture's response to demands for increased supplies of food and raw materials seems to have been more elastic than Malthusian theories and depictions might

[83] An excellent survey is Holton, *Transition from feudalism.* The most recent debate on property rights and class struggle in the European countryside is Aston and Philpin, eds., *Brenner debate.*

[84] Macfarlane, *Origins of English individualism.*

[85] Wade, *Institutional determinants;* Hoffman, 'Institutions and agriculture'; **Grantham,** reproduced as article 17.

[86] Heywood, 'Role of the peasantry'; O'Brien, Heath, and Keyder, 'Efficiency of agriculture'.

[87] Shiel, 'Improving soil productivity'; Chorley, 'Agricultural revolution in northern Europe'.

[88] Galassi, 'Reassessing Mediterranean agriculture'.

[89] **Wrigley,** reproduced as article 13.

[90] Campbell and Overton, eds., *Land, labour and livestock.*

suggest.[91] Examples of more technologically progressive and productive agricultures can often be located in the hinterlands of towns and cities, or on farmland well connected by waterways to urban markets. In these histories, geographical and institutional constraints upon the levels of agricultural productivity, required to sustain more rapid rates of structural change long before the eighteenth century, are not regarded as fundamental or even as highly significant for variations in the pace and pattern of industrialization down to 1914.[92] On the contrary, some historians wish to begin their explanations for long-run development from the early concentration of populations in towns and industrial occupations, which in their view actively promoted institutional changes and rising land and labour productivity within agriculture.[93] Progressive agricultures thus tended to emerge as the outcome of demographic pressures and of investment in towns, transportation networks, and earlier industrialization of the workforce.[94] Perhaps such interconnections are better represented as symbiotic, but they are certainly interesting to investigate country by country—even region by region—over the centuries before 1750.[95]

V Patterns and typologies

At the opposite pole from the accumulation of detail embodied in histories of European regions, towns, and localities, are the macro typologies associated with writings by Gerschenkron, Landes, Rostow, and others who have offered usable frameworks for the organization of research and teaching in European economic history.[96] Largely because he perceived industrialization of the post-1750 period to be a process of diversity rather than uniformity, Gerschenkron's enduring contribution has been to focus attention upon variations in the *way* European economies passed through an initial but *decisive* cycle of accelerated industrial growth, which he referred to as a discontinuity or great spurt.

[91] Historians of French agriculture have tended to adopt a Malthusian model to explain the static nature of the agrarian economy before 1750. See Le Roy Ladurie, *French peasantry,* but that view for France and for the rest of European agriculture is now being questioned. See Campbell and Overton, eds., *Land, labour and livestock;* Clark, 'Productivity growth'; Cleary, 'French agrarian history'; Goldsmith, 'Agrarian history'; Komlos, 'Thinking about the industrial revolution'.

[92] Grantham, 'Agricultural supply'; Bairoch, 'Urbanization and the economy'.

[93] Grantham, 'Urban provisioning zones'; Hoffman, 'Land rents and productivity'.

[94] The theory that positive effects flow from demographic pressure was developed by Boserup, *Conditions of agricultural growth.* For the industrial revolution period see Anderson, *Population change.*

[95] For an excellent survey of the interconnections see Anderson, *Explaining long-term economic change.* Persson has formulated a model to test the main relationships: Persson, *Preindustrial economic growth.*

[96] **Gerschenkron,** reproduced as article 5, and see *idem, Continuity in history,* and a collection of essays which discuss Gerschenkron's seminal hypothesis: Sylla and Toniolo, eds., *Patterns of European industrialization.* On the Landes approach, see Landes, *Unbound Prometheus,* and the critique in O'Brien and Keyder, *Economic growth in Britain and France,* ch. 1; **Rostow,** reproduced as article 12, and *idem, Stages of economic growth,* and two collections of critical appraisals of the stages in Rostow, *Economics of the take off,* and Kindleberger and di Tella, *Economics in the long view.* Other recent attempts to construct and criticize typologies include **Crafts,** reproduced as article 3; Pollard, *Typology of the industrialization processes;* O'Brien, 'Do we have a typology?'.

Gerschenkron expected that the study of carefully delineated contrasts in the methods used to build up modern industry within several European countries would help historians to explain differences in the time taken by national economies to converge to the highest levels of per caput income. In his view, Britain had forged ahead because conditions for 'spontaneous industrialization' developed early—and the problem for relatively 'backward economies' on the mainland was how to find effective substitutes for missing (i.e. British) preconditions for growth. Thus, the delays exhibited by continental economies in their endeavours to catch up with Britain are imputed, first, to the absence of a range of more or less important 'prerequisites' for early and successful industrialization available to the British economy; and secondly, to the time taken by 'follower countries' to develop their own substitutes for these agrarian, institutional, and technological deficiencies.

Gerschenkron concentrated upon an important but still restricted number of the components of modern industrialization, which he hypothesized would vary systematically, in form and significance, with the timing and degree of backwardness exhibited by different European economies as they passed through a first 'great spurt' in industrial production. Thus, he predicted that when historians examined the build-up of modern industry in Europe country by country they would observe that the more backward or late the economy was when it began that process, the more rapid would be the rate of acceleration in the growth of industrial production, and the more pronounced would be the reliance on capital as distinct from consumer goods; that among 'latecomers' new industrial enterprises would emerge as more technically sophisticated and capital intensive, larger in scale, and more prone to monopoly; that such enterprises would also rely to a greater degree upon external sources (banks and the state) for investible funds and for entrepreneurial initiatives; that, finally, for their markets and supplies of cheap inputs late developers would depend far less than their British precursors upon rising productivity in domestic agriculture.[97]

Backwardness was never clearly defined, but statistical tests (based on small samples of European countries) and using income per head, shares of the workforce in agriculture, and times of arrival at a threshold level of per caput income as proxies for backwardness, do not correlate well with several dependent elements of the industrialization process singled out by Gerschenkron as ways of surmounting retardation.[98] Furthermore, recent historical research (often inspired by his hypothesis) has not merely widened the sample of countries to include deviant cases from the Austrian Empire, the Balkans, and Scandinavia, but has deepened perceptions of British, Dutch, French, Italian, and even Russian industrialization in several ways that appear inimical to the survival of Gerschenkron's typology as an heuristic device.[99]

[97] **Gerschenkron,** reproduced as article 5, and for an excellent summary of his views, Sylla and Toniolo, eds., 'Introduction: patterns of European industrialization'.

[98] Barsby, 'Economic backwardness'.

[99] Sylla and Toniolo, *Patterns of European industrialization,* supplies the modern bibliography. See also **Good,** reproduced as article 22; **Komlos,** reproduced as article 23; **Lampe,** reproduced as article 24.

First (and this criticism applies even more forcibly to Rostow's attempt to periodize Europe's long-run development into demarcated *stages* of growth), the imperfect state of the indices available for the study of nineteenth-century European economic history means that the statistical problems of delineating phases of 'take off', 'trend acceleration', or any other euphemism for locatable and important discontinuities, especially for national, but even for industrial outputs, are probably insurmountable. This point was initially established by Spree and other statisticians who used classical time series analysis on German production data for the period 1820–1913.[100] Mathematically the case for the timing of decisive discontinuities proposed by Rostow, Gerschenkron, and other historians has also been seriously undermined by Crafts, Leybourne, and Mills's application of a 'non observed components model with stochastic trend and cycle and no prior selection of structural breaks' to a dataset for seven European countries.[101] It remains unclear what this recently revived interest in testing imperfect statistical series for underlying trends might add to our understanding of European economic growth. Using more mundane techniques, historians have traditionally charted industrialization as a cyclical process.[102] They were, moreover, aware that transitions from household and local to nationally integrated systems of production pose grave difficulties for the measurement of total or industrial outputs.[103] Thus the segmentation of any time series, however well made, into stages, spurts, long cycles, or objective trends for purposes of analysis cannot be other than crude and impressionistic. Tilly's brave attempt to distinguish a German 'take off' from swings (1845–61, 1861–79), shorter cycles, and international booms in railway construction is an elegant example of just how difficult it is to periodize a national record in ways that might reveal causal factors at work.[104]

After extensive discussion, Gerschenkron's notion of a 'decisive spurt', and Rostow's famous 'take off' have been abandoned by historians of the Habsburg dominions. In the Bohemian and Moravian provinces of that empire modern industry emerged early and developed over a long time span from the late eighteenth century through to the 1870s.[105] For Austria itself the beginnings of modern industrialization have been identified with the period 1826–47; and more or less rapid industrial growth occurred in 1841–54, 1861–71, and 1898–1907.[106] Hungary experienced appreciable industrial growth from a low initial level, but two cycles of acceleration (1874–84 and 1906–12) mark the curve and neither phase coincided with the rapid reallocation of the workforce from

[100] This literature in German is cited by O'Brien, 'Do we have a typology?', p. 307.

[101] Crafts, Leybourne, and Mills, 'Britain'.

[102] For an excellent discussion of the conceptual and empirical difficulties of periodizing growth into cycles, trends, swings, long waves, etc., see Solomou, *Phases of economic growth.*

[103] Marketed output is easier to measure, but may simply represent the replacement of household production. See Berg, *Markets and manufacture,* and Hudson and Lee, eds., *Women's work.*

[104] Tilly, 'The take off in Germany'.

[105] Ashworth, 'Typologies and evidence'.

[106] The periodization is not yet settled. See Good, *Economic rise;* Komlos, *Habsburg monarchy;* Huetas, *Economic growth and economic policy;* Rudolph, *Banking and industrialization;* Good, 'Austria'.

agricultural to industrial employment.[107] Similarly the Norwegian and Dutch economies passed through not one but two periods of extraordinarily rapid industrial growth.[108] Just before the First World War, the Balkans (Greece, Serbia, Romania, and Bulgaria) also witnessed remarkable growth in the corporate and larger-scale sectors of industry, but Lampe derogates these 'stirrings' because they 'fell short of a sustained spurt'.[109] How economies growing rapidly during the international boom from 1899 to 1914 might have continued to grow in the radically altered conditions after 1914 has, however, not been considered in discussions of 'spurts that failed'.[110] French and Italian historians reject take offs or crucial spurts as visible or analytically useful concepts to pursue.[111] Research on the economic history of Belgium, Switzerland, Spain, Denmark, and Sweden also seems disinclined to single out phases or cycles of industrial growth for particular emphasis.[112] Even the industrialization of Tsarist Russia (that paradigm case for discontinuity) begins to emerge in the wider and longer-term perspectives of Blackwell, Crisp, Gregory, and Gatrell as a gradual movement on a far broader front than the breakthroughs posited by Gerschenkron.[113]

When historians turned away from imperfect indices of industrial and national production to investigate the agrarian, legal and pre-modern foundations of industry, the drama of 'take offs', 'spurts', and 'discontinuities' in European economic history faded from the scene. Unwilling to accept the Rostovian notion that after a certain key stage (or take off) growth became 'self-sustaining', they also realized that discussions of upswings in industrial production or in national output can degenerate into semantics because neither history nor economics offers acceptable criteria for ranking particular stages or conjunctures in the growth process as 'decisive'.[114] More significantly, they also perceived that all the available typologies offer no models or mechanisms which explain how and why economies or industrial sectors passed from one stage to another. They simply set an agenda for comparative history to address.[115]

Gerschenkron also expected that national historical records would reveal distinctive patterns of industrial organization, but this element of his typology has not been tested because the statistics required to compare his predictions as to the prevalence of larger-scale, more capital intensive, and monopolistic firms

107 **Komlos,** reproduced as article 23; *idem, Economic development.*

108 **Hodne,** reproduced as article 26; *idem, Economic history of Norway; idem, Norwegian economy;* **de Vries,** reproduced as article 20; Griffiths, 'Creating of a national Dutch economy'.

109 **Lampe,** reproduced as article 24; Lampe and Jackson, *Balkan economic history.*

110 Palairet, 'Land, labour and industrial progress'; Lampe, 'Debating Balkan potential'.

111 Federico and Toniolo, 'Italy'; Lévy-Leboyer and Lescure, 'France'; **Crouzet,** reproduced as article 16.

112 Lebrun *et al., Histoire quantitative;* Bergier, *Naissance et croissance;* **Jorberg,** reproduced as article 27; Kranz, 'New estimates of Swedish historical GDP'; Harrison, 'Economic history of Spain'; Sanchez-Albornoz, *Economic modernization;* Kergard, 'Industrial development of Denmark'.

113 Gatrell, *Tsarist economy;* Crisp, *Studies in the Russian economy; idem,* 'Russia'; Gregory, 'Role of the state'.

114 Supple, 'Revisiting Rostow'.

115 But see Rostow's long reply to his critics in the second edition of *Stages of economic growth.*

suggest.[91] Examples of more technologically progressive and productive agricultures can often be located in the hinterlands of towns and cities, or on farmland well connected by waterways to urban markets. In these histories, geographical and institutional constraints upon the levels of agricultural productivity, required to sustain more rapid rates of structural change long before the eighteenth century, are not regarded as fundamental or even as highly significant for variations in the pace and pattern of industrialization down to 1914.[92] On the contrary, some historians wish to begin their explanations for long-run development from the early concentration of populations in towns and industrial occupations, which in their view actively promoted institutional changes and rising land and labour productivity within agriculture.[93] Progressive agricultures thus tended to emerge as the outcome of demographic pressures and of investment in towns, transportation networks, and earlier industrialization of the workforce.[94] Perhaps such interconnections are better represented as symbiotic, but they are certainly interesting to investigate country by country—even region by region—over the centuries before 1750.[95]

V Patterns and typologies

At the opposite pole from the accumulation of detail embodied in histories of European regions, towns, and localities, are the macro typologies associated with writings by Gerschenkron, Landes, Rostow, and others who have offered usable frameworks for the organization of research and teaching in European economic history.[96] Largely because he perceived industrialization of the post-1750 period to be a process of diversity rather than uniformity, Gerschenkron's enduring contribution has been to focus attention upon variations in the *way* European economies passed through an initial but *decisive* cycle of accelerated industrial growth, which he referred to as a discontinuity or great spurt.

[91] Historians of French agriculture have tended to adopt a Malthusian model to explain the static nature of the agrarian economy before 1750. See Le Roy Ladurie, *French peasantry,* but that view for France and for the rest of European agriculture is now being questioned. See Campbell and Overton, eds., *Land, labour and livestock;* Clark, 'Productivity growth'; Cleary, 'French agrarian history'; Goldsmith, 'Agrarian history'; Komlos, 'Thinking about the industrial revolution'.

[92] Grantham, 'Agricultural supply'; Bairoch, 'Urbanization and the economy'.

[93] Grantham, 'Urban provisioning zones'; Hoffman, 'Land rents and productivity'.

[94] The theory that positive effects flow from demographic pressure was developed by Boserup, *Conditions of agricultural growth.* For the industrial revolution period see Anderson, *Population change.*

[95] For an excellent survey of the interconnections see Anderson, *Explaining long-term economic change.* Persson has formulated a model to test the main relationships: Persson, *Preindustrial economic growth.*

[96] **Gerschenkron,** reproduced as article 5, and see *idem, Continuity in history,* and a collection of essays which discuss Gerschenkron's seminal hypothesis: Sylla and Toniolo, eds., *Patterns of European industrialization.* On the Landes approach, see Landes, *Unbound Prometheus,* and the critique in O'Brien and Keyder, *Economic growth in Britain and France,* ch. 1; **Rostow,** reproduced as article 12, and *idem, Stages of economic growth,* and two collections of critical appraisals of the stages in Rostow, *Economics of the take off,* and Kindleberger and di Tella, *Economics in the long view.* Other recent attempts to construct and criticize typologies include **Crafts,** reproduced as article 3; Pollard, *Typology of the industrialization processes;* O'Brien, 'Do we have a typology?'.

Gerschenkron expected that the study of carefully delineated contrasts in the methods used to build up modern industry within several European countries would help historians to explain differences in the time taken by national economies to converge to the highest levels of per caput income. In his view, Britain had forged ahead because conditions for 'spontaneous industrialization' developed early—and the problem for relatively 'backward economies' on the mainland was how to find effective substitutes for missing (i.e. British) preconditions for growth. Thus, the delays exhibited by continental economies in their endeavours to catch up with Britain are imputed, first, to the absence of a range of more or less important 'prerequisites' for early and successful industrialization available to the British economy; and secondly, to the time taken by 'follower countries' to develop their own substitutes for these agrarian, institutional, and technological deficiencies.

Gerschenkron concentrated upon an important but still restricted number of the components of modern industrialization, which he hypothesized would vary systematically, in form and significance, with the timing and degree of backwardness exhibited by different European economies as they passed through a first 'great spurt' in industrial production. Thus, he predicted that when historians examined the build-up of modern industry in Europe country by country they would observe that the more backward or late the economy was when it began that process, the more rapid would be the rate of acceleration in the growth of industrial production, and the more pronounced would be the reliance on capital as distinct from consumer goods; that among 'latecomers' new industrial enterprises would emerge as more technically sophisticated and capital intensive, larger in scale, and more prone to monopoly; that such enterprises would also rely to a greater degree upon external sources (banks and the state) for investible funds and for entrepreneurial initiatives; that, finally, for their markets and supplies of cheap inputs late developers would depend far less than their British precursors upon rising productivity in domestic agriculture.[97]

Backwardness was never clearly defined, but statistical tests (based on small samples of European countries) and using income per head, shares of the workforce in agriculture, and times of arrival at a threshold level of per caput income as proxies for backwardness, do not correlate well with several dependent elements of the industrialization process singled out by Gerschenkron as ways of surmounting retardation.[98] Furthermore, recent historical research (often inspired by his hypothesis) has not merely widened the sample of countries to include deviant cases from the Austrian Empire, the Balkans, and Scandinavia, but has deepened perceptions of British, Dutch, French, Italian, and even Russian industrialization in several ways that appear inimical to the survival of Gerschenkron's typology as an heuristic device.[99]

[97] **Gerschenkron,** reproduced as article 5, and for an excellent summary of his views, Sylla and Toniolo, eds., 'Introduction: patterns of European industrialization'.

[98] Barsby, 'Economic backwardness'.

[99] Sylla and Toniolo, *Patterns of European industrialization,* supplies the modern bibliography. See also **Good,** reproduced as article 22; **Komlos,** reproduced as article 23; **Lampe,** reproduced as article 24.

among latecomers are not available.[116] Once again the typology seems designed to juxtapose continental deviations against a misperceived version of British industrial organization. Recent research reveals variance in scale, capital intensity, and departure from competitive norms across industries throughout the first industrial revolution.[117] Furthermore, it might be misleading to accept the suggestion of conspicuous contrasts across countries when accessible data and the majority of business histories are (inevitably but unfortunately) confined to the incorporated sector of industry. Variations in plant size, competitiveness and factor intensity are, moreover, unlikely to be systematically correlated to backwardness and are probably related to legal rules for incorporation, the development of markets for stocks and shares, changing technology and, above all, to variations in product mix.[118]

Early statistical investigations by Hoffman into the evolution of industrial sectors led historians to expect that industrial growth would proceed as a systematic and explicable process of diversification from consumer goods through intermediate products to capital goods industries.[119] Although the empirical foundations for Hoffman's attempt to establish stages of industrial growth have been rightly criticized by Kuznets, there are good reasons to expect European industrialization to proceed more or less in the way that Hoffman suggested.[120] First and in general terms, Hoffman's hypothesis accords with descriptions of orderly transitions from simple to complex processes of manufacturing activity. For example, the initial mechanization and standardization of several consumer goods industries tended to be located within well-defined regions of specialization, utilizing the investible funds, skilled labour, and entrepreneurial talents of traditional proto-industrial bases. Factories emerged on a small scale with relatively low capital-labour ratios to engage in the linear transformation of raw materials into familiar products sold on well-established markets. In many cases such resources could be mobilized from within the traditional economy in order to replace imported manufactures or the products of handicraft industries.[121]

Building and construction apart, the manufacture of producers' or capital goods (such as metals, machines, tools, transport equipment, steam and hydraulic engines, and chemicals) required more skills, organizational capacity, mobile capital, and steady demand than seems to have been available to most economies in the initial phases of development.[122] As manufacturing expanded, as the agricultural sector embarked on mechanization and farmed with

[116] Kocka has respecified this hypothesis in a wider context: **Kocka,** reproduced as article 14. A model econometric study testing the possible connections between scale and industrial growth is Nye, 'Firm size and economic backwardness'. See also Freedman, *Joint stock enterprise.*

[117] The British case is surveyed in Berg, 'Industrial organization'.

[118] Von Tunzelmann, 'Technological and organizational change'; Tilly, 'Mergers, external growth and finance'; and Payne; 'Industrial entrepreneurship'.

[119] Hoffman, *Growth of industrial economies.*

[120] Kuznets, *Modern economic growth.*

[121] See discussion of proto-industrialization and references cited above, nn. 54–8; **Heikkinan and Hjerppe,** reproduced as article 25; **Hodne,** reproduced as article 26; Knapp, 'Europe's textile artisans'; Sheridan, 'Political economy of artisan industry'.

[122] Klima, 'Beginning of the machine building industry'; Fremdling, 'Development of the iron industry'.

chemicals, and as transport networks began to cover the terrain, demand and supply constraints on the expansion of a domestic capital goods sector gradually diminished and at that stage a second wave of industrialization carried the economy forward.[123]

Gerschenkron's insistence on discontinuities leads, however, to a rejection of Hoffman's orderly stages of industrial evolution and to a derogation of industrialization before the 'great spurt'. But statistical sources for the century before 1914 are nowhere near good enough to test his hypothesis that backward nations forced their way up the ranks of European economies by abrupt and unusual changes in the commodity mix of industrial output. By taking the longer view and a wider focus, recently published analyses of industrial growth (for the Habsburg Empire, Scandinavia, Italy, Portugal, Spain, and even for Tsarist Russia) seem disinclined to single out the rise of producers' goods as either a sudden or significant element in the industrialization of latecomers.[124]

At the level of particular industries and commodities, proceeding from low initial levels of domestic output, breakthroughs certainly occurred in the manufacture of some types of producers' goods, including steel, heavy chemicals, steam engines, agricultural tools, and simple machinery.[125] On examination, these examples usually represent the exploitation of natural resources rendered valuable by technical change and investment in transport, often attracted by subsidies, tariffs, and other governmental concessions.[126]

In general, their history before 1914 suggests that the relatively backward European countries did not exploit strategies for rapid industrialization which were unrelated either to prior internal developments in the manufacture of consumers' goods, or to natural resource bases, or to opportunities open to them on international or imperial markets. For example, Hungarian industrialization, which rested in large part on the processing of grain into flour and spirits, can be linked to the soils of the Danubian plain, to its favoured position within the Habsburg customs union, and access to German markets.[127] Both the timing and the pattern of industrial growth in Sweden from 1860 to 1914 have been linked to exports and to natural advantages based on timber and iron ore as well as the prior accumulation of a stock of literate manpower.[128] Norway's two upswings are inseparable from exports of timber in the 1840s and the diffusion of hydroelectricity in the early 1900s,[129] while the failure to expand exports of

[123] Hoselitz, 'Some problems in the quantitative studies of industrialization'; Sylla and Toniolo, eds., *Patterns of European industrialization.*

[124] Good, *Economic rise of the Habsburg empire;* Komlos, *Economic development;* Hansen, *Early industrialization;* **Heikkinan and Hjerppe,** reproduced as article 25; **Hodne,** reproduced as article 26; **Reis,** reproduced as article 21; Carreras de Odriozola, 'Producción industrial'; Crisp, *Studies in the Russian economy.*

[125] Klima, 'Beginnings of the machine building industry'; Westebbe, 'State entrepreneurship'; Day, 'Making of mechanical engines'; Tilly, 'Capital formation'; Fenoaltea, 'Extractive industries'.

[126] Trebilcock, *Industrialization of the continental powers;* Ville, *Transport and the development of the European economy;* Crisp, *Studies in the Russian economy;* Falkus, *Industrialization of Russia;* Hertner and Jones, *Multinational theory.*

[127] Komlos, *Habsburg monarchy.*

[128] Sandberg, 'Banking and economic growth'; *idem,* 'Ignorance, poverty and economic backwardness'.

[129] **Hodne,** reproduced as article 26; *idem, Economic history of Norway.*

manufactured goods apparently constrained the upswing in Czech and Austrian production from 1895 to 1914.[130] After Waterloo, French industry found its markets in an international economy dominated by Britain by concentrating upon high-quality products and the finishing section of manufacturing activity.[131]

Typologies of European industrialization which do not link its pace and character to geography and foreign trade neglect essential elements of the process.[132] Europe developed as a continent of large and small political units, with a skewed distribution of natural resources. When possibilities for foreign trade expanded rapidly, in the wake of investment in transportation, that promoted very different rates and patterns of industrial growth from country to country.[133] For example, at comparable levels of per caput income very few European economies achieved anything like Britain's comparative advantages in the export of manufactures. As late as 1913 the shares of 'manufactured commodities' to total exports varied from under 10 per cent for Greece, Denmark, Norway, and Portugal, to over 75 per cent for Britain, Belgium, Switzerland, and Germany, with the rest of Europe somewhere in between.[134] Estimates of 'revealed comparative advantage' for manufactured exports alone expose a variety of national patterns. For example, in 1900 Belgium's comparative advantages resided in glass, non-ferrous metals, rails, ships, iron, and steel, while Italy's advantages can be ranked to place wines, tobacco, textiles, and fancy goods at the top of the list.[135]

Systematic comparisons of trade data across countries and studies of long-term trends in the composition of net imports (exports minus imports) for major manufactured commodities offer historians a promising approach to the statistical delineation and comprehension of the varieties of industrialization displayed by different European economies before 1914.[136] From textbooks with a European frame of reference they have already been made aware of small and large country patterns.[137] Studies of foreign trade will help students to appreciate the significance of factor supplies, endowments, and locations, to understand variations promoted by different tariff regimes, and to realize the complexities of forward and backward linkages between trade and industrial development.[138] They will also realize that the cycles in production and capital formation so carefully measured for several European economies can be properly understood in relation to broader movements only across an evolving international economy.[139] For Britain, upswings and downswings in economic activity have

130 Rudolph, *Banking and industrialization.*
131 O'Brien and Keyder, *Economic growth.*
132 Foreman-Peck, *History of the world economy.*
133 Bairoch, 'European foreign trade'; *idem,* 'Geographical structure and trade balance'.
134 **Crafts,** reproduced as article 3.
135 Tyszinski, 'World trade'.
136 **Berend and Ranki,** reproduced as article 1; Berend and Ranki, *European periphery.*
137 Especially Milward and Saul, *Development of economies.*
138 Crafts, 'Revealed comparative advantage'; Bairoch, *Commerce extérieur et développement économique de l'Europe au 19me siècle* (Paris, 1976); *idem,* 'European trade policy'.
139 Lewis, 'Rate of growth'; Rostow, *World economy.*

traditionally been analysed within the boundaries of an Atlantic economy. It should prove illuminating to place the considerable body of knowledge and insights that economic historians now possess on trade and growth for individual countries within the international economy as a whole in order to observe and analyse trends, cycles, and long-term changes in comparative advantage before 1914.[140]

VI Capital formation, investment, and banks

Rates at which different countries accumulated productive assets (such as the tools, machines, buildings, transport equipment, communications networks, and inventories) required to assist workers to raise output per unit of labour input continue to be of central concern for European economic history. Impressive amounts of empirical research have gone into attempts to measure and compare changes in the volume and structure of capital stocks available to national economies and their workforces over time.[141] As countries industrialized, their capital stocks accumulated more rapidly than their workforces, and usually changed in composition in fairly predictable ways because the shares of land, mercantile inventories, and residential housing declined as a proportion of national wealth relative to producers' goods (machinery, transportation equipment) and communications networks, especially railways.[142] Rostow and Lewis posited that European industrialization had typically been accompanied by rather abrupt jumps in the shares of national income invested in the formation of capital, but the general evidence for that particular discontinuity seems to be as tenuous and theoretically problematical as the inferences and evidence for decisive spurts and take offs in national outputs.[143]

For example, this premature and uncritical extrapolation of Keynesian growth models into economic history in order to account for higher rates of growth in terms of upswings in rates of investment and capital accumulation is now recognized to be taxonomic at best and empirically misconceived at worst.[144] Relations between investment and output, however, turned out to be less simple and direct than Harrod-Domar models might suggest. Increments to outputs over any period of time obtained from observed changes in rates of national investment depended upon the allocation of capital among particular sectors and industries; the size and shape of the initial stock of capital; the accessibility of a country's natural resources; and the availability of complementary resources, particularly skilled labour.[145]

[140] **Prados de la Escosura,** reproduced as article 29; Lains, 'Growth exports'.

[141] Mathias and Postan, eds., *Cambridge economic history of Europe,* 7.

[142] Solow and Temin, 'Introduction: the inputs of growth'.

[143] Rostow, *Stages of economic growth;* Lewis, *Theory of economic growth.*

[144] Rostow, *Economics of take off* contains useful critiques of this simple notion by Cairncross and Kuznets.

[145] Solow and Temin, 'Introduction: inputs for growth'.

Such connections have not, moreover, proved at all easy to quantify. In theory, and *if* the formidable demands for the data required to measure long-term changes in output and in major inputs (including land, labour, and capital) utilized to generate growth can be met, then it might become possible: (a) to decompose observed changes in output into the shares 'imputable' to the use of more land, labour, and capital by an economy or an industry; (b) to measure the size of residual which usually emerges from these exercises as the share of the increment to output unexplained by changes in the *quantities* of land, labour, and capital; and (c) to begin to break down the often large residual into technical changes, organizational improvements, education and training, economies of scale, government policies, and other elements of explanations for economic growth long familiar to economic historians.[146]

Alas, 'accounting' for the observed growth of output in terms of changes in inputs rests upon several theoretical assumptions that have been contested by economists and demands sets of data that are rarely available to historians of European industrialization.[147] Capital formation series have been published for but a few national economies and industries.[148] The ostensibly more manageable task of constructing indices to measure increases in the size of workforces (even inputs of cultivable and other land) employed in production is possible only within margins of error that render this otherwise promising and well specified approach to comparative economic history very difficult to bring on stream.[149] Nevertheless it would be illuminating to appreciate the relative scale and the timing of contributions made by labour, capital formation, and the diffusion of new technologies (and other elements of the residual) to the pace and pattern of development across European countries.[150] Although the evidence for rapid transitions from lower to higher rates of gross and net investment has not been established, economic historians expect that accelerated rates of capital accumulation, and increasing ratios of capital to labour accompanied industrialization. Meanwhile, they have turned their attention to more manageable lines of research involved in the analysis of the costs and benefits of capital formation in particular sectors of national economies, including housing, roads, canals, and above all railways. They have concentrated even greater attention upon banks, state policies, and other institutional arrangements designed to encourage and to mobilize the domestic and foreign savings required to fund the investments required for structural change.

146 Model studies are Matthews, Feinstein, and Odling-Smee, *British economic growth;* Crafts, *British economic growth.*

147 The debate over the use of growth accounting methods to decompose explanations for British economic growth into shares imputable to capital, labour, and a residual seem to have no analogues for other European economies. References to the British discussion may be found in Berg and Hudson, 'Rehabilitating the industrial revolution'.

148 Mathias and Postan, eds., *Cambridge economic history of Europe,* 7.

149 Maddison, *Economic growth in the west* is excellent on problems of defining and measuring labour input, but see O'Brien, *International productivity comparisons;* Bairoch, *Working population.*

150 Lévy-Leboyer and Bourgignon, *French economy in the nineteenth century* offers an alternative approach to macro production function analysis. Maddison, *Dynamic forces* prefers partial productivity comparisons.

Investment in social overhead capital has merited the widespread and rigorous attention it has received because that kind of capital formation: (a) created the framework for market integration and modern manufacturing industry concentrated in towns and regions across Europe; (b) promoted backward linkages which stimulated the development of important industries such as iron and steel, engineering and construction; and (c) promised (following the publication of seminal books by Fogel and Fishlow on railways in America) to allow economic historians to measure the benefits that accrued from investment in European railways and other forms of transportation.[151]

As a research programme in comparative economic history, the large volume of publications on European communications are illuminating to read because they have exposed nearly all the interconnections between investment in transportation and economic growth. Extensions to and improvements in communication networks and equipment clearly made a positive but very variable contribution to national rates of development across Europe and we now understand why such interesting differences occurred. Although social savings (as developed by Fogel and Fishlow) provided a powerful heuristic and organizational device for modern research, historians now know why they cannot measure the contribution made to European economic growth by investment in transportation. They and their American mentors have not quantified, and perhaps cannot quantify, possible declines in the rates of growth of national products which might counterfactually have occurred *if* economic efficiency in several modes of transportation (including roads, canals, and shipping, as well as railways) had not improved over time.[152] Nevertheless when similar degrees of rigour and quantification are brought to bear upon the comparative economic analysis of investment and technical change in, say, steel, textiles, coal, electricity, and other major industries, the comprehension of European industrialization may take another leap forward.[153]

Gerschenkron suggested that when they built up modern industry Europe's latecomers depended far more on banks and the state for investible funds and entrepreneurial stimulus than British and French industrialists had at comparable phases of their earlier development. This particular hypothesis (despite an apparent consensus that the availability of investible resources rarely constituted a serious constraint upon capital formation in the nineteenth century) probably inspired more research across more countries than all other elements of his typology put together.[154] For example, Cameron and a team of distinguished collaborators published two seminal volumes on the role of banks and banking in the early stages of industrialization in England, Scotland,

[151] Ville, *Transport and the development of the European economy;* Szostak, *Role of transportation;* Caron, 'French railroad investment'.

[152] **O'Brien,** reproduced as article 9, and *idem*, ed., *Railways and the economic development of western Europe,* which includes chapters on France, Britain, Germany, Italy, Belgium, and Spain, and an extensive bibliography.

[153] Williamson, 'Greasing the wheels', sums up the controversy.

[154] A good bibliography is in Sylla, 'Role of banks'.

Belgium, France, Germany, Russia, Spain, Italy, and Austria.[155] More recently the suggestion has occupied a central place in studies of Italian, Czech, Hungarian, Austrian, Swedish, Norwegian, and Balkan industrialization.[156]

Gerschenkron's general propositions about banks are difficult to translate into a form that makes them amenable to statistical testing. And no estimates have appeared to measure the relative contributions of several forms of financial intermediaries, the state and other classifiable sources of finance for capital invested and credit deployed by industrial sectors, as they developed over time.[157] Such macro data for flows of funds are unlikely to emerge from financial records, although business history does come up with the odd significant case study for particular industries.[158]

Some inferences and comparisons can be drawn about potential changes through time in the *availability* of bank loans by using ambiguous indicators, such as number of banks per square kilometre of territory and the (more relevant) ratios of bank assets to GNP now published for several European economies.[159] These ratios lend little support to the notion of any systematic connections across Europe between per caput incomes or industrial output per head on the one side and relative levels of bank assets on the other; a finding that prompted Sandberg to describe Sweden in 1850 as an 'impoverished sophisticate'.[160] Furthermore, no observable threshold levels of bank lending can be observed prior to upswings in industrial production. According to Good's correlations, banks neither led nor (with the possible exception of Germany and Belgium) followed upon the growth of industrial production.[161] As national histories illustrate, the diffusion of banks and expansion of bank loans proceeded across Europe at rather disparate rates that can be related to a variety of local conditions, including legal regulations, government deficits, literacy and education, the scale of internal and external trade, regional concentrations of economic activity, urbanization, social attitudes (including antipathy to banks), and even (as Gerschenkron once remarked) standards of honesty. Until bank assets are broken down between loans to industry and to other sectors of the economy such as transport, the state, house-building, and utilities, the scale of the contribution made by banks to capital formation in industry *per se* cannot be assessed.

155 Cameron, *Banking in the early stages of industrialization; idem, Banking and development.*

156 Cohen, *Finance and industrialization;* Rudolph, *Banking and industrialization;* **Komlos,** reproduced as article 23; Sandberg, 'Banking and economic growth'; *idem,* 'Case of the impoverished sophisticate'; Eyge, 'Transformation of bank structures'; Nygren, 'Transformation of bank structures'; Nielsen, 'Aspects of industrial financing'; **Lampe,** reproduced as article 24.

157 Kindleberger, *Financial history of western Europe,* makes no attempt to quantify financial flows.

158 Freedman, 'Joint stock business organization'. For the difficulties see Zamagni, *Finance and the enterprise.*

159 Goldsmith, *Financial structure;* Cameron, *Banking in the early stages;* **Cameron,** reproduced as article 2.

160 Sandberg, 'Case of the impoverished sophisticate', but Kindleberger, 'Financial institutions', regards bank assets as a reasonable proxy for levels of development.

161 Good, 'Backwardness and the role of banking'.

Modern research on financial intermediation has, however, shifted attention away from tenuous macro correlations to the analysis of the actual responses, positive and negative, by banks to industrialization as it proceeded in particular national or regional contexts. Cameron, when summarizing the results of his international comparisons, candidly admitted that he lacked criteria to measure their success or efficiency, although he was prepared to laud banks in Scotland and Sweden, and to deplore their meagre contribution in France—an economy which somehow managed to achieve respectable levels of per caput income and industrial output with little help from its banks.[162] For Germany (once the paradigm example for the positive role of banks), there is now a disputed counter suggestion that the over-praised Kreditbanken misallocated capital and slowed up overall growth.[163] In the Czech Crownlands (Bohemia, Moravia, and Upper Silesia) risk averse banks did little except supply a modicum of working capital.[164] After 1873 Austrian banks 'lost most of their interest in industrial enterprises for years to come'. For Hungary it was not banks but 'unencumbered access to the Austrian capital market' that appeared crucial to the beginnings of industrialization.[165] Norwegian commercial banks showed little initiative and failed to play any leading role in the process of economic growth.[166] Belgium, Italy, and Germany are contested cases, but in general it seems difficult to find clear examples where banks evidently played the progressive and entrepreneurial role in industrialization marked out for them by Gerschenkron.[167]

VII The state

Both the decisions to promote and the finance required for the advance of European industry before 1914 continued to be overwhelmingly private in origin. Although governments certainly played discernibly different roles within their respective national and international spheres, their civil expenditures (which usually represented a small proportion of central budgets) seemed basically designed to create conditions within which private enterprise could operate more efficiently; while their fiscal, commercial, and monetary policies sought to encourage higher rates of investment, both domestic and foreign.[168] Investments undertaken by central governments were pretty well confined to organizing and funding extensions and improvements to national networks of communication—roads, canals, rivers, ports, and especially

162 **Cameron,** reproduced as article 2; but see Kindleberger, *Financial history of western Europe.*

163 Neuberger and Stokes, 'German banks and German growth'; Fremdling and Tilly, 'German banks and German growth'; see **Tilly,** reproduced as article 15; Tilly, 'German banking'.

164 Rudolph, *Banking and industrialization.*

165 **Komlos,** reproduced as article 23.

166 Eyge, 'Transformation of bank structures'.

167 Sylla, 'Role of banks'; Federico and Toniolo, 'Italy'; Morrison, 'Financial intermediaries'.

168 Tilly, *Formation of states; idem, Capital, coercion.*

railways, often in collaboration with private business.[169] Outside the manufacture of armaments, direct investment by governments in industry was uncommon and on a small scale even in Tsarist Russia.[170] Economic historians still need to provide proper estimates of the significance of government funds (local, regional, and central) for national capital formation before 1914. Thus the concentration of recent research which analyses (and occasionally compares) the possible economic effects of labour, fiscal, tariff, monetary, and exchange rate policies certainly places the emphasis on the role of European states between 1750 and 1914 where it belongs.[171]

Linked to political history (and to related studies in national traditions in economic thought), the literature on economic policy now brings increasingly sophisticated applications of modern macroeconomic theory (both Keynesian and classical) to the appreciation of the states' role in European development.[172] The illuminating debate on Russian monetary policy and its connection with industrial growth from 1861 to 1914 is but one excellent example.[173] Other studies have revealed how increasing budget deficits and an inept exchange rate policy created a climate of uncertainty which crowded out and depressed private capital formation in Austria from 1848 to 1866.[174] For Hungary, Komlos observed that changes in 'governmental fiscal policy correlate well with the industrial production index'. Thus in the 1870s and 1880s the ability of that government to attract foreign investments shifted the burden of the debt on to foreigners and enabled the state to pursue an active fiscal policy without harming the private sector through excessive taxation.[175] German agriculture 'made impressive technological strides' behind the walls of protection it received after 1879, and the compromises of the Meline tariff helped to create the stable political conditions required for the growth of larger scale industry in France during the *belle époque*.[176] If the Italian tariff of 1887 succeeded in fostering the development of some industries, the price paid by the whole economy in terms of the misallocation of resources seems to have been high.[177] Research into connections between public policy and the industrialization of Europe could soon become sophisticated enough to create a basis for comparisons about the scope and effectiveness of governmental intervention.[178]

169 Supple, 'The state and the industrial revolution'; Freedman, 'Cartels and the law'; Smith, 'Free trade versus protectionism'.

170 Kahan, *Russian economic history;* Gregory, 'Role of the state'.

171 Mathias and Pollard, eds., *Cambridge economic history of Europe,* 8.

172 Harling and Mandler, 'From fiscal military state'; Poggi, *The state, its nature.*

173 Barkai, 'Macro-economics of Tsarist Russia'; Drummond, 'Russian gold standard'; Gregory and Sailors, 'Russian monetary policy'; Gregory, 'Russian balance of payments'.

174 Huetas, *Economic growth and economic policy.*

175 **Komlos,** reproduced as article 23; Eddie, 'Economic policy'.

176 Webb, 'Agricultural protection'; Smith, *Tariff reform;* Lebovics, *Alliance of iron and wheat.* On trade policies see Bairoch, 'European trade policy'.

177 Federico and Toniolo, 'Italy'; Toniolo, *Economic history of liberal Italy.*

178 For the evaluation of tariff policies, see esp. above, n. 176, and the debate between Irwin and Nye, 'Free trade and protectionism', and Nye, 'Myth of free trade'.

But to confine investigations into the role of the state to periods before and during discontinuities (*pace* Gerschenkron and Rostow) or even to the entire period 1750 to 1914 would derogate its most important historical function: namely, the creation and destruction of institutions and legal rules which, over the long run, sustained and restrained the operation and spread of markets. Older traditions in economic history (represented by the writings of the German historical school and their followers elsewhere in Europe) were fully aware that markets for commodities, labour, capital, land, and technology had not evolved 'naturally'—even in England.[179] Quantitative analyses of the 'causes' of European economic growth, of the kind exemplified in approaches based upon social accounting and macro production functions, often take for granted the frameworks of law, property rights, institutions, as well as security from internal and external aggression, which had taken centuries to build up.[180] Recent critiques of these methods and exercises in quantification seem well founded.[181] Historians have long recognized that such approaches offer more or less acceptable statistical outlines and analyses of the rates and patterns of economic growth which occurred country by country, *once* efficient institutions for the exchange of commodities and the allocation of factors of production were in place.[182]

Yet such institutions usually emerged at locatable points in history and often before manufactured goods and farm products became cheaper to produce. When they appeared, the costs of undertaking exchanges over distance and through time began to decline. Costs of 'transacting business' fell in response to the construction of recognized and secure property rights which found tangible embodiment in political constitutions, legal systems, central regulations, and more widely accepted codes of business behaviour.[183]

In short, a long period of 'institutional development' and commercial experience allowed Adam Smith to observe in 1776 that the pursuit of private profit led inexorably through the operation of the invisible hand to economic growth and gains for everybody. That development, which is essentially the history of the market, was neither costless nor continuous. Even within the cradle of capitalism, in western Europe, it proceeded at different speeds in several locations at various times. Venice led, Holland followed, and eventually the rulers of England's maritime realm steered their national economy into calmer waters whence economic progress became steadier, then more rapid, and eventually irreversible.[184]

Monarchical and aristocratic states of the *ancien régime* operated as the principal 'architects' of the institutions and rules which eventually brought about some real measure of congruence between the pursuit of private profit and social

[179] Hoselitz, 'Theories of the stages'; Hartwell, *Industrial revolution and economic growth.*

[180] See above, section II, and Maddison, *Dynamic forces* basically begins and ends with numbers.

[181] **North and Thomas,** reproduced as article 8; North, *Structure and change.*

[182] Macfarlane, 'Capitalism', and *idem,* 'Individualism reconsidered' provide a good summary of the argument and bibliography.

[183] North and Thomas, *Rise of the western world;* Lamb, *Property markets.*

[184] O'Brien, 'Government and the economy'; North, 'Government and the costs of exchange'.

rates of return.[185] States, with their constitutions and behaviour, are the stuff of politics—something economic historians have eschewed. They prefer to leave the study of governments to political and social history. But since the institutional foundations of national and international markets need to be compared, explained, and not assumed, then the state is clearly too important to be left entirely to students of constitutions and politics, who are often too absorbed with 'process' and with the short run to analyse implications for structural change. In general terms, we can track the rise of states through data on revenues and expenditures.[186] The economic historian's interest in the state is bounded and focused upon the provision of security for business and upon the design of rules which operated to reinforce codes of functional behaviour within which exchange and specialization evolved.[187] Economic history needs to analyse when, why, and how well different states provided security for private investors and enforced rules for competition and cooperation conducive to the extension of markets. Economic historians will, however, wish to avoid leaving the impression that advance towards industrial market economies depended upon the autonomous actions of enlightened states. As *architects* of institutional development, Europe's monarchical states could not be autonomous. They responded to demands from clients, operated within budgetary constraints, worked through recalcitrant aristocrats and bureaucrats.[188] Nevertheless, the depiction may be heuristic because there are effective and incompetent architects. They may not be constrained adequately by clients and often go their own way in pursuit of personal and aesthetic ends. Finally, as Hegel remarked, 'the state stands on earth and so is in the sphere of caprice, chance and error'.[189]

Forays by neoclassical economists into the history of European political economy espouse the opposite assumption that seems to leave little room for autonomous state action. Writers of this persuasion attempt to convince us that European governments provided *public goods* as and when it became profitable for rulers to do so.[190] Now tensions between opportunities for profit and restraints on the realization of substantial private gains clearly operated as one major force for the promotion of politically inspired change reacting to pressure from below. Nevertheless, that kind of analysis of the state not only begs every conceivable question but ignores centuries of scholarship in political, institutional, and administrative history.[191] Political historians are unlikely to accept a theory of the state formulating and enforcing rules and institutions conducive to the spread of markets as a response to opportunities for private profit

[185] Mann, *Sources of social power;* Dyson, *State tradition.*

[186] Bonney, 'State and its revenues'; O'Brien and Hunt, 'Rise of a fiscal state'.

[187] Lane, 'Role of governments'.

[188] An excellent modern text is Bonney, *European dynastic states.*

[189] Evans *et al., Bringing back the state.*

[190] Summaries and cogent theoretical critques of this neo-institutional approach to the development of economically functional states are provided by Field, 'Problem with neoclassical institutional economics', and by Jones, 'Institutional determinism'.

[191] Contrast Akelund and Tollison, *Mercantilism as a rent seeking society* with Dyson, *State tradition.*

emanating from a 'natural' (and in some accounts distinctively English) propensity to truck and barter. Alas, most enquiries into the origins of the 'Wealth of European Nations' seem to generate circular propositions. Governments created and safeguarded markets; markets in turn affected the behaviour of kings and statesmen. In relation to the development of markets, states acted as architects, as demolition contractors, and often as the principal impediment to their diffusion and integration. In European history they often found themselves playing all three roles at one and the same time.[192]

VIII The way ahead

For the study of industrialization, a European frame of reference has been employed by several generations of historians. In the 1930s, 1940s, and 1950s British and American students were introduced to varieties of 'continental' experience (usually French and German) and Europeans could hardly avoid the first industrial revolution. Read today, the texts of that period appear to lack structure and seem impossible to assimilate. They parade scholarship country by country, language by language. Their readers are left to impose order, to build bridges across frontiers, and to 'induce' whatever generalizations they can from the immense variety of national, regional, and sectoral histories set down on hundreds of printed pages. In their wisdom, the scholars of yesteryear (as well as those who continue to write in this way) leave us with an impression of a European economy that is almost as uniquely local and multifarious as the languages, cultures, and peoples of our complex continent.[193]

Shortly after mid-century historians began to formulate and to use models and typologies in order to compare and generalize about Europe's modern industrial development. Perhaps the most accessible heuristic device available at that time was to refer continental industrialization to a paradigm case, the British industrial revolution, as it was perceived to have proceeded from the 1750s to the 1840s. On the assumption that Britain's lead down to 1914 could be ascribed to superior industrial technology, it followed that the economic history of the rest of the continent could be understood as the spread of techniques of production and new forms of industrial and commercial organization from the first industrial nation to more backward economies on the mainland. British methods constituted not merely initial but best practice with almost universal application; diffusion then became a major concept and the central problem for European economic history, at least for the nineteenth century, was to explain Britain's early start and the rate at which other economies adopted British techniques and modes of economic organization.[194]

[192] Tilly, *Coercion capital;* Mann, *States, war and capitalism.*

[193] See above, n. 12.

[194] The classic study is Landes, *Unbound Prometheus,* which first appeared as 'Technological change and developments in western Europe, 1750–1914'. See also Lilley, 'Technological progress'; Hughes, *Industrialization and economic history;* Kemp, *Industrialization in nineteenth-century Europe;* Rostow, *How it all began.*

Yet it is not clear that European statesmen, businessmen and other contemporary observers of the economic scene conceived of their national or industrial goals in terms of a need to import British methods of production.[195] Indeed many continental travellers to Victorian England expressed doubts about the 'British way'. European historians have recognized the illumination embodied in the idea of diffusion but seldom allowed their national case studies of industrialization to be dominated by such a simplistic model.[196] Today few economic historians are prepared to accept the notion that for several decades after 1750 there was an observable and optimal path to higher levels of industrial output which could be identified with stylized versions of how the industrial revolution had proceeded in Britain.[197]

For a start, diffusion is now understood as a complex process of creative adaptation rather than a synonym for emulation.[198] The transfer of technology succeeded only if and when a whole matrix of related capacities (such as energy supplies, skilled labour, and marketing methods) into which machinery can fit had been built up and proved receptive to change.[199] Once in place, European firms spawned their own problems, solutions, and trajectories for future innovation.[200]

Furthermore, recent historical writing has restored traditional perceptions of the first industrial revolution as a drawn-out European process in which the accelerations of the eighteenth century were not especially dramatic or unique. New mechanized techniques of production now seem less novel, less British, and diffused slowly even within the Hanoverian kingdom. They eventually succeeded because they rested upon the prior spread of a market economy throughout the British Empire over several generations before the American War of Independence.

Although the success of the Victorian economy represented an 'example' for western Europe of what could be achieved, Britain's industrial revolution can no longer be reified into a 'paradigm' which rival economies emulated. Except in terms of the high level and now commonplace generalizations associated with taxonomies of economic growth constructed by Kuznets, Rostow, and Maddison, the ways in which follower countries industrialized and caught up with Britain might now be perceived to be more interesting for their contrasts than for their similarities with the experience of the first industrial nation.[201] Comparative European history has revealed several 'paths' to the twentieth century

[195] Milward and Saul, *Economic development of continental Europe;* Lévy-Leboyer, *Les banques européennes;* O'Brien and Keyder, *Economic growth.*

[196] Pollard remains an exception. Pollard, *Peaceful conquest,* but see Cameron, 'New view of European industrialization'; Crouzet, 'Western Europe and Great Britain'; Saul, 'Nature and diffusion of technology'; Cameron, 'International diffusion of technology'.

[197] Mokyr, *Lever of riches;* Rosenberg and Birdzell, *How the west grew rich.*

[198] Bruland, *Technology transfer;* Fremdling, 'Foreign trade patterns'; *idem,* 'Foreign competition and technological change'.

[199] Bruland, *British technology;* Davis, 'Technology and innovation'.

[200] Pinedo, 'From bloomery to blast furnace'; Anderson and Pilling, 'Spanish entrepreneurs'; Redlich, 'Leaders of the German steam engine industry'; Pierenkemper, 'Entrepreneurs in heavy industry'.

[201] Pollard, *Typology of industrialization processes.*

that serve to highlight many salient but peculiar features of the British case, including excellent resource endowments (especially abundant supplies of coal), an exceptionally productive agriculture, a secure system of private property rights, and an effective state. These advantages could not be replicated easily or quickly by follower countries on the mainland, a fact which undermines the idea of a British-style industrial revolution as a model strategy for those who wished to catch up.[202] Britain certainly prompted other states to support industrialization by its politically challenging economic and contingent fiscal success.[203] Its contribution to their economic growth resided in the export of a restricted range of mechanical technologies and skills, and in maintaining a liberal international economic order. As a 'do it yourself' package, the first industrial revolution seems too British to be of widespread and profound transnational significance.[204] In any case, many of its technical, scientific, and organizational elements were international property before 1750.[205] While the kingdom's early lead was not random or accidental, there seems no reason to argue that the industrialization of other European economies would have been long delayed without an earlier British example of success before them.[206] Meanwhile Britain moved ahead by building up and exploiting its comparative advantages (especially cheap energy and naval power) and experienced a pattern of long-run development based upon balanced productivity growth between industry, agriculture, and transportation, and rapid, if confined, advance in a few staple manufactures, sold to foreigners.[207]

It is now more than 30 years since three distinguished American scholars (Gerschenkron, Landes, and Rostow) offered students and teachers of European economic history concepts, approaches, and typologies designed to help them organize and clarify the accumulation of historical knowledge in their rapidly expanding subject. Today, when the first industrial revolution is seen as more of a special case, the potential for histories of European industrialization constructed around the concept of technological diffusion seems diminished, even disabled.

From philosophers of the Enlightenment, through Marx to the German historical school, historians of European development have found it difficult to organize their thoughts without recourse to stages of growth. Rostow certainly publicized a new and memorable vocabulary from his stages of growth but (after long debate and for familiar reasons) his schema is not considered to be a particularly illuminating way to write Europe's industrial history. Conceptually, the whole notion of sharp decisive discontinuities no longer attracts historians of economic growth, inclined to analyse that process with biological rather than aeronautical metaphors and to explain accelerations (if and when they can be

[202] Guha, *Evolutionary view of economic growth;* Adams, *Paradoxical harvest.*
[203] O'Brien, 'Political preconditions'.
[204] Crafts, 'British industrialization'.
[205] Inkster, 'Mental capital transfers'.
[206] Crafts, 'Industrial revolution in Britain'.
[207] O'Brien, 'Modern conceptions'.

measured) as the product of fortuitous political events, or the final outcome of a protracted process of institutional change, and the prior accumulation of physical and human capital. Apart from insurmountable problems with statistics, discussion usually runs into that hoary old debate on the relative importance of necessary and sufficient conditions. Gerschenkron's typology inspired two generations of scholars and survives as a viable framework for economic historians who are concerned to place their research within those wider and potentially deeper perspectives that emanate from international comparisons. More often than not, its heuristic value seems to reside in supplying explanations and predictions that simply fail to fit the country or case under investigation. Meanwhile, and as research upon economies, regions, towns, sectors, and factors of production proliferates, no scholar with Gerschenkron's awesome learning and capacity for generalization has emerged to reconstruct an alternative typology which could serve to draw economic history into a European framework for comparative history.[208]

Apart from the unsatisfactory mode of confounding Gerschenkron, the majority of Europe's economic historians continue to pursue their subject within the legal and institutional framework of nation states, and to offer their research as a necessary basis for comparative history. Indeed that programme is exemplified by most English-language textbooks in the field, which tend to be surveys of foreign scholarship arranged country by country or input by input.[209] Readers of these scholarly and analytically sophisticated books in European economic history are implicitly asked to make the necessary intellectual leaps from national histories moving through a century or more of industrialization to comparative history unfolding across the continent. Students, not yet scholars, are being left to draw out illuminating comparisons and to discover missing connections in the circuits that generated economic progress throughout Europe before 1914. Of course, interesting and potentially significant contrasts are sometimes delineated between agricultural institutions, techniques of production, transport facilities, and types of financial intermediation, but they are rarely investigated or specified in ways that suggest how variations might be tested for significance.

It is certainly not difficult for editors of collections lost in the thickets of national, regional, and local scholarship critically to review the paths marked out by their predecessors and to call for new signposts for a comparative economic history of Europe. But three new approaches seem to offer prospects of escape from merely adding more and more printed information to the volume of historical research bounded by national frontiers. All three call for the continued accumulation and further improvement of national statistics. First, there is the task of mapping out and explaining the long-term evolution of revealed comparative advantage country by country from the bodies of trade data at our

[208] Sylla and Toniolo, eds., *Patterns of European industrialization.*

[209] See above, n. 7; Mathias and Postan, eds., *Cambridge economic history of Europe,* 7 (pts. 1 and 2); and Mathias and Pollard, eds., *Cambridge economic history of Europe,* 8.

disposal. Secondly, scholars who wish to explore the heuristic power of the inductive methods (pioneered by Chenery and introduced to historians by Crafts) to expose paths or patterns of industrialization among developing economies, need to be supplied with improved and wider sets of data.[210]

Thirdly, European economic history might find a way forward if we are prepared to follow the hard road trodden by our colleagues in historical demography. In that field a truly massive assault on parish registers, population censuses, and other primary sources over the past four decades makes it possible systematically to compare long-run demographic experience across Europe and also with other continents.[211] Demographers have now reached the more difficult stage of a research programme inaugurated by Louis Henry in 1954, and they begin to explain measured and settled facts dealing with variations in fertility, nuptiality, mortality, and illegitimacy across countries and regions. Their database excites envy from historians who try to account for economic change within the same continent but with only a fraction of the body of macro and micro statistics demographers now command.[212]

Since mid-century European economic history has certainly been concerned with the measurement and explanation of differences in levels of per caput income and productivity attained by its constituent national economies from the late eighteenth century down to the Great War. Perceptions of advanced and backward economies are shared by most scholars who expect to find Britain at the top of European league tables and Russia towards the bottom.[213] In between some kind of rank order among countries has apparently been accepted. Alas, the fact is that economic historians are simply not yet in a position to appreciate the scale of variations among European nations in terms of per caput income and consumption levels: in labour productivities for major industries, or for agriculture, in yields per hectare for important field crops, or indeed for any of the other indicators of development and relative economic efficiency which constitute standard indicators for economists concerned with the measurement of comparative levels of economic development among European economies since the Second World War. Thus the most urgent task for the subject, and a precondition for a general understanding of European industrialization, is to reconstitute the data now available on a national or industry-wide basis into a form that would command acceptance and facilitate comparisons across countries. That, as Marc Bloch and Simon Kuznets suggested a long time ago, is the most illuminating way to write European economic history.

210 Chenery and Syrquin, *Patterns of development;* Crafts, 'Patterns of development'; Adelman and Morris, *Comparative patterns.* For a model attempt to apply these models to Spain, see Molinos and Prados de la Escosura, 'Was Spain different?'.

211 Flinn, *European demographic system;* Tilly, *Historical studies in changing fertility;* Lee, *Population patterns.*

212 Lee, *European demography;* Anderson, *Population changes.*

213 Senghaas, *European experience.*

Bibliography

Abramovitz, M., 'Catching up, forging ahead, falling behind', *J. Econ. Hist.*, XLVI (1986), pp. 385–406.
Adams, R. N., *Paradoxical harvest: energy and explanation in British history, 1870–1914* (Cambridge, 1982).
Adelman, I. and Morris, C. T., *Comparative patterns of economic development* (Baltimore, 1988).
Aldcroft, D. and Rodger, R., *Bibliography of European economic and social history* (Manchester, 1984).
Aldrich, R., 'Latecomer or early starter? New views of French economic history', *J. Eur. Econ. Hist.*, 16 (1987), pp. 89–100.
Almquist, E., 'Pre-Famine Ireland and the theory of European proto-industrialization', *J. Econ. Hist.*, XXXIX (1979), pp. 699–718.
Amin, S., *Accumulation on a world scale: a critique of the theory of underdevelopment*, 2 vols. (New York, 1974).
Anderson, B. L. and Pilling, P. W., 'Spanish entrepreneurs and British technology in early nineteenth-century Andalucia', *J. Eur. Econ. Hist.*, 19 (1990), pp. 35–72.
Anderson, J. L., *Explaining long-term economic change* (1991).
Anderson, M., *Population change in north western Europe, 1750–1850* (1988).
Arnez, J. A., *Slovenian lands and their economies, 1848–73* (New York, 1983).
Ashworth, W., 'Industrialization and the economic integration of Europe', *Eur. Stud. Rev.*, 4 (1974), pp. 291–316.
Ashworth, W., 'Typologies and evidence: has nineteenth-century Europe a guide to economic growth?', *Econ. Hist. Rev.*, 2nd ser., XXX (1977), pp. 140–58.
Aston, T. H. and Philpin, C. H. E., eds., *The Brenner debate: agrarian class structure and economic development in pre-industrial Europe* (Cambridge, 1985).
Baechler, J., Hall, J. A., and Mann, M., *Europe and the rise of capitalism* (1987).
Bairoch, P., *The working population and its structure* (Brussels, 1968).
Bairoch, P., 'European foreign trade in the nineteenth century', *J. Eur. Econ. Hist.*, 2 (1973), pp. 5–36.
Bairoch, P., 'Geographical structure and trade balance of European foreign trade', *J. Eur. Econ. Hist.*, 3 (1974), pp. 557–608.
Bairoch, P., 'Europe's gross national product, 1500–1975', *J. Eur. Econ. Hist.*, 5 (1976), pp. 273–340.
Bairoch, P., 'International industrialization levels from 1750–1980', *J. Eur. Econ. Hist.*, 11 (1982), pp. 269–333.
Bairoch, P., *Cities and economic development from the dawn of history to the present* (1988).
Bairoch, P., 'European trade policy, 1815–1914', in P. Mathias and S. Pollard, eds., *Cambridge economic history of Europe*, 8 (Cambridge, 1989), pp. 1–160.
Bairoch, P., 'Urbanization and the economy in pre-industrial societies: the findings of two decades of research', *J. Eur. Econ. Hist.*, 18 (1989), pp. 239–90.
Barkai, H., 'The macro-economics of Tsarist Russia in the industrialization era', *J. Econ. Hist.*, XXXIII (1973), pp. 339–71.
Barsby, S., 'Economic backwardness and characteristics of development', *J. Econ. Hist.*, XXIX (1969), pp. 449–72.
Berend, I. T. and Ranki, G., 'Foreign trade and the industrialization of the European periphery in the nineteenth century', *J. Eur. Econ. Hist.*, 9 (1980), pp. 539–84.

Berend, I. T. and Ranki, G., *The European periphery and industrialization, 1780–1914* (Cambridge, 1982).
Berg, M., 'Industrial organization', in R. Floud and D. N. McCloskey, eds., *The economic history of modern Britain since 1700* (2nd edn., Cambridge, 1994).
Berg, M., *Markets and manufacture in early industrial Europe* (1991).
Berg, M., 'Industrial organization: transition from pre-industrial to industrial forms', *Warwick Econ. Res. Pap.*, 380 (1991), pp. 1–40.
Berg, M., Hudson, P., and Sonenscher, M., eds., *Manufacture in town and country before the factory* (Cambridge, 1983).
Berg, M. and Hudson, P., 'Rehabilitating the industrial revolution', *Econ. Hist. Rev.*, XLV (1992), pp. 24–50.
Bergier, J. F., *Naissance et croissance de la Suisse industrielle* (Berne, 1974).
Blackwell, W. L., *The beginnings of Russian industrialization, 1800–60* (Princeton, 1968).
Boltho, A., ed., *The European economy: growth and crisis* (Oxford, 1981).
Bonney, R., *The European dynastic states* (Oxford, 1991).
Bonney, R., 'The state and its revenues in ancien régime France', *Hist. Res.*, 65 (1992), pp. 156–76.
Boserup, E., *The conditions of agricultural growth: the economies of agrarian change under population pressure* (1965).
Bowden, W., Karpovich, M., and Usher, A. P., *An economic history of Europe* (New York, 1937).
Braudel, F., *Civilization and capitalism, 15th-18th century,* 3 vols. (1981, 1982, 1984).
Bruland, K., *British technology and European industrialization: the Norwegian textile industry in the mid nineteenth century* (Cambridge, 1989).
Bruland, K., ed., *Technology transfer and Scandinavian industrialization* (Oxford, 1991).
Burke, P., ed., *New perspectives on historical writing* (1991).
Campbell, B. M. S. and Overton, M., *Land, labour, livestock: historical studies in European agricultural productivity* (Manchester, 1991).
Cameron, R., 'The international diffusion of technology and economic development', in *idem*, ed., *Papers presented to the 6th International Economic History Association Conference* (Copenhagen, 1974).
Cameron, R., 'A new view of European industrialization', *Econ. Hist. Rev.*, 2nd ser., XXXVIII (1985), pp. 1–23.
Cameron, R., *A concise economic history of the world* (Oxford, 1993).
Cameron, R. ed., *Banking in the early stages of industrialization* (New York, 1967).
Cameron, R., 'Conclusion', in *idem,* ed., *Banking in the early stages of industrialization* (New York, 1967), pp. 290–321.
Cameron, R. *et al.*, eds., *Banking and development: some lessons of history* (New York, 1972).
Cameron R. and Freedman, C. E., 'French economic growth: a radical revision', *Soc. Sci. Hist.*, 7 (1983), pp. 3–30.
Caron, F., 'French railroad investment, 1850–1914', in R. Cameron, ed., *Essays in French economic history* (Homewood, Ala., 1970), pp. 315–40.
Carreras de Odriozola, A., 'La producción industrial española, 1842–1981', *Revista de Historia Economica,* 2 (1984), pp. 127–57.
Chenery, H. and Syrquin, M., *Patterns of development, 1950–1970* (Oxford, 1975).

Chorley, P., 'The agricultural revolution in northern Europe, 1750–1880', *Econ. Hist. Rev.,* 2nd ser., XXXIV (1981), pp. 71–93.
Cipolla, C. M., *Before the industrial revolution* (1976).
Clapham, J. H., *Economic development of France and Germany* (Cambridge, 1936).
Clark, G., 'Productivity growth without technical change in European agriculture before 1850', *J. Econ. Hist.,* XLVII (1987), pp. 419–32.
Cleary, M. C., 'French agrarian history after 1750: a review and a bibliography', *Agric. Hist. Rev.,* 37 (1989), pp. 65–75.
Clough, S. B. and Cole, C. W., *Economic history of Europe* (Boston, 1952).
Cohen, J., *Finance and industrialization in Italy* (New York, 1977).
Coleman, D. C., 'Proto-industrialization: a concept too many', *Econ. Hist. Rev.,* 2nd ser., XXXVI (1983), pp. 435–48.
Crafts, N. F. R., 'Industrial revolution in Britain and France: some thoughts on the question why was England first?' *Econ. Hist. Rev.,* 2nd ser., XXX (1977), pp. 429–41.
Crafts, N. F. R., 'Gross national product in Europe, 1879–1910: some new estimates', *Exp. Econ. Hist.,* 20 (1983), pp. 387–401.
Crafts, N. F. R., 'Patterns of development in nineteenth-century Europe', *Oxf. Econ. Pap.,* 36 (1984), pp. 438–58.
Crafts, N. F. R., *British economic growth during the industrial revolution* (Oxford, 1985).
Crafts, N. F. R., 'British industrialization in an international context', *J. Interdisc. Hist.,* 19 (1989), pp. 415–28.
Crafts, N. F. R., 'Revealed comparative advantage in manufacturing, 1899–1950', *J. Eur. Econ. Hist.,* 18 (1989), pp. 127–38.
Crafts, N. F. R. and Broadberry, S. N., eds., *European productivity in the twentieth century, Oxf. Bull. Econ. & Stat.,* 52(4) (Special issue) (1990).
Crafts, N. F. R., Leybourne, S. J., and Mills, T. C., 'Britain', in R. Sylla and G. Toniolo, eds., *Patterns of European industrialization in the nineteenth century* (1991), pp. 124–52.
Crisp, O., *Studies in the Russian economy before 1914* (1976).
Crisp, O., 'Russia', in R. Sylla and G. Toniolo, eds., *Patterns of European industrialization in the nineteenth century* (1991), pp. 201–62.
Crouzet, F., 'Wars, blockades and economic change in Europe, 1792–1815', *J. Econ. Hist.,* XXIV (1964), pp. 567–90.
Crouzet, F., 'Western Europe and Great Britain: "catching up" in the first half of the nineteenth century', in A. J. Youngson, ed., *Economic development in the long run* (1972), pp. 98–125.
Crouzet, F., 'Britain and France in the eighteenth century: criticisms and replies to the comparison', in *idem, Britain ascendant: comparative studies in Franco-British economic history* (Cambridge, 1990), pp. 12–104.
Davis, J. A., 'Technology and innovation in an industrial latecomer: Italy in the nineteenth century', in P. Mathias and J. Davis, eds., *Innovation and technology in Europe from the eighteenth century to the present day* (Oxford, 1991), pp. 83–106.
Davis, L. and Huttenback, R. A., *Mammon and the pursuit of empire* (Cambridge, 1986).
Day, C. R., 'The making of mechanical engines in France, 1803–1914', *French Hist. Stud.,* 10 (1977–8), pp. 439–60.
De Rosa, L., 'Urbanization and industrialization in Italy, 1861–1921', *J. Eur. Econ. Hist.,* 17 (1988), pp. 467–90.

De Vries, J., *The economy of Europe in an age of crisis* (Cambridge, 1976).
De Vries, J., *European urbanization, 1500–1800* (1984).
Dowrick, S. and Gemmell, N., 'Industrialization, catching up and economic growth: a comparative study across the world's capitalist economies', *Econ. J.*, 101 (1975), pp. 263–75.
Drummond, I., 'The Russian gold standard, 1899–1914', *J. Econ. Hist.*, XXXVI (1976), pp. 663–88.
Dumke, R., 'The political economy of German unification: tariffs, trade and politics of the Zollverein era', *J. Econ. Hist.*, XXXVIII (1978), pp. 277–8.
Dyson, K. H. F., *The state tradition in western Europe: a study of an idea and an institution* (Oxford, 1980).
Eddie, Scott M., 'Economic policy in Austria-Hungary, 1867–1913', in P. Mathias and S. Pollard, eds., *Cambridge economic history of Europe*, 8 (1989), pp. 814–86.
Egge, A., 'Transformation of bank structures in the industrial period: the case of Sweden, 1820–1913', *J. Eur. Econ. Hist.*, 12 (1983), pp. 271–94.
Ekelund, R. B. and Tollison, R. D., *Mercantilism as a rent seeking society* (College Station, Tex., 1981).
Evans, P., Dietrich, R., and Skocpol, T., *Bringing the state back* (1985).
Falkus, M., *The industrialization of Russia, 1700–1914* (1972).
Federico, G. and Toniolo, G., 'Italy', in R. Sylla and G. Toniolo, eds., *Patterns of European industrialization in the nineteenth century* (1991), pp. 197–217.
Feinstein, C. H., 'Economic growth since 1870: Britain's performance in international perspective', *Oxf. Rev. Econ. Pol.*, 4 (1988), pp. 1–13.
Fenoaltea, S., 'The extractive industries in Italy, 1861–1913', *J. Eur. Econ. Hist.*, 17 (1988), pp. 117–26.
Field, A. J., 'The problem with neoclassical institutional economics', *Exp. Econ. Hist.*, 18 (1981), pp. 174–98.
Fischer, W., McInnis, R., and Schneider, J., *The emergence of a world economy, 1500–1914* (1986)
Flinn, M., *European demographic system, 1500–1800* (Brighton, 1981).
Foreman-Peck, J., *A history of the world economy: international economic relations since 1850* (Brighton, 1983).
Fraser, D. and Sutcliffe, A., eds., *The pursuit of urban history* (1983).
Frank, G., *World accumulation, 1492–1789* (1978).
Freedman, C., 'Joint stock business organization in France, 1807–67', *Bus. Hist. Rev.*, 39 (1965), pp. 184–204.
Freedman, C., *Joint stock enterprise in France, 1807–67* (Chapel Hill, N.C., 1979).
Freedman, C., 'Cartels and the law in France before 1914', *French Hist. Stud.*, 15 (1988), pp. 462–78.
Fremdling, R., 'The development of the iron industry in western Europe, 1820–60: a comparative view of the adaptation of coke smelting and puddling in Belgium, France, and Germany', in L. Jorberg and N. Rosenberg, eds., *Technical changes, employment and investment* (Budapest, 1982), pp. 111–21.
Fremdling, R., 'Foreign trade patterns, technical change, cost and productivity in west European iron industries, 1820–70', in R. Fremdling and P. K. O'Brien, eds., *Productivity in the economies of Europe* (Stuttgart, 1983), pp. 152–74.
Fremdling, R., 'Productivity comparison between Great Britain and Germany', *Scand. Econ. Hist. Rev.*, 39 (1991), pp. 28–42.

Fremdling, R., 'Foreign competition and technological change: British exports and the modernization of the German iron industry', in W. R. Lee, ed., *German industry and German industrialization* (1991), pp. 47–76.

Fremdling, R. and O'Brien, P. K., eds., *Productivity in the economies of Europe* (Stuttgart, 1983).

Fremdling, R. and Tilly, R., 'German banks and German growth and economic history', *J. Econ. Hist.*, XXXVI (1976), pp. 416–24.

Galassi, F. L., 'Reassessing Mediterranean agriculture: retardation and growth in Tuscany, 1870–1914', *Rev. di Storia Econ.*, 3 (1986), pp. 91–121.

Gatrell, P., *The Tsarist economy, 1850–1917* (1986).

Gerschenkron, A., 'Economic backwardness in historical perspective', in *idem, Economic backwardness in historical perspective: a book of essays* (Cambridge, Mass., 1962), pp. 5–30.

Gerschenkron, A., *Continuity in history and other essays* (Cambridge, Mass., 1968).

Goldsmith, J., 'The agrarian history of pre-industrial France', *J. Eur. Econ. Hist.*, 13 (1984), pp. 175–98.

Goldsmith, R., *Financial structure and economic development* (New Haven, 1969).

Good, D., 'Backwardness and the role of banking in nineteenth century European industrialization', *J. Econ. Hist.*, XXXIII (1973), pp. 845–50.

Good, D. F., *The economic rise of the Habsburg empire* (Berkeley, 1984).

Good, D. F., 'Uneven development: the Habsburg empire and the United States', *J. Econ. Hist.*, XLVI (1986), pp. 137–51.

Good, D. F., 'Austria', in R. Sylla and G. Toniolo, eds., *Patterns of European industrialization in the nineteenth century* (1991), pp. 197–217.

Goodman, J. and Honeyman, K., *Gainful pursuits* (1988).

Goubert, P., *The French peasantry in the seventeenth century* (Cambridge, 1989).

Graff, H. J., *The legacies of literacy: continuities and contradictions in western culture* (Bloomington, 1987).

Grantham, G., 'Agricultural supply during the industrial revolution: French evidence and European implications', *J. Econ. Hist.*, XLIX (1989), pp. 43–72.

Grantham, G., 'The persistence of open field farming in France', *J. Econ. Hist.*, XL (1980), pp. 515–31.

Grantham, G., 'Urban provisioning zones before the industrial revolution' (unpub. paper, McGill Univ., 1992).

Gregory, P., 'The Russian balance of payments, the gold standard and monetary policy', *J. Econ. Hist.*, XXXIX (1979), pp. 379–99.

Gregory, P., *Russian national income, 1885–1913* (Cambridge, 1982).

Gregory, P. R., 'The role of the state in promoting economic development: the Russian case and its general implications', in R. Sylla and G. Toniolo, eds., *Patterns of European industrialization in the nineteenth century* (1991), pp. 64–79.

Gregory, P. and Sailors, J., 'Russian monetary policy and industrialization', *J. Econ. Hist.*, XXXVI (1976), pp. 836–51.

Griffiths, R., 'The creating of a national Dutch economy, 1795–1909', *Tijdschrift voor Geschiedenis*, 95 (1982), pp. 513–37.

Guha, A. S., *An evolutionary view of economic growth* (Oxford, 1981).

Gutmann, M. P., *Towards the modern economy: early industrialization in Europe* (Philadelphia, 1988).

Hacken, R. D., *Central European economic history, 1815–75: a bibliography* (New York, 1987).

Hanssen, S. A., *Early industrialization in Denmark* (Copenhagen, 1970).
Harling, P. and Mandler, P., 'From fiscal military state to laissez-faire state, 1760–1850', *J. Brit. Stud.*, 32 (1993), pp. 44–70.
Harris, J. R., 'Michael Alcock and the transfer of Birmingham technology to France before the industrial revolution', *J. Eur. Econ. Hist.*, 15 (1986), pp. 7–58.
Harrison, J., 'The economic history of Spain since 1800', *Econ. Hist. Rev.*, 2nd ser., XLII (1990), pp. 79–89.
Hartwell, R. M., *The industrial revolution and economic growth* (1971).
Heaton, H., *Economic history of Europe* (New York, 1948).
Heikkinan, S. and Hjerppe, R., 'The growth of Finnish industry in 1860–1913: causes and linkages', *J. Eur. Econ. Hist.*, 16 (1987), pp. 227–44.
Henderson, W. O., *The industrial revolution on the continent: Germany, France and Russia, 1800–1914* (1961).
Hertner, P. and Jones, G., eds., *Multinational theory and history* (1986).
Heywood, C., 'The role of the peasantry in French industrialization', *Econ. Hist. Rev.*, 2nd ser., XXXIV (1981), pp. 359–76.
Heywood, C., *The development of the French economy, 1750–1914* (1992).
Hodne, F., 'Growth in a dual economy: the Norwegian experience, 1814–1914', *Econ. & Hist.*, 16 (1973), pp. 81–110.
Hodne, F., *An economic history of Norway, 1815–70* (Bergen, 1975).
Hodne, F., *The Norwegian economy, 1920–80* (1983).
Hoffman, P. T., 'Land rents and productivity in the Paris basin, 1450–1789', *J. Econ. Hist.*, LI (1991), pp. 771–806.
Hoffman, P. T., 'Institutions and agriculture in old regime France,' *J. Theoret. & Instit. Econ.*, 145 (1989), pp. 166–88.
Hoffman, W., *The growth of industrial economies* (Manchester, 1958).
Hohenberg, P. and Lees, L., *The making of urban Europe, 1000–1950* (Cambridge, Mass., 1985).
Holton, R. J., *The transition from feudalism to capitalism* (1985).
Holtfrerich, C. L., 'The growth of net domestic product in Germany, 1850–1913', in R. Fremdling and P. K. O'Brien, eds., *Productivity in the economies of Europe* (Stuttgart, 1983), pp. 124–32.
Honeyman, K. and Goodman, J., 'Women's work, gender conflict and labour markets in Europe, 1500–1900', *Econ. Hist. Rev.*, XLIV (1991), pp. 608–28.
Hoselitz, B., 'Theories of the stages of economic growth', in *idem*, ed., *Theories of economic growth* (New York, 1960), pp. 193–238.
Hoselitz, B., 'Some problems in the quantitative studies of industrialization', *Econ. Dev. & Cult. Change*, 9 (1961), pp. 437–49.
Hudson, P. and Lee, R., eds., *Women's work and family economy in historical perspective: women in the workforce* (Manchester, 1990).
Huetas, T., *Economic growth and economic policy in a multi-national setting: the Habsburg monarchy, 1841–65* (New York, 1977).
Hughes, J., *Industrialization and economic history* (New York, 1970).
Inkster, I., 'Mental capital, transfers of knowledge and technique in eighteenth century Europe', *J. Eur. Econ. Hist.*, 19 (1990), pp. 403–41.
Irwin, D. A. and Nye, J. V., 'Free trade and protectionism in nineteenth century Britain and France', *J. Econ. Hist.*, LIII (1993), pp. 146–53.
Isacson, M. and Magnusson, L., *Proto-industrialization in Scandinavian craft skills in the industrial revolution* (1987).

Israel, J., *Dutch primacy in world trade, 1585–1740* (Oxford, 1989).
Jones, E. L., 'Institutional determinism and the rise of the western world', *Econ. Inquiry,* 12 (1974), pp. 114–24.
Jones, E. L., *The European miracle* (Cambridge, 1981).
Jones, E. L., *Growth recurring: economic change in world history* (Oxford, 1988).
Jorberg, L., 'Structural change and economic growth: Sweden in the nineteenth century', *Econ. & Hist.,* 8 (1965), pp. 3–46.
Kaelble, H., 'Was Prometheus most unbound in Europe? The labour force in Europe during the late nineteenth and twentieth centuries', *J. Eur. Econ. Hist.,* 18 (1989), pp. 65–104.
Kahan, A., *Russian economic history in the nineteenth century* (Chicago, 1989).
Koergärd, N., 'The industrial development of Denmark, 1840–1913', *J. Eur. Econ. Hist.,* 19 (1990), pp. 271–92.
Kemp, T., *Industrialization in nineteenth century Europe* (1969).
Kendrick, J. and Vacara, B., eds., *New developments in productivity measurements and analysis* (Chicago, 1980).
Kindleberger, C., 'Commercial expansion and the industrial revolution', *J. Eur. Econ. Hist.,* 4 (1975), pp. 613–54.
Kindleberger, C., 'Financial institutions and economic development: a comparison of Great Britain and France in the eighteenth and nineteenth centuries', *Exp. Econ. Hist.,* 21 (1984), pp. 103–24.
Kindleberger, C., *A financial history of western Europe* (1993).
Kindleberger, C. and di Tella, G., *Economics in the long view: essays in honour of W. W. Rostow,* 3 vols. (1982).
Kisch, H., *From domestic manufactures to industrial revolution* (Oxford, 1989).
Klima, A., 'The beginning of the machine building industry in the Czech lands in the first half of the nineteenth century', *J. Eur. Econ. Hist.,* 4 (1975), pp. 49–78.
Knapp, V. J., 'Europe's textile artisans in the early stages of the industrial revolution', *Stud. Mod. Eur. Hist. & Cult.,* 3 (1977), pp. 62–78.
Knowles, L. C., *Economic development in the nineteenth century: France, Germany, Russia, and the United States of America* (1932).
Kocka, J., 'Capitalism and bureaucracy in German industrialization before 1914', *Econ. Hist. Rev.,* 2nd ser., XXXIV (1981), pp. 453–68.
Komlos, J., 'Economic growth and industrialization in Hungary, 1830–1913', *J. Eur. Econ. Hist.,* 10 (1981), pp. 5–46.
Komlos, J., *The Habsburg monarchy as customs union: economic development in Austria-Hungary in the nineteenth century* (Princeton, 1983).
Komlos, J., *Economic development in the Habsburg monarchy in the nineteenth century* (Boulder, 1983).
Komlos, J., 'Thinking about the industrial revolution', *J. Eur. Econ. Hist.,* 18 (1989), pp. 191–203.
Kooy, P., 'Urbanization: what's in a name', in H. Schmal, ed., *Patterns of European urbanization since 1500* (1981), pp. 62–85.
Krantz, O., 'New estimates of Swedish historical GDP since the beginning of the nineteenth century', *Rev. Income & Wealth,* 34 (1988), pp. 215–42.
Kriedte, P., *Peasants, landlords and merchant capitalists* (Warwick, 1983).
Kriedte, P., Medick, H., and Schlumbohm, J., *Industrialization before industrialization* (Cambridge, 1981).
Kuznets, S., *Modern economic growth rate structure and spread* (New Haven, 1966).

Kuznets, S., *The economic growth of nations* (Cambridge, Mass., 1971).
Lains, P., 'Growth exports and causality in western Europe' (unpub. paper, Universidad Nova de Lisboa, 1991).
Lamb, R. B., *Property markets and the state in Adam Smith's system* (New York, 1987).
Lampe, J., 'Varieties of unsuccessful industrialization in the Balkan states before 1914', *J. Econ. Hist.*, XXXV (1975), pp. 56–85.
Lampe, J., 'Debating Balkan potential for pre-1914 development', *J. Eur. Econ. Hist.*, 12 (1983), pp. 187–96.
Lampe, J. and Jackson, M. R., *Balkan economic history* (Bloomington, 1982).
Landes, D., 'Technological change and industrial development in western Europe, 1750–1914', in H. J. Habakkuk and M. M. Postan, eds., *Cambridge economic history of Europe,* 6 (Cambridge, 1966), pp. 274–310.
Landes, D., *The unbound Prometheus* (Cambridge, 1970).
Lane, F. C., 'The role of governments in economic growth in early modern times', *J. Econ. Hist.*, XXXV (1975), pp. 8–17.
Lebovics, H., *The alliance of iron and wheat in the third French republic* (Baton Rouge, 1988).
Lebrun, P. *et al.*, eds., *Histoire quantitative et développement de la Belgique,* 11 vols. (Brussels, 1979).
Lee, C., *Regional economic growth in the U.K.* (1971).
Lee, R. D., ed., *Population patterns in the past* (New York, 1977).
Lee, W. R., *European demography and economic growth* (1979).
Leon, P. *et al.*, eds., *L'industrialization en Europe au 19me siècle: cartographie et typologie* (Paris, 1972).
Le Roy Ladurie, E., *The French peasantry, 1450–1660* (Cambridge, 1987).
Lévy-Leboyer, M., *Les banques européennes et l'industrialization internationale* (Paris, 1964).
Lévy-Leboyer, M. and Bourgignon, F., *The French economy in the nineteenth century: an essay in econometric analysis* (Cambridge, 1990).
Lévy-Leboyer, M. and Lescure, M., 'France', in R. Sylla and G. Toniolo, eds., *Patterns of European industrialization in the nineteenth century* (1991), pp. 153–4.
Lewis, A., *The theory of economic growth* (1955).
Lewis, A., 'The rate of growth of world trade', in S. Grassman and E. Lundberg, eds., *The world economic order* (1981), pp. 11–74.
Lilley, S., 'Technological progress and the industrial revolution', in C. M. Cipolla, ed., *The Fontana economic history of Europe,* 3 (1973), pp. 187–254.
Liss, P. K., *Atlantic empires: the network of trade and revolution* (Baltimore, 1983).
McCloskey, D. N., *Econometric history* (1987).
Macfarlane, A., *The origins of English individualism* (1978).
Macfarlane, A., 'Capitalism and individualism reconsidered', in *idem, The culture of capitalism* (1987), pp. 170–241.
Maddison, A., *Economic growth in the west* (1964).
Maddison, A., 'A comparison of levels of G.D.P. in developed and developing countries, 1700–1980', *J. Econ. Hist.*, XLIII (1983), pp. 27–43.
Maddison, A., *Dynamic forces in capitalist development* (Oxford, 1991).
Mann, M., *The sources of social power: a history of power from the beginning to A.D. 1760* (Cambridge, 1986).
Mann, M., ed., *States, war and capitalism* (Oxford, 1988).

Marczewski, J., 'Le produit physique de l'économie française de 1789 à 1913 (comparisons avec le Grande Bretagne)', *Cahiers de l'Institut Statistique Economique Appliqué,* (1965).

Mathias, P. and Pollard, S., eds., *Cambridge economic history of Europe,* 8: *The industrial economies* (Cambridge, 1989).

Mathias, P. and Postan, M. M., eds., *Cambridge economic history of Europe,* 7, pts. 1 and 2 (Cambridge, 1978).

Matthews, R. C. O., ed., *Slower growth in the western world* (1982).

Matthews, R. C. O., Feinstein, C. H., and Odling-Smee, J. C., *British economic growth, 1856-73* (Oxford, 1973).

Mendels, F., 'Proto-industrialization: theory and reality. General report', A themes, Eighth International Economic History Congress, Budapest (1982), pp. 69-107.

Meyer, J. *et al., Etudes sur les villes en Europe occidentale (milieu du 17me siècle à la veille de la Révolution Française,* 2 vols. (Paris, 1983).

Milward, A. and Saul, S. B., *The development of the economies of continental Europe, 1850-1914* (1977).

Milward, A. and Saul, S. B., *The economic development of continental Europe* (1979).

Mokyr, J., 'The industrial revolution in the Low Countries in the first half of the nineteenth century: a comparative case study', *J. Econ. Hist.,* XXXIV (1974), pp. 365-91.

Mokyr, J., 'Growing up and the industrial revolution in Europe', *Exp. Econ. Hist.,* 13 (1976), pp. 371-96.

Mokyr, J., 'Industrialization in two languages', *Econ. Hist. Rev.,* 2nd ser., XXXIV (1981), pp. 143-9.

Mokyr, J., *The lever of riches: technological continuity and economic progress* (Oxford, 1990).

Molinas, C. and Prados de la Escosura, L., 'Was Spain different? Spanish historical backwardness revisited', *Exp. Econ. Hist.,* 26 (1989), pp. 385-402.

Morrison, R. J., 'Financial intermediaries and economic development: the Belgian case', *Scand. Econ. Hist. Rev.,* 15 (1967), pp. 56-70.

Munzer, S. R., *A theory of property* (Cambridge, 1990).

Neuberger, H. and Stokes, H., 'German banks and German growth, 1883-1913', and 'Reply', *J. Econ. Hist.,* XXXV (1975), pp. 710-31.

Nielsen, P., 'Aspects of industrial financing in Denmark, 1840-1914', *Scand. Econ. Hist. Rev.,* 31 (1983), pp. 79-108.

North, D. C., 'Government and the cost of exchange in history', *J. Econ. Hist.,* XLIV (1984), pp. 255-64.

North, D. C. and Thomas, R. P., 'An economic theory of the growth of the western world', *Econ. Hist. Rev.,* 2nd ser., XXIII (1970), pp. 1-17.

North, D. C. and Thomas, R. P., *The rise of the western world: a new economic history* (1973).

Nunes, A. B., Mata, E., and Valerio, N., 'Portuguese economic growth, 1855-1913', *J. Eur. Econ. Hist.,* 18 (1989), pp. 291-330.

Nye, J., 'Firm size and economic backwardness: a new look at the French industrialization debate', *J. Econ. Hist.,* XLVII (1987), pp. 649-69.

Nye, J. V., 'The myth of free trade Britain and fortress France: tariffs and trade in the nineteenth century', *J. Econ. Hist.,* LI (1991), pp. 23-46.

Nygren, A., 'Transformation of bank structures in the industrial period: the case of Norway, 1820-1914', *J. Eur. Econ. Hist.,* 12 (1983), pp. 29-68.

O'Brien, P. K., 'Agriculture and the industrial revolution', *Econ. Hist. Rev.*, 2nd ser., XXX (1977), pp. 166–81.
O'Brien, P. K., 'European economic development: the contribution of the periphery', *Econ. Hist. Rev.*, 2nd ser., XXXV (1982), pp. 1–18.
O'Brien, P. K., *Railways and the economic development of western Europe, 1830–1914* (1983).
O'Brien, P. K., 'Transport and economic development in Europe, 1789-1914', in *idem, ed., Railways and the economic development of western Europe, 1830–1914* (1983), pp. 1–27.
O'Brien, P. K., 'Do we have a typology for the study of European industrialization in the nineteenth century?', *J. Eur. Econ. Hist.*, 15 (1986), pp. 291–333.
O'Brien, P. K., ed., *International productivity comparisons and problems of measurement* (Berne, 1986).
O'Brien, P. K., 'The costs and benefits of British imperialism', *P. & P.*, 120 (1988), pp. 163–200.
O'Brien, P. K., 'Modern conceptions of the industrial revolution', in P. K. O'Brien and R. E. Quinault, eds., *The industrial revolution and British society* (Cambridge, 1993), pp. 1–30.
O'Brien, P. K., 'Political preconditions for the industrial revolution', in P. K. O'Brien and R. E. Quinault, eds., *The industrial revolution and British society* (Cambridge, 1993), pp. 124–55.
O'Brien, P. K., 'Government and the economy, 1688–1815', in R. C. Floud and D. N. McCloskey, eds., *The economic history of modern Britain since 1700* (2nd edn., Cambridge, 1994).
O'Brien, P. K., Heath, D., and Keyder, C., 'The efficiency of agriculture in Britain and France', *J. Eur. Econ. Hist.*, 6 (1977), pp. 339–92.
O'Brien, P. K. and Keyder, C., *Economic growth in Britain and France, 1780–1914* (1978).
O'Brien, P. K. and Engerman, S. L., 'Exports and the growth of the British economy from the Glorious Revolution to the Peace of Amiens', in B. L. Solow, ed., *Slavery and the rise of the Atlantic system* (Cambridge, 1991), pp. 177–209.
O'Brien, P. K. and Hunt, P. A., 'The rise of a fiscal state in England, 1485–1815', *Hist. Res.*, 66 (1993), pp. 129–76.
O'Brien, P. K. and Prados de la Escosura, L., 'Agricultural productivity and European industrialization, 1890–1980', *Econ. Hist. Rev.*, XLV (1992), pp. 514–36.
O'Brien, P. K. and Toniolo, G., 'The poverty of Italy and the backwardness of its agriculture before 1914', in B. M. S. Campbell and M. Overton, eds., *Land, labour and livestock: historical studies in European agricultural productivity* (Manchester, 1991), pp. 144–82.
Palairet, M., 'Land, labour and industrial progress in Bulgaria and Serbia before 1914', *J. Eur. Econ. Hist.*, 12 (1983), pp. 163–86.
Parker, W. N. and Jones, E. L., eds., *European peasants and their markets* (1975).
Parker, W. N., 'The pre-history of the nineteenth century', in *idem, Europe, America and the wider world: essays on the economic history of western capitalism*, 1 (Cambridge, 1984), pp. 15–54.
Payne, P. L., 'Industrial entrepreneurship, and management in Great Britain', in P. Mathias and M. M. Postan, eds., *Cambridge economic history of Europe*, 7 (Cambridge, 1978), pp. 180–230.
Persson, K. G., *Preindustrial economic growth, social organization and technical progress in Europe* (Oxford, 1988).

Phillips, C. R., 'The growth and composition of trade in the Iberian empires, 1450–1750', in J. Tracy, ed., *The rise of merchant empires* (Cambridge, 1990), pp. 34–101.
Pierenkemper, T., 'Entrepreneurs in heavy industry: Upper Silesia and the Westphalian Ruhr region, 1852–1913', *Bus. Hist. Rev.*, 53 (1979), pp. 65–78.
de Pinedo, E., 'From bloomery to blast furnace: technical change in Spanish iron machinery, 1658–1822', *J. Eur. Econ. Hist.*, 17 (1988), pp. 7–32.
Poggi, G., *The state: its nature, development and prospects* (Oxford, 1991).
Pohl, H., ed., *The European discovery of the world and its economic effects on pre-industrial society* (Stuttgart, 1990).
Pollard, S., 'Industrialization and the European economy', *Econ. Hist. Rev.*, 2nd ser., XXVI (1973), pp. 636–48.
Pollard, S., *Typology of industrialization processes in the nineteenth century* (1990).
Pollard, S., *Peaceful conquest: the industrialization of Europe, 1760–1970* (Oxford, 1981).
Pounds, N. J. G., *An historical geography of Europe, 1500–1840* (Cambridge, 1979).
Pounds, N. J. G., *An historical geography of Europe, 1800–1914* (Cambridge, 1985).
Prados de la Escosura, L., 'Foreign trade and the Spanish economy during the nineteenth century', in N. Sanchez-Albornoz, ed., *The economic modernization of Spain, 1830–1930* (New York, 1987), pp. 128–50.
Prados de la Escosura, L., 'Spain's gross domestic product, 1850–1990: a new series' (Madrid: Ministerio de Economía y Hacienda, *Documentos de Trabajo*, D93002, 1993).
Reddy, W. M., *The rise of market culture: the textile trade and French society, 1750–1900* (Cambridge, 1984).
Redlich, F., 'The leaders of the German steam engine industry during the first hundred years', *J. Econ. Hist.*, IV (1944), pp. 121–48.
Reis, J., 'Industrial development in a late and slow developer: Portugal, 1870–1913', *Rivista di Storia Economica*, 3 (1986), pp. 67–90.
Riley, J., 'The Dutch economy after 1650', *J. Eur. Econ. Hist.*, 13 (1984), pp. 521–69.
Rosenberg, N. and Birdzell, L. E., *How the west grew rich* (1986).
Rostow, W. W., ed., *The economics of the take off into self sustained growth* (1963).
Rostow, W. W., 'The take-off into self-sustained growth', in B. Supple, ed., *The experience of economic growth* (1963), pp. 81–110.
Rostow, W. W., *The stages of economic growth* (Cambridge, 1965; 2nd edn. 1971).
Rowstow, W. W., *How it all began: origins of the modern economy* (New York, 1975).
Rostow, W. W., *The world economy: history and prospect* (1980).
Rudolph, R. L., *Banking and industrialization in Austria-Hungary* (Cambridge, 1976).
Sanchez-Albornoz, N., 'Convergence among Spanish economic regions in the nineteenth century', *J. Eur. Econ. Hist.*, 3 (1974), pp. 725–46.
Sanchez-Albornoz, N., ed., *The economic modernization of Spain, 1830–1930* (New York, 1987).
Sandberg, L., 'Banking and economic growth in Sweden before World War I', *J. Econ. Hist.*, XXXVIII (1978), pp. 22–41, 650–80.
Sandberg, L., 'The case of the impoverished sophisticate: human capital and Swedish economic growth before World War I', *J. Econ. Hist.*, XXXIX (1979), pp. 225–42.
Sandberg, L., 'Ignorance, poverty and economic backwardness', *J. Eur. Econ. Hist.*, 11 (1982), pp. 675–98.

Saul, S. B., 'The nature and diffusion of technology', in A. J. Youngson, ed., *Economic development in the long run* (1972), pp. 61–85.
Scandinavian Economic History Review. Special issue on proto-industrialization, vol. 30 (1982).
Scholliers, P., ed., *Real wages in nineteenth and twentieth century Europe* (Oxford, 1989).
Schremmer, E., 'Proto industrialization: a step towards industrialization', *J. Eur. Econ. Hist.,* 10 (1981), pp. 653–70.
Senghaas, D., *The European experience: a historical critique of development theory* (Leamington Spa, 1982).
Sewell, W. H., 'Marc Bloch and the logic of comparative history', *Hist. & Theory,* 6 (1967), pp. 208–18.
Sheridan, G., 'The political economy of artisan industry: government and people in the silk trade at Lyons, 1830–70', *French Hist. Stud.,* 11 (1979–80), pp. 215–38.
Shiel, R., 'Improving soil productivity in the pre-fertilizer era', in B. M. S. Campbell and M. Overton, eds., *Land, labour, livestock: historical studies in European agricultural productivity* (Manchester, 1991), pp. 51–78.
Skocpol, T., ed., *Vision and method in historical sociology* (Cambridge, Mass., 1984).
Skocpol, T. and Somers, M., 'The use of comparative history in macro social inquiry', *Comp. Stud. Soc. & Hist.,* 22 (1980), pp. 174–97.
Smith, M., 'Free trade versus protectionism in the early Third Republic', *French Hist. Stud.,* 10 (1977–8), pp. 293–314.
Soderberg, J., 'A long term perspective on regional economic development in Sweden *c.* 1550–1914', *Scand. Econ. Hist. Rev.,* 32 (1984), pp. 1–16.
Soderberg, J., 'Regional economic disparity and dynamics, 1840–1914: a comparison between France, Great Britain, Prussia and Sweden', *J. Eur. Econ. Hist.,* 14 (1985), pp. 273–98.
Solomou, S., *Phases of economic growth: Kondratieff waves and Kuznets swings* (Cambridge, 1987).
Solow, B. L., ed., *Slavery and the rise of the Atlantic system* (Cambridge, 1991).
Solow, B. L., 'Slavery and colonization', in *idem,* ed., *Slavery and the rise of the Atlantic system* (Cambridge, 1991), pp. 21–42.
Solow, B. L. and Engerman, S. L., eds., *British capitalism and Caribbean slavery: the legacy of Eric Williams* (Cambridge, 1987).
Solow, R. M. and Temin, P., 'Introduction: the inputs for growth', in P. Mathias and M. M. Postan, eds., *Cambridge economic history of Europe,* 7 (Cambridge, 1978), pp. 1–27.
Stoianovich, T., *French historical methods: the Annales paradigm* (Ithaca, 1976).
Supple, B. E., 'The state and the industrial revolution, 1700–1914', in C. M. Cipolla, ed., *The Fontana economic history of Europe,* 3 (1973), pp. 301–57.
Supple, B. E., 'Revisiting Rostow', *Econ. Hist. Rev.,* 2nd ser., XXXVII (1984), pp. 107–14.
Sylla, R., 'The role of banks', in R. Sylla and G. Toniolo, eds., *Patterns of European industrialization in the nineteenth century* (1991), pp. 45–63.
Sylla, R. and Toniolo, G., eds., *Patterns of European industrialization in the nineteenth century* (1991).
Sylla, R. and Toniolo, G., 'Introduction: patterns of European industrialization in the nineteenth century', in R. Sylla and G. Toniolo, eds., *Patterns of European industrialization in the nineteenth century* (1991), pp. 1–26.

Szostak, R., *The role of transportation in the industrial revolution: a comparison of England and France* (Montreal, 1991).
Tann, J. and Breckin, M. J., 'The international diffusion of the Watt engine, 1775–1820', *Econ. Hist. Rev.*, 2nd ser., XXXI (1978), pp. 541–64.
Tilly, C., ed., *The formation of states in western Europe* (Princeton, 1975).
Tilly, C., ed., *Historical studies in changing fertility* (Princeton, 1978).
Tilly, C., *Capital, coercion and states, A.D. 900–1900* (Oxford, 1990).
Tilly, R., 'Capital formation in Germany in the nineteenth century', in P. Mathias and M. M. Postan, eds., *Cambridge economic history of Europe,* 7, pt. 1 (Cambridge, 1978), pp. 382–441.
Tilly, R., 'The take off in Germany' (unpub. paper, Institüt für Wirtschafts und Sozialgeschichte, Münster, 1980).
Tilly, R., 'Mergers, external growth and finance in the development of large scale finance in Germany', *J. Econ. Hist.*, XLII (1982), pp. 629–58.
Tilly, R., 'German banking, 1830–1914: development assistance for the strong', *J. Eur. Econ. Hist.,* 15 (1986), pp. 113–52.
Tilly, R., 'German industrialization and Gerschenkronian backwardness', *Rivista di Storia Economica,* 6 (1989), pp. 139–64.
Tipton, F., *Regional variations in the economic development of Germany* (Middleton, Conn., 1976).
Toniolo, G., *An economic history of liberal Italy, 1850–1918* (1990).
Tortella, G., ed., *Education and economic development since the industrial revolution* (Valencia, 1990).
Tortella, G. and Prados de la Escosura, L., 'The pattern of economic modernization in southern Europe' (unpub. paper, Dept. of Econ. Hist., Universidad de Alcalá de Henares, 1984).
Toutain, J. C., 'Le produit intérieur brut de France de 1789 à 1982', *Economies et Sociétés,* 15 (1987), pp. 66–87.
Tracy, J. D., ed., *The rise of merchant empires* (Cambridge, 1990).
Tracy, J. D., ed., *The political economy of merchant empires* (Cambridge, 1991).
Trebilcock, C., *The industrialization of the continental powers, 1780–1914* (1981).
Tyszinski, H., 'World trade in manufactured commodities, 1899–1950', *Man. School,* 19 (1951), pp. 272–304.
Van Ark, B., 'Comparative levels of manufacturing labour productivity in postwar Europe', *Oxf. Bull. Econ. & Stat.*, 52 (1990), pp. 343–74.
Vandenbroeke, C., 'The regional economy of Flanders and industrial modernization in the eighteenth century', *J. Eur. Econ. Hist.,* 16 (1987), pp. 149–70.
Van der Wee, H., *The rise and decline of urban industries in Italy and the Low Countries* (Leuven, 1988).
Van der Woude, A., de Vries, J., and Hayami, A., eds., *Urbanization in history: a process of dynamic interaction* (Oxford, 1990).
Van Zanden, J. L., 'The first green revolution: the growth of productivity in European agriculture, 1870–1914', *Econ. Hist. Rev.,* XLIV (1991), pp. 215–39.
Ville, S. P., *Transport and the development of the European economy, 1750–1918* (Basingstoke, 1990).
Von Tunzelmann, G. N., 'Technological and organizational change in industry', in P. K. O'Brien and R. E. Quinault, eds., *The industrial revolution and British society* (Cambridge, 1993), pp. 254–82.

de Vries, J., 'The decline and rise of the Dutch economy, 1675–1900', *Res. Econ. Hist.*, supplement 3 (1984), pp. 149–89.
Wade, W. W., *Institutional determinants of technical change and agricultural productivity growth* (New York, 1981).
Wallerstein, I., *The modern world system,* 3 vols. (New York, 1974, London, 1980, San Diego, 1989).
Webb, S. B., 'Agricultural protection in Wilhelmian Germany', *J. Econ. Hist.,* XLII (1982), pp. 309–26.
Westebbe, R., 'State entrepreneurship: King Willem I, John Cockerill and the Seraing engineering works', *Exp. Entrepren. Hist.,* 8 (1955), pp. 205–32.
Williamson, J. G., 'Regional inequality and the process of national development', *Econ. Dev. & Cult. Change* (Special issue, July 1965).
Williamson, J. G., 'Greasing the wheels of sputtering export engines, mid-western grains and American growth', *Exp. Econ. Hist.,* 17 (1980), pp. 211–7.
Woolf, E., *Europe and the people without a history* (Berkeley, 1982).
Wrigley, E. A., *Industrial growth and population change* (Cambridge, 1960).
Wrigley, E. A., 'Urban growth and agricultural change: England and the continent in the early modern period', *J. Interdisc. Hist.,* XV (1985), pp. 683–728.
Zamagni, V., *Industrializione esquilibri regionali in Italia* (Bologna, 1978).
Zamagni, V., ed., *Finance and the enterprise* (San Diego, 1992).

The Industrial Revolution in Europe

ARTICLES

Foreign Trade and the Industrialization of the European Periphery in the XIXth Century

I.T. Berend
University of Budapest

G. Ranki
Historical Institute of the
Hungarian Academy of Sciences, Budapest

Analysis of foreign trade and the role of export sectors reveals both the manner and the degree to which the more backward regions became absorbed into the world economy, as well as giving some clearer picture of the nature of this process. As Marx said: ' Modern capitalism started with world trade and the world market '.[1] And in terms of the theories of comparative benefits and international specialisation, which were fundamental assumptions of XIXth century economic liberalism, it was foreign trade that provided the principal guarantee of harmonious development. In Marx's economic theory considerable emphasis was laid on the advantages of specialization and the benefits of foreign trade,[2] but the notion of a harmony of interests was firmly rejected and Marx pointed to the contradictions inherent in a unified world capitalist system and the unevenness of the levels of actual and possible integration within it.

More recently the economists of the New Left, in dealing

1 KARL MARX, *Capital*, Vol. I, Progress Publishers, Moscow, 1977, p. 145.

2 " It is... impossible to conceive a capitalist nation without foreign trade...", *Lenin Collected Works*, Vol. 3, Progress Publishers, Moscow, 1972, p. 65.

with the predicament of the Third World, have tended to reject both liberal economics and Marx's interpretation, arguing that the absorption of backward countries into the world market does nothing other than form a peripheral zone for the world economy. This perpetuates a wholly one-sided process of accumulation, in which foreign trade becomes the means whereby the economies of the backward, or belatedly developing, countries are exploited and subordinated.[3]

The situation of the backward regions of XIXth century Europe provides us with a particularly valuable historical case against which to test these theories. In fact, the case of the European periphery provides one of the very rare examples of the entry of relatively backward zones into the world market that can be studied historically. It allows us both to examine the impact of contact with the capitalist world economy on a backward region and to set this process in historical perspective. Although the volumes and the composition of the foreign trade of the countries we shall be examining varied, there can be no doubt that they all shared the features generally characteristic of the trade relations between underdeveloped and developed countries. In all these countries, the products exported were predominantly raw materials, and agricultural or food products. Their imports, in turn, were overwhelmingly manufactured goods.

What was the effect of this type of foreign trade on the more backward areas? In seeking to answer this question, the first thing we must bear in mind is that economic change and the growing specialization of labour in both the domestic and international contexts in the course of the XIXth century meant that foreign trade had a rate of growth that far surpassed industrial production and national income.

---

[3] See, among others, SAMIR AMIN, *L'accumulation à l'échelle mondiale*, Paris, 1970; A. G. FRANK, "Multilateral Merchandise Trade Inbalances and Uneven Economic Development", *The Journal of European Economic History*, Vol. 5, No. 2, 1976.

Paul Bairoch's latest calculations give the following picture:

Europe's exports as % of GNP

1830	4.0
1860	9.4
1890	12.6
1913	14.0

This meant that the value of exports — calculated at current prices — had jumped from 645 million dollars to 10,550 million dollars, that is, had grown twenty-six-fold. These data alone indicate that foreign trade was one of the chief means of expansion and chief indicators of the development of a capitalist world economy. Looking at the foreign trade figures, we might well regard the economic development of the countries of the European periphery as an example of demand-oriented growth; something that even Adam Smith had considered to be a very important factor of economic growth. A number of economists do consider the economic development of even the leading industrial nations to have been due primarily to the rapid growth of international trade. Such a view is mainly supported by data indicating that in England, too, it was in the so-called export sectors that production growth was the most rapid.[4]

Most historians dealing with the subject, however, argue that it was the growth of the domestic market that played a decisive role in the transformation of the leading industrial nations, all the more so because foreign trade — since it depends on certain determinate natural preconditions — is a function of a specific level of domestic production. And we could hardly account for the coincidence of demand and supply without examining other production factor trends, such as capital and labour, for instance.

---

[4] See, M.R. HARTWELL, "The Causes of the Industrial Revolution", *The Economic History Review*, 1969.

I. T. Berend - G. Ranki

But the first question we must tackle in examining the reactions of the countries of the European periphery to those riding on the crest of the first wave of industrialization, is how they had come to enter the stream of international trade, a stream swelled by the industrial revolution. None of the countries of the periphery was cut off from European trade. The Italian city-states had been the centers of world trade in the Middle Ages; Spain and Portugal had dominated it in the XVIth century. The Balkan countries, on the other hand, had only been marginally involved in foreign trade; but involved they were, particularly between the XVIth and XVIIIth centuries. The volume of their trade, however, had been relatively small, not only because the volume of world trade during that period was generally smaller, but also because the Balkans were countries with low productivity, minimal surpluses, and no economic specialization. The goods exported were those typical of the countries of the periphery from the XVIth century on (the time when the exchange of agricultural products for mass industrial goods first started as did the relatively large-scale export of certain industrial raw materials).

The countries of the European periphery that we have been dealing with were, for the most part, exporters of agricultural products to Western Europe. Sweden and Russia also exported significant quantities of brass, iron and wood. However, prior to the XIXth century, their trade cannot, or can but barely, be considered examples of international specialization and as far as the domestic economies were concerned, foreign trade remained supplementary to rather than determinative of, economic development. The fragmentary, and by no means totally reliable, data at our disposal indicate that Russia's share of European trade in the 1830s was relatively large, nearly 8 %. During the same period, the share of the Habsburg Monarchy was nearly 5 %; of Spain, more than 2 %; of Portugal, 1.5 %; of Sweden, more than 1 %. Portugal led the countries of Europe in terms of per capita foreign trade at this time, probably on account of her intensive colonial trade, and her special trade relations witl. England.[5]

The English Industrial Revolution, and its diffusion throughout the western half of the continent created new conditions for European trade. The countries that were less developed but had close economic, political and cultural ties with Western Europe were all influenced by the new circumstances, though both the form this took varied from country to country as did the consequences. Everywhere, however, the way was being cleared for the spread of capitalism and economic specialization. It was this that was largely to determine the development of foreign trade throughout the XIXth century. This had two main features. The first was the growth of trade among the developed countries, though England continued to keep her lead over the rest; the reason here was the growing specialization that more developed production gave rise to, and the exploitation of the advantages each country derived from its own special factors of production. The second characteristic of XIXth century European trade was that the trade between the developed and the underdeveloped nations grew by leaps and bounds. The incentives here were the nearly insatiable need of the former for agricultural products and raw materials, and the fact that trade with the nearest underdeveloped agricultural nations — those of Europe — seemed the most natural way to satisfy these needs.

It was, thus, a natural first step in the world-wide expansion of capitalism to spread the fruits and demands of the industrial revolution to the European periphery. The countries of this area were not only easily accessible geographically, they were themselves anxious to establish contact with the industrialized

---

[5] In the mid-XVIIIth century, 15% of English exports were sold on the Portuguese market. Bairoch calculates that around 1800 Portugal accounted for 5-7% of all European exports-a percentage that seems to us exaggerated. See P. BAIROCH, *Commerce exterieur et developpement économique de l'Europe*, Paris, 1976, p. 266.

nations; and, with the socio-political changes taking place and the bourgeois institutional systems coming into being, they were becoming particularly well equipped to do so. The railways built with foreign investments provided the technical preconditions of their participation in world trade. Europe's backward countries thus became at once the market for Western Europe's growing store of industrial products, and the providers of its supplies of food and raw materials. At a time when the other continents were just starting to join in the new system of world trade initiated by the industrial revolution, except for a few special tropical products, it was mainly South-Eastern Europe, and to a certain degree Northern Europe, which functioned as a peripheral area to the industrial European centre.[6]

There can be no doubt, therefore, that the needs of the developed industrialized countries stimulated an export orientation in these hitherto self-sufficient, and consumption-oriented economies. The scale and consequences of this change, however, were functions of external and domestic factors combined. The geographical proximity of the European periphery — even though the revolution in transport in the XIXth century, the building of the railway network and the consequent drop in transportation costs was to make this a secondary consideration — was in itself a potential advantage. The scope and the effects of foreign trade, however, depended primarily on how far the developed countries' import demands coincided with what the backward countries had to offer. The exporters of crucially important raw materials of high demand elasticity had an advantage — in the given phase of development — over the exporters of agricultural

[6] It is not our task to give a detailed analysis of the ever growing literature dealing with the relationship of the centre and periphery. We would note, however, that one of its weakest points seems to be the failure to account adequately for the relationship between the developed and underdeveloped parts of Europe. Content to regard the underdeveloped parts of Europe as "semi-peripheral" or a belatedly developing adjuncts of the core, most authors avoid the question of why it was that these semi-peripheral areas could later become parts of the core.

products, whose demand elasticity was lower. The domestic repercussions of foreign trade were no less important. The process through which capitalism drew ever newer areas into the international economy, which reached Scandinavia before it did the Balkans, can, indeed, be regarded as being unified.[7] Nonetheless, we can hardly consider the relationship of the first industrialized and the second-comer countries simply in terms of a time-lag, nor see their functional relationships simply as an example of natural specialization based on comparative production costs.

As we have already mentioned in speaking of the pull of the centre, the XIXth century industrial revolution led to a phenomenal rise in import needs. The new machines were insatiable in their consumption of cotton, wool, coal and metals, and domestic agricultural production was unable to meet the needs of the growing industrial population, even with standards of living as low as they were. The industrialization of the centre, therefore, meant extraordinary opportunities for the more backward countries of Europe to increase their exports of agricultural goods and raw materials. Accordingly, agriculture became the chief and most common export branch of the countries of the periphery, these being the countries most directly affected by Western Europe's ever growing demand for imports of foodstuffs. Calculated at constant 1913 prices, Western Europe's per capita food imports grew from an annual $ 6.00 in 1872, to $ 12.80 in 1900, and to $ 13.90 by 1913[8] There was a corresponding growth in the exports of the European periphery: from Scandinavia, from Italy, from Greece and from Spain ever larger quantities of foodstuffs and raw materials were being shipped to England, Germany and Austria.

---

[7] Cf. S. Pollard, "Industrialization and the European Economy", *The Economic History Review*, No. 4, 1973.

[8] Ch. Kindleberger, *Foreign Trade and National Economics*, New Haven, 1962.

6

I. T. Berend - G. Ranki

TABLE 1

RATE OF EXPORT GROWTH,[9]
Annual Average, in %

Country	1860–80	1880–90	1890–1910	1860–1910
Hungary	—	1,6	3,6	2,7 *
Bulgaria	—	9.4	3.3	5.3
Greece	2.1	6.7	2.9	3.3
Romania	—	2.1	3.9	3.3 *
Serbia	—	1.9	4.2	3.4 *
Russia	4.1	3.5	3.6	3.8
Sweden	4.9	3.7	3.2	4.0
Norway	3.3	1.8	3.8	3.2
Finland	6.6	1.6	5.5	5.1
Denmark	7.2	1.8	4.8	5.1
Italy	3.4	— 2.0	4.2	2.6
Spain	4.4	2.9	0.9	2.7
Portugal	2.0	0.5	2.1	1.7
Europe	3.2	1.3	3.2	2.8

* 1880–1910.

The first conclusion we can draw from Table 1 is that in most of the countries concerned, the rate of export growth was faster than the European average. The countries with the fastest growing export rates were Finland and Denmark, both with an annual growth rate of over 5 per cent. At the other pole, we find the countries of the Iberian Peninsula and Italy, with an export growth that was below the European average. These latter countries, then, were unable to join fully in the export boom induced by the Western-Europe center.

The value of the periphery's exports (Table 2) is another indication of the part they played in European trade.

[9] Calculations based on P. BAIROCH, " European Foreign Trade in the XIXth Century: The Development of the Value and Volume of Exports ", *The Journal of European Economic History*, 1973. For the figures on Hungary, see SCOTT EDDIE, " Terms and Patterns of Hungarian Foreign Trade ", *The Journal of Economic History*, June, 1977.

TABLE 2

THE PERIPHERY'S CONTRIBUTION TO EUROPEAN TRADE,[10] Value of Exports, in %

Country	1860	1890	1910
Hungary	—	4,4	3.8
Bulgaria	—	0.1	0.3
Greece	0.2	0.3	0.3
Romania	—	1.1	1.3
Serbia	—	0.2	0.2
Russia	5.6	8.3	8.9
Sweden	1.0	1.8	1.8
Norway	0.7	0.7	0.8
Finland	0.2	0.4	0.6
Denmark	0.5	1.0	1.5
Italy	5.1	3.8	4.5
Spain	2.3	3.4	2.2
Portugal	0.7	0.5	0.4

Here, too, the growth trend is obvious, though in a number of cases — as we shall see — the volume of trade was so small that even a faster rate of growth led to no essential change in the ratios. And here again, it is the countries of Southern Europe that show a declining trend. Portugal's contribution was ever smaller. In Spain we find signs of an export boom between 1860 and 1890, but the 20 years following were much less auspicious. For Italy, on the other hand, the period between 1860 and 1880 was the low point; the 1890s and the first decade of the century brought a gradual upswing.

All this indicates the period as a whole to have been one of

[10] Bairoch did not include Hungary in his calculations, and even when calculating the total volume of European foreign trade, he included the figures referring to the Monarchy as a whole. When we also take into consideration the trade that occurred between the two halves of the Monarchy, we obtain considerably higher figures for the Monarchy's share in European trade than those that he gave. Naturally, we included all of Hungary's foreign trade (i.e. the exports going to the Austrian half of the Monarchy) too.

relatively rapid export growth, with a corresponding rise in the demand for import goods, all of which served to accelerate the development of the economy and its capitalist transformation. Demonstrating the growth of exports and foreign trade, however, is by no means tantamount to establishing how far it contributed to economic development as a whole, or how far it led to the growth through savings and investments. Nor can the social ramifications of these questions be fully discussed. In this connection, we shall merely mention that in many countries the bourgeois transformation was at that time an incomplete, and still ongoing process; the power of the old ruling classes survived in a number of countries, as did the feudal system of big estates. All this served to concentrate a great deal of the income from exports in the hands of a disproportionately small sector of the population; and one which was by no means unequivocally committed to fostering capitalist values.

But to return to the economic aspects of the role of foreign trade, let us examine first of all the " axiom " that a foreign trade structure based on the export of raw materials is, in some sense, indicative of backwardness.[11] Just as common a point of departure is the assumption that in any exchange of food-stuffs and raw materials for manufactured goods, those selling the former are bound to be at a disadvantage, while those selling the latter are bound to profit from the deal. With terms of trade thus favouring the sellers of manufactured goods — the countries of the industrialized centre — the rich were bound to get richer, and the backward were bound to fall even farther behind. There can be no doubt that the XIXth century exchange of agricultural goods and raw materials for industrial products seemed to confirm this assumed regularity. For instance, the world market price for machine products was on average 16 per cent higher in 1900 than it had been in 1872, and rose another 31 per cent

[11] See SAMIR AMIN, *op. cit.*, p. 472.

by 1913. At the same time, grain prices, which had been rising in the third quarter of the XIXth century, fell drastically during the last quarter. Between 1855-64 and 1885-94, the price of wheat on the Copenhagen market fell by a third, the price of rye by nearly a quarter, the price of oats by 10 per cent, while barley prices more or less stagnated. Between 1872 and 1913, the price of wood fell by a sixth, as did the prices of both coal and cake. Raw material prices as a whole fell by an average of one-third between 1872 and 1900. (We must note, however, that it would be mistaken to consider this the inevitable result of some hard and fast " law ". Grain prices, as we have seen, did rise for a time, and the price of animal products rose throughout the period.) The facts indicate that it was not *generally* true that trade terms were to the advantage of the countries of the centre, and disadvantageous for the countries of the periphery. Who benefited most was always the result of a combination of factors, the nature of the actual goods being imported and exported, and of actual world market price trends.

There can be no doubt, however, that between 1860 and 1914, a decisive half century from the point of view of the European periphery, England's terms of trade improved by 20%: that is, for the same volume of exports, England at the end of the period was getting a fifth more imports than she had fifty years earlier. Belgium was another country of the centre whose trade showed similar trends. (Yet even here, we should note that in the previous forty years between 1820 and 1860, when England's industrial monopoly was much more decisive, her terms of trade had followed an unfavourable trend, and, by 1860, England had to export 50% more than in 1820 for the same volume of import goods.[12]) For France and Germany, the prewar decades brought poor trade terms. Between 1880 and 1913, for instance, Germany's terms of trade deteriorated by no less than 20%.

[12] Br. Mitchell-Ph. Deane, *British Historical Statistics*, Cambridge, 1952, pp. 331-332.

(The fact that chemical product prices in 1900 were only 25% of 1872 prices, and by 1913 only 45% of the 1900 prices obviously had a great deal to do with this).

During the same period, the terms of trade of Sweden, an exporter of foodstuffs and raw materials, improved by 50%, and Denmark's too became more favourable; Norway's terms of trade, however, were damaged by the fall in revenues lost as a consequence of the fall in shipping costs. Lãszló Katus' calculations for Hungary indicate that the Monarchy's protectionist agricultural policy after the turn of the century brought similar improvements in her terms of trade.[13] However, the situation was quite different in the case of the countries of the Iberian Peninsula.

Certain general trends notwithstanding, therefore, a country's profits or losses from fôreign trade were a function of its particular circumstances. A wheat exporting country was obviously worse off selling at world market prices than a country exporting meat and other animal products (Hungary, which sold her wheat to the Monarchy, was a special case); yet, both were " exporters of agricultural products ". In the same way, machine exporters were at an advantage over exporters of chemical products — for which the terms of trade were unfavourable indeed — though both were cases of " industrial export ".

All this indicates that the losses or gains sustained in the course of trade did not simply follow from a country's central or peripheral position, but were the consequence of the given country's particular production and export activity. Just what these were, however, was very much determined by a country's

---

[13] J. PEDERSEN-O. PEDERSEN, *An Analysis of Price Behaviour*, Copenhagen, 1938; W. HOFFMANN, *Das Wachstum der deutschen Wirtschaft seit der Mitte des XIX. Jarhhunderts*, Berlin, 1965; C. KINDLEBERGER, *The Terms of Trade*, New York, 1956; L. KATUS, the chapter on economic development for Vol. VII of the 10-volume history of Hungary, Budapest, 1978; SCOTT EDDIE, " The Terms and Patterns of Hungarian Foreign Trade 1882-1913 ", *The Journal of Economic History*, No. 2. 1977.

ability — or inability — to adapt to the demands of the market. This element of flexibility was, in fact, a crucial factor in determining whether foreign trade became the means of a country's subordination to a more developed economy, or whether conversely it became the incentive for the country itself to develop a more advanced economic structure. What was the case in any given country can be determined only through a complex analysis of all the factors of both the international and the internal economic system.[14] In this respect, we must agree with H. Mynt: " The traditional concept of « terms of trade » is no adequate measure of the advantages of international trade, any more than is the distribution of the income derived from it. We must also take into consideration the growth of economic activity, the changes in employment and level of technological knowhow induced by derivative investments, and all the other dynamic incentives originating from the growth of the country's volume of trade." [15]

That this is indeed so is demonstrated by the fact that a great many of the developed countries exported foodstuffs and raw materials in the initial phase of their development, and often

---

[14] This is why we find the concept of " dependence " as it is used in a great many works treating the relationship of core and periphery too general. The factors making for dependence are seen by D. Senghaas to be the following:

a) the inequalities already present at the take-off point,

b) the inequalities in structure (in raw materials, industry, trade; whether the export structure is one-sided or diversified),

c) the one-sided effects of financial successes,

d) unequal development opportunities.

D. SENGHAAS, *Imperialismus und strukturelle Gewalt. Analysen über abhängige Reproduktion*, Frankfurt, 1972, p. 23.

[15] H. MYINT: *An Interpretation of Economic Backwardness in the Economics of Underdevelopment*, Ed. by A.N. Agarwala - S.P. Singh Oxford, 1958. S. KUZNETS, (*Modern Economic Growth*, New Haven-London, 1966, p. 302) adds that dependence on foreign trade is a function not only of the foreign trade-national income ratio, but also of the relative weight of the key sectors, or rather, the degree to which foreign trade is diversified. The advantages to be derived from foreign trade depend to a great extent on how far the goods imported can be substituted for, and on the mobility of the productive factors invested in export goods.

I. T. Berend - G. Ranki

much later too, in the effort to speed up capital accumulation. In a number of countries, then, it was the development of the raw-material based export sectors that laid the groundwork for the structural transformation of industry, and for an overall economic boom. The question, therefore, is not whether the export of foodstuffs and raw materials *in itself* leads to favourable or unfavourable terms of trade for a given area, but rather whether the countries of an area become trapped in the role of raw material exporters, or else are able to go on from there to build up a suitably developed economic structure.[16]

The ability to build up such a developed industrial structure, to respond to the Western stimulus, and to compete successfully with the more developed countries was, naturally, the result of the complex interaction of a number of factors. The wealth of a country's natural and geographic resources, its accessibility, its communications network, the antecedents of its proto-industrialization, its chances of capital accumulation and of adopting foreign technology, and last but not least, government policy, were all decisive factors of the response to the West.[17] However, we

---

[16] It is instructive, in this respect, to look at the export structure of Canada and the United States:

	Food, raw material			Food and semi-finished products		Industrial goods	
	USA	Canada		Canada	USA	USA	Canada
1850	68	—		—	19	13	—
1868–70 *	54		95		29	16	5
1914 **	42		87		33	26	13
1926–1929	31	47		30	24	45	23
1936–1939	23	32		40	25	52	28

* In USA 1861-1870; ** In USA 1901-1910.

In: CH. KINDLEBERGER: *Foreign Trade and National Economics*, New Haven, 1962. 41 and 43 p.

[17] For these as factors of European industrialization, see: S. POLLARD, "European Industrialization", *The Economic History Review*, No. 4, 1973.

must take care to distinguish within the too general category of foodstuffs and raw materials those products which were particularly conducive to the development of industrialization from those which were less likely to lead to technological improvements, or else did not do so at all. We find it very significant that it was the latter group of products that dominated the exports of those countries where the rate of export growth was less than, or only just on, the average. In Greece, for instance, agricultural products were the main export item. In 1887, 75% of Greece's exports consisted of agricultural primary products. We find little change by 1912, when agricultural products amounted to 78 per cent of all exports. But just what agricultural products were these? In 1887, the three most important export products were raisins (comprising 56.5% of all exports), wine (5.9%) and olives and olive oil (4.5%). By 1912, raisins had fallen to 28.8% of the total export; tobacco comprised 14.1% of the total export, olives and olive oil 14.9%, and wine, 11.8%. None of these products, however, though they needed some minimal processing before export (the tobacco had to be dried and graded, the oil pressed from the olives, the grapes dried for raisins, etc.) necessitated the development of a genuine food industry. The technology needed for processing tobacco, for instance, was minimal, for the tobacco factories of the period were closer to the traditional manufactories than to the industrial plants of today. Nor did the production of raisins, oil or wine set new industrial tasks.

Here, therefore, the dominant export branches were by no means organically related to the first possible steps to industrialization.

However, this was not the only impediment which prevented Greece's exports becoming the starting point of her economic prosperity and industrial transformation. No less important was the fact that the principal Greek exports tied in with no key sector of Western European economic development. Raisins, oil and tobacco all had a low price elasticity of demand, and showed price trends that were less than favourable. The lack of economic impetus was reflected also in the import trends, aggravated by the fact that Greece needed to import foodstuffs as well. In 1887, cereals comprised 38% of her imports; in 1912, cereals accounted for 19%; sugar comprised a further 5%. Coal, too, was a significant import product, accounting for 12% of Greece's imports throughout the period. The relatively small volume of industrial imports, too, was in keeping with the conventional picture of a backward economy: the most important industrial imports were the textiles brought in for mass consumption; machinery and investment goods came to a minimal 1.5% of total imports. In the case of Greece, then, the annual export growth of 3% and the Greek economy's heavy export orientation — foreign trade contributed 26% of the national income — tended, rather, to conserve the given economic structure, rather than to renew or to transform it. In fact, there was only one industry which was directly related to foreign trade, the shipbuilding industry. A trading ship's capacity during these decades grew ten-fold; however, for lack of capital, technology and skilled labour, the really big, modern ships were unlikely to have been built in Greece.[18]

Portugal during these years presents an even sadder picture. For generations, Portugal had been one of England's chief trading partners: by the Methuen Agreement of the XVIIIth century, Portugal had become a free trade area for English textiles and in return Portuguese wines received significant tariff concessions in English ports. This trade of English textiles for Portuguese wines has provided the great example for free-trade and comparative-costs theoreticians since Ricardo. This is not the place to give a general critique of the comparative costs theory; the

---

[18] For the data, see B. STERJOS, Industrial Revolution in Greece (Hungarian) in BEREND T. IVÁN-RÁNKI GYÖRGY: Gazdasági elmaradottság (Budapest, 1979.)

European, and particularly the Portuguese experience, however, goes a long way toward refuting it. For the growth in the demand for wine was very far from keeping pace with the growth in the demand for the more important raw products or consumer products of the period.

Bairoch attributes the stagnation of the Portuguese export sector to three factors: 1) the country's loss of the Brazilian trade monopoly; 2) the stagnation of wine exports; 3) the extraordinary one-sidedness of Portugal's exports (up to 1890, wines accounted for about 50% of her export trade).[19]

The slow growth of Portugal's foreign trade indicates first of all that what she had to offer did not sufficiently attract the new, dynamic world market. (Sardine exports did, however, grow by leaps and bounds after the turn of the century.[20]) This being so, Portugal had a constant balance of payments deficit. Her export industries, though they could hardly be called colonial in character, were in sectors that could only minimally stimulate her internal economic transformation. They also failed to contribute to the partial modernization of agriculture, or even temporarily to become the leading economic sector. When, after the turn of the century, the textile industry and textile exports did show an upward trend, this was due in no small part to the salutary influence of Portugal's colonial market.

S. Kuznets and W. Rostow[21] are absolutely right in saying that a more than average rapid growth rate is by no means all that is needed for a sector to become a "leading sector". We can speak of genuine transformation, of genuine "take-off", only when there emerges an economic sector or branch which is sufficiently widely and deeply embedded in the national eco-

---

19 P. Bairoch, *op. cit.* (1976), p. 267. The wine exported in 1800 reached 200,000 hecto-liters, and was never higher until 1870.

20 A. Castro, *A revolucao industrial em Portugal no seculo XIX*, Lisbon, 1972, p. 73.

21 W. Rostow, "The Leading Sectors and the Takeoff", in *The Economics of Take-off into Sustained Growth*, London, 1963.

nomy as a whole to create a mass demand for labour, raw materials, and other industrial goods, and to stimulate the creation of further industries (either through narrow cross-sections, or through cutting investment costs). The criterion, in short, is to set off a general chain-reaction. It is not enough for export industries to start developing, or export production to start growing, to stimulate the economy's capitalist transformation. What is needed is a whole series of dynamic changes, in which the setting up of new substitution and complementary industries leads to the transformation of the whole of the country's economic structure.[22]

This being so, we can hardly be surprised to find that the huge foreign loans received failed to act as a stimulus to the Portuguese economy. For one thing, with her considerable import surplus, the interest payments and principal repayments that had to be made on the foreign loans became an ever heavier — practically insuperable — burden on the Portuguese economy.[23] For another, foreign investment was predominantly in sectors which had only minimal spin-off effect, and which produced no surplus value capable of supporting a new more developed economic structure such as might tilt the balance of assets and deficits in favour of the former. Thus, though foreign investments did accelerate economic development and expand the domestic market, they did so only slightly, and inadequately for industrialization. Portugal had the biggest per capita debt in Europe, a very low level of national income, and a very slowly changing economic structure. Trapped at a very incipient stage of development, her dependence on the developed industrial nations, particularly England, followed an almost classical pattern. Once the external impetus for change slowed down, it was more and more these distorting and inhibiting influences which came to pre-

[22] Cf. A. Hirschmann, *Strategy of Economic Development*, Yale, 1958.

[23] A. Castro, *op. cit.*, p. 146. Between 1898 and 1910, the value imported was generally twice the value exported.

dominate. We have here what was, in fact, a typical, and by no means a unique situation for a country on the European periphery: foreign influences could accelerate to a degree, but could by no means solve the question of economic development. In Portugal's case, the preconditions determining the direction and effectiveness of foreign investment and of international trade relations were, for the most part, highly unfavourable. Neither the sectors invested in, nor their effect on the national income, were conducive to the transformation of the economy; and the country's economic structure, which determined both the investment and the export possibilities, was even less so.

This was true in spite of the indubitable growth in the volume of Portugal's exports, partly because the country's export sector was characterized by low productivity. Both export growth and foreign investments boosted trade rather than industry, and had only a minimal accelerator and multiplier effect on production. There was no significant growth in domestic demand, investment remained dispersed, capital profits tended to trickle out of the country, and the effectiveness of the impetus the economy had received proved, in the long run, to have been severely limited.[24]

It seems quite clear that the export sectors of the Balkan countries — which started to develop much later than those of Greece and Portugal, and showed extraordinary dynamism (between 1850 and 1913, Serbia's export grew from 4.7 million to 23.4 million dollars, and Romania's from 9.4 million to 135.9 million dollars, while Bulgaria's grew from 4.4 million dollars in 1870 to 36.9 million in 1913) — followed a line of development quite similar to that of the latter two countries. Five leading export items — all of them agricultural products — comprised between three-quarters to four-fifths of the Balkan exports

[24] For a theoretical treatment of the subject, see G.M. Meyer, "The Problem of Limited Economic Development", in Agarwala Singh (ed.), *The Economics of Underdevelopment*, New York, 1963.

throughout the period. The income received from these exports was only sufficient to increase agricultural production slightly, and to start building up a more modern, XIXth century infrastructure. Though a number of the Balkan countries, expecially Romania, also offered export products for which there was constantly a great demand on the European market and at prices that were far from being disadvantageous, nevertheless here, too, we can detect only the first steps toward capitalism during this period. However, the centre's need for exports did give a boost to these countries' previously stagnant economies, and accelerated their development. No radical structural transformation or rise in economic levels took place, however; for while the centre did serve as a stimulus to the economies of the countries of the periphery, the contradictions inherent in this kind of economic relationship were, in the case of the Balkans, particularly acute. In a number of cases, the pull of the centre did not operate simply through the automatic mechanism of the world market. Not infrequently, it took the administrative and political intervention of the great industrial powers to remove the legal and institutional impediments that tradition had placed in the way of progress. The domestic conditions, however, remained so rudimentary that it necessitated foreign help to build up even the minimal infrastructure that was achieved.

Under the circumstances, the limitations and distortions of the pattern of economic growth induced by the centre were particularly obvious. The developed countries not only insisted on exports of foodstuffs and raw materials, but also flooded the underdeveloped Balkan markets with their mass-produced industrial products. The small local handicraft industries that had been prosperous earlier were soon ruined by this kind of competition; no genuine big industrial development, however, came to take their place. Foreign trade, too, proved but a limited incentive to development at this time, though there was growth in branches of mining, and in the industrial processing of certain foodstuffs. Incipient urbanization, and a start at building up the economic basis of a modern transport system were two other effects of foreign trade. But the incentive provided by the centre only served to integrate these countries too in the world economic system, in a way that tended to keep them as underdeveloped and structurally backward as they had been. The exogeneous forces dominated these countries' early efforts at economic modernization to such an extent that they shaped the endogeneous forces too to suit their own priorities. There can be no doubt, however, that the former could not have been so decisive had the latter been stronger, had the domestic conditions of take-off been more conducive to economic growth. The region was by no means ripe for industrialization. Though it would be unfair to say that it was free trade which prevented industrialization, free trade both reinforced the area's economic shortcomings, and created new impediments to development. At the same time, it did provide a limited and onesided incentive to growth in these branches which the availability of raw materials and the conditions of agricultural production made optimal, and which best satisfied the centre's industrial needs. The predominance of these sectors made for a certain backwardness in the structure of production; but since the domestic incentives for industrialization were as weak as they were, they were all there was to accelerate economic growth.

Nonetheless, the Balkans were seldom able to copy the developmental pattern of the leading industrial nations; we find few instances of "imitative" economic development here. In the case of the smaller Balkan nations, it was rather a matter of developing economies "complementary" to those of the advanced centre, where domestic conditions and responsiveness to the demands of the international market enabled them to do so.[25]

There are no totally reliable data on the trends in terms of

---

[25] Iván T. Berend–György Ránki, *The Economic Development of East Central Europe*, New York, (1974).

trade during this period; generally speaking, however, it is by no means proven that it was a bad time for exporters of agricultural products and raw materials.[26] But this is not the crux of the matter. The important point, rather, is that for these countries, joining the capitalist world market meant nothing more than acquiring a bigger and better version of the economic structure they had had before. Practically speaking, all that happened was exploitation of the auxiliary factors of production (land and labour) in order to increase production, without this really leading to the restructuring or more efficient use of the factors of production themselves — without, in other words, this leading to industrial take-off. Much of the augmented income from exports was, thus, wasted on imports or on unproductive budget items, or was simply squandered on luxuries for the dominant classes and upper strata of society. Both the enormous production and capital superiority of the developed countries, and the trade policies pursued by the Balkan governments served but to compound the adverse effects of these trends.

In these countries, then, we can hardly speak of spin-off effects. With the accelerated industrial growth of the developed countries putting them ever farther ahead socially and economically, two very unequal groups of countries met as partners in international trade: the economically strong, technologically and structurally developed centre; and the weak, structurally backward periphery, its economy determined by foreign influences. All this was not only automatic guarantee of the superiority of the former over the latter, but was also a chance for the more developed region to shape — or rather, deform — the less ad-

[26] Since the price of wheat and corn, the decisive export products, rose on the world market from the turn of the century, we have no reason to suppose that the exporters of agricultural products suffered from worse terms of trade. See: J. LAMPE, " Varieties of Unsuccessful Industrialization. The Balkan States before 1914 ", *The Journal of Economic History*, No. 1, 1975, p. 65. We do not find convincing or applicable the theory of unequal terms of trade presented by ARGHIRI EMMANUEL, in *L'échange inégal*, Paris, 1969.

vanced to suit its own needs.[27] Thus, though foreign investment did encourage economic growth, this by no means compensated for the extraordinarily low proportion of the productive investments, nor for the disastrous competition that the industrial imports concomitant with capital import meant for the underdeveloped domestic industries.

The massive loans — which led to indebtedness which later resulting in the formal curbing of the economic sovereignty of these countries — did, in fact, initiate a certain process of economic transformation, but this never led to self-sustained economic growth, nor to the development of a modern economic structure. The change was that these economies, consolidated in their backwardness, now conformed to the needs of the developed economies. The centre thus proved to be only a very tendentious incentive to development. For one thing — as our data in the chapter on the Balkans indicate — many of the loans contracted were used to build up new armies and state apparatus. These largely unproductive investments were a great burden on the economy without bringing any direct profits.[28] What is more, given that there were no returns on these investments, increased taxation remained the only way to repay the debts.

This is well reflected in the growing importance of excise taxes in the national budget: 50% of Romania's national revenues, and 42% of Bulgaria's came from this source. Bulgarian peasants paid out 20% of their incomes in excise taxes.[29] All

[27] For a general treatment of the problem, see D. SENGHAAS, *op. cit.*, p. 19, as well as J. GALTUNG, *Eine strukturelle Theorie des Imperialismus*, in D. SENGHAAS: *Kritische Friedenforschung* Frankfurt, 1971. Naturally, we can hardly accept a view which explains the economic backwardness of the Balkan countries exclusively, or even primarily, in terms of their relationship to the developed ones.

[28] J. K. Galbraith aptly calls such investments of the newly independent states instances of " symbolic modernization ". " Yet it will be clear that the economic well-being of the people is not much advanced by symbolic modernization." J.K. GALBRAITH, *Economic Development*, Boston, 1964, 51.

[29] A. MILWARD-S.B. SAUL, *The Development of the Economies of Continental Europe 1850-1914*. London, 1977, p. 461.

this, of course, served only to further enervate the already weak domestic market, and made it even more unlikely that the sectors benefitting from foreign investments could stimulate the creation of spin-off industries.[30]

Disregarding both the largely unproductively invested government loans, and the technology borrowed from the centre, the positive influence of the developed countries consisted mainly in stimulating agricultural exports, in helping to build up the Balkan railway networks, and in capital investments in certain extractive industries. (Except for Romanian oil after the turn of the century, the export of Balkan mineral resources was, at this time, negligible.)

In the absence of a spin-off, and with its share in world market as small as it was, the Balkan agricultural exports could not become the basis of general economic prosperity. (Romania was an exception in this respect, and supplied 8% of the world's wheat exports.)

The Balkan railway network — though far less extensive than the railways of Western Europe, or even those of Italy or Hungary — did, nevertheless, bring a certain unity to the national market, and served to connect it to the network of international trade. It did not, however, become the chief factor of economic growth as it had in Western Europe, or Italy, or Hungary, or even in Russia, where the economic processes induced by railway building became the basis of lasting economic prosperity.

Foreign investors, who held the lion's share in railway construction, naturally first of all wanted to see the international lines completed. Neither political nor strategic considerations dictated the construction of a network joining the various areas of the domestic market, nor did it promise to be particularly profitable to do so. The economic effects of foreign railway

[30] The interest due was very high as well. Romania had to pay an interest rate of 5 1/2 - 5 3/4%. Serbia and Bulgaria paid a rate of 6 1/2% in the early 1900s, and 5 1/2% just before the war. J. LAMPE, *op. cit.*, p. 76.

investment were rather one-sided and external, and essentially, railways became a factor of international specialization, and, as such, tended to help keep the area in its state of backwardness.[31]

Even Romanian oil exports suffered all the drawbacks typical of an export-oriented economy set in motion by foreign investments. The rapid growth of a few industries had relatively little effect on the development and transformation of the economy as a whole. Though the extractive industries prospered, this hardly influenced the processing industries. No adequate sectoral structure developed; investment goods were still largely imported, and even the crude oil that was extracted was, for the most part, processed abroad. Before the war, Romania was still importing refined oil. Unlike other countries, in Romania oil gave rise to no industrial development. Her industrial sectoral structure was only minimally altered by the increase in oil production, for the profits were either invested in expanding the extractive sector, or else were exported abroad.

Foreign capital investment did, then, speed up the process through which more developed production methods came to the Balkans; it served to increase productivity and to expand the market, and helped establish technologically more advanced factories and even some new industries. However, given the system of international specialization, and the absence of domestic incentives, this was not enough for a breakthrough. At the same time foreign investment also had aspects which tended to preserve the structural backwardness of these economies. The export sectors formed isolated " enclaves " within the economy as a whole,[32] and gave rise to a dualism which, though it differed from the classical colonial model, (where the industries set up with the help of foreign investments differed from the backward,

[31] IVÁN T. BEREND-GYÖRGY RÁNKI, *op. cit.* (1974).

[32] For the " enclave " type of export sectors in general, see TAMÁS SZENTES, *Az elmaradottság és a fejlettség dialektikája a tőkés világgazdaságban* (The dialectics of backwardness).

traditional sector to the point of conflict) was yet a sign that the Balkan economies too were " dual " economies.

Very similar to the Greek, Portuguese and Balkan developmental pattern that we have been examining was that of Spain. Spanish export growth was by no means rapid; it fell somewhat short of the European average, and, what was particularly interesting, showed a downward trend. While between 1860 and 1880 export growth was 4.4% per annun, in the 1880s it was only 2.9%, falling to only 0.9% by between 1890 and 1910. This decline in export growth was accompanied by changes in the structure of Spanish exports which occurred between 1850 and the turn of the century. In 1850, wine and brandy made up 28% of Spain's exports. Of the ten leading exports, nine were agricultural products, with lead being the only industrial raw material. At this time, Spain was even exporting wheat.[33] Initially, then, trade with the industrialized nations here too meant a growth in agricultural production. Western Europe's growing industrialization, however, demanded more and more metals and ores. From the 1870s on, it was from the countries of the European periphery that were already established trading partners that the West strove to satisfy its demand for raw material. Foreign capital, as we have seen, flowed into the periphery precisely to exploit its mineral wealth. It is, then, hardly surprising to find that by 1913 it was no longer wine, but metals that headed the list of Spanish exports, and accounted for 12%. Wine took second place, with ores (11.6%) coming a very close third. Among the ten leading exports we now also find textiles. Characteristic of the change in the export structure was the decline in grain exports. In fact, in the quarter of century preceding World War I Spain was more often an importer of grain. At the same time, fresh fruit became the fourth most important

---

[33] For the question of Spanish grain exports in the XIXth century, see N. SANCHEZ ALBORNOZ, *Los crises de subsistencias de España en el siglo XIX*, Rosario, 1963.

export (8.4%), much of it coming from the abundant citrus plantations of Valencia.

Foreign capital investments in Spain were less diffused, therefore, concentrating as they did on producing for the world market. It was for this reason that they also served to induce a certain degree of autonomous investment. Spain's income from her exports was considerably higher than Portugal's. A great deal of foreign investment went into railways and mining, and a developed infrastructure was the result. The basis of an iron industry was also laid; and, with the help of native investments, an expanding domestic market and the existence of a traditional textile manufacture helped lay the basis of a strong, modern textile industry in Catalonia.[34] But even in Spain it was not primarily as a stimulus to the economy that foreign capital investment functioned. Foreign investment failed to lead to self-sustained economic growth, though there were some signs that it might. In the long run, however, the structural changes that did take place stopped short of a genuine transformation, and the principal characteristic of the economy remained foreign dependence.

Although Spain's entry into the world market had much in common with the analogous attempts of the previously discussed underdeveloped countries, the process was much more fraught with contradictions. Spain's was not merely a case of quantitative development within the framework of the old economic structure, but a case of genuine change in the factors that made for economic development and sustained it. Economic development started in response to external stimuli, but came to encompass certain elements indicative of self-sustained growth. The old economic structure underwent some change, but there was no radical transformation. The conflicting forces of stagnation

---

[34] H. Vasquez de Roda, " El processo de l'industrializacion en España ", in *l'Industrialization en Europe*, Lyon, 1974.

and change waged a tug of war to determine the economy's future.

It was much the same process that took place in Italy and Hungary, but the changes which occurred were more decisive and far-reaching. Here, too, foreign demand provided the decisive impetus for change, with the export sectors doing a great deal to speed up economic development and to create a capitalist economy.

Between 1850 and 1913, Hungary's exports jumped from 30 million dollars to 368 million. In the 40 years preceding World War I, exports increased more than three-fold, which meant an annual growth of over 3%.

The chief export item was grain, which accounted for over 50% of all exports. Grain was a fortunate export item in all respects, at least to the turn of the century. The demand for it grew constantly as the western areas of the Austro-Hungarian Monarchy became more industrialized, and as per capita consumption increased. Prices — until the flood of cheap American grain to Europe brought them down — showed an upward trend, and long continued to be advantageous within the tariff wall protecting the Monarchy. Hungary — primarily because of her position within the Habsburg Monarchy — felt the pull of the Western economies much earlier than her neighbours, and was also able to respond to it about three decades earlier than they.

Naturally, the speed of this reaction had a great deal to do with the strength of the domestic factors of development. At the same time, this earlier start meant that Hungary fell relatively less far behind, and had a greater variety of options open to her as to the form her participation in world trade would take. The volume of Hungary's grain exports was such that it led to an extraordinarily fast growth rate in agricultural production: an annual average of 2 per cent during the half century of the Dual Monarchy. But it was not merely a matter of using new production techniques to increase productivity. As early as the late 1860s, Hungary began to exploit the industrial potential of being an exporter of grain. Unlike a great many other agricultural products, grain was eminently suited for industrial processing; what was more, the conditions of its production, and the technological and manpower demands so generated promised to affect the country's economic structure as a whole. Although the manpower needs — and so the influence on social formation — of the milling industry were rather limited, the organic composition of capital that it required was high: in other words, it required large-scale technological investments, a great deal of capital and a skilled although not very large labour force. Though there were local milling industries in all the countries of Europe, Hungary was practically the only European country to be able to transform a part of her agricultural exports into exports of processed food products. All this indicates intense economic activity, and a responsiveness to the stimulus of the industrial revolution; it also indicates that certain structural changes were taking place in the country's economy.

Hungary's milling industry soon became an exporter on a world scale. Budapest's milling capacity was next only to that of Minneapolis, U.S.A. How can we account for this extraordinary boom? Primarily, because Hungary had the advantage of relative costs, not only in grain production but also in the milling industry, both because of cheaper raw materials, and also because of better technology and higher productivity. Though the machinery in the mills was partly of foreign import, during the critical phase of the boom it was a Hungarian invention, the roller mill, which guaranteed the country's milling industry its technological superiority, and its lead in quality and productivity.

This clearly was a case of an economy more ready to respond to external stimuli. Had Hungary not already had a certain technological expertise, she could hardly have achieved such superior productivity; nor can we doubt that there already was a great deal of domestic capital accumulation, and the requisite capitalist stratum, since Hungary's milling industry was almost entirely the product of domestic investments. As a consequence, after the turn of the century, the ratio of wheat to flour in Hungary's exports was 33: 67, while it was 92: 8 in Romania, and 98: 2 in Russia.[35]

It was, however, not merely in becoming an exporter of foodstuffs rather than of unprocessed agricultural products that Hungary differed from the countries we have been examining. Just as initially the profits from the export of food products were invested in the milling industry, so the later capital accumulation fostered by the milling industry became a basis for industrialization, and of a partial transformation of the country's economic structure. This was reflected in the changes in Hungary's export structure, too: just before World War I, agricultural products no longer comprised 76-80% of all exports as they had earlier, (and continued to do in many neighbouring countries), but only about a half. The effects of the initial phase of industrialization were reflected in the import structure as well: while among exports it is the food and other industrial products that deserve our notice, we find that among imports, investment goods were beginning to compete with the prime import category, industrial consumer products. In Spain, for instance, more money was spent in 1850 on the import of cinnamon than on the import of machinery, and in 1913, too, machinery accounted for only 8.7% of all imports; in Hungary, on the other hand, machinery comprised 12% of all imports during the prewar boom years.[36]

The changeover from wheat exports to the export of flour and its consequences for capital accumulation and industrial and technological development therefore removed Hungary from the ranks of those peripheral countries where the pull of the centre

[35] Iván T. Berend-György Ránki, *op. cit.* (1974).

[36] Iván T. Berend-György Ránki, *Magyarország gyáripara 1900-1914* (Hungary's manufacturing industry 1900-1914), Budapest, 1955, pp. 190 ff.

induced only the accelerated development of sectors geared to satisfy the centre's own needs. This, however, was by no means enough for the radical transformation of the economic structure. We must also examine the peculiarities of foreign investment activity in Hungary in order to place her within the complicated system of interrelationships linking the industrial centre and the countries of the European periphery.

We have already had occasion to note the potential for inducing autonomous development inherent in the railway building, that accounted for least half of all foreign investment activity. We have also noted that its spin-off effect for both demand and supply was considerable — though not Western-European in scope — and that this, in some measure, acted as a multiplier. As early as the 1890s, there were signs of the effect of the active foreign capital invested, too. It is at this point that the intensive structural transformation of the Hungarian economy, industrialization, really takes place, and foreign capital investors continued to play a very important role in this process: about a third of all industrial shares were in the hands of foreign capitalist groups.

In Hungary foreign capital investment also drained the national income, and led to onesidedness, to distortions in the economic structure that developed. But, on balance, in Hungary's case, it was the impetus that foreign investment gave to economic development that must be considered the more significant: it got the economy over its period of stagnation, and helped start a period of lasting economic growth.[37] After the turn of the century — one of the most significant phases of Hungary's economic and industrial development — three-quarters of the nation's capital needs were satisfied from domestic sources, a telling

[37] See. L. KATUS, " A Kelet-Európai iparasodás és az önálló tökés fejlödés kérdéséhez " (On the problems of Eastern European industrialization and independent capitalist development), *Történelmi Szemle* (Historical Review), No. 1, 1976; and L. KATUS, *Economic Growth in Hungary during the Age of Dualism*, *Studia Historica*, 62, Budapest, 1970, p. 52.

indication of the fact that the autonomous economic forces at work in Hungary were by then sufficient to exclude the possibility of her economy becoming dependent on external factors.

Essentially, it was the domestic forces that enabled external influences have a spin-off effect. It was for this reason that these influences tended to reinforce the various — rather weak — domestic developmental trends. Hungary had a unique and intermediate position as a country which responded to the stimulus of Western European industrialization and was able also to adapt and take part in the processes initiated by industrialization, without, however, herself undergoing economic transformation. For a long time, she remained a country with a specialized and linked economy, while integrating as a dependent area in a world economy dominated by the developed centre.

Italy's development was similar in a number of respects. Initially, silk was Italy's primary export product, the traditional silk producers of Northern Italy contributing a third of all the country's export goods in the late 1860s. Sicilian sulphur and marble stood second and third on the export list. Ores, expecially iron ore, were next, with agricultural products, especially wine, olive oil, and fruits following.[38] As in Greece, Spain and Portugal, in Italy, too, the agricultural exports failed to stimulate the development of spin-off industries. Next to wine and olive oil, citrus fruit was the most significant export item. (In 1862, 454 quintals of citrus fruit were exported: in 1900, 2 million quintals, and in 1913, 4.3 million quintals.)[39] The technology of the age, however, did not permit the industrial processing of these products or the development of a food industry.

Quite a different picture is provided by the silk industry, built up as it was around already existing domestic industries

[38] G. LUZATTO, *L'economia Italiana del 1861 el 1914*, Vol. 1, Milan, 1963, pp. 193–194.

[39] F.J COPPA, "The Italian Tariff and the Conflict between Agriculture and Industry: The Commercial Policy of Liberal Italy", *The Journal of Economic History*, No. 4, 1970, pp. 755, 764.

and small manufactories. Significant domestic capital accumulation, urbanization, and more developed trade and manufacturing had laid the groundwork for the changes induced in Italy's foreign trade by contact with the developed industrial nations.

Though economic development during the decades following Unification was very slow, it soon became obvious that Italy had joined the world market through more than simply increasing her raw material exports, and that other factors besides a more active and more profitable foreign trade [40] were working to stimulate the economy. In fact, the structural transformation of the economy practically coincided with the growth of foreign trade and of the profits from it. Already by the turn of the century we can see signs of this: silk fabrics came increasingly to replace raw silk as an export product (by 1913, silk materials comprised 17% of all exports); the export of cotton yarn grew ten-fold, that of woollen cloth, twenty-fold.

Italy, then, very soon ceased to be dependent on imported textiles — an unmistakable sign of a backward economy — and became herself a textile exporter. More and more imported machinery improved the Italian textile industry, especially during the boom years of Italian heavy industry after the turn of the century.[41] Foreign trade, however, was a relatively less important factor of the Italian economy than of most of the more backward economies, exports accounting for only 11% of Italy's national income.[42] That this was indeed the case can be seen also from examining the role of foreign investments in the Italian economy. Foreign investments, as we have seen, were neither particularly big, nor of long duration. But foreign capital pro-

---

[40] This important point was made by J. VINER, (*The Economics of Development in the Economics of Underdevelopment*, New York, 1963). Though a number of his conclusions could be questioned, we can only wholeheartedly agree with him that the question of the benefits of foreign trade is not merely a matter of exchange rates.

[41] L. CAFAGNA, Italy 1830–1914, *The Fontana Economic History of Europe*, Vol. 4, Part 1, Glasgow, pp. 300 ff.

[42] A. MADISON, *Economic Growth in The West*, New York, 1964, p. 67.

moted Italy's economic development at crucial periods and in crucial areas, giving it the necessary boost at the critical moments. The fact, however, that Italy's quickly growing economy did not remain an export-oriented dependency of foreign capital was due, obviously, to her internal development, and her receptivity to external influences.

Here, then, the spin-off effects of foreign investment did not stop at the export-oriented extractive industries, but — with the internal preconditions as favorable as they were — could reinforce the internal forces promoting the overall development of all branches of the economy; under these circumstances, it was the positive, rather than the negative effects of foreign investment that predominated.

Obviously, this could not be the case wherever there was a traditional economy of the Balkan type. We could hardly make sense of Italy's development in the XIXth century if we overlook that before the XVIth century Italy had been in the vanguard of European social and economic development. Her extensive trade, her flourishing handicrafts, her advanced urbanization had far surpassed that of the rest of Europe. Although she later fell behind in all these respects and entered on a period of decline, exchanging her position of leadership for one that became ever more "semi-peripheral" in nature, nevertheless the industrial revolution found her — especially the northern areas — with domestic preconditions more conducive to development than any of the other countries we have been examining.

Naturally, the role and effect of external factors was nowhere confined solely to foreign trade. Foreign capital investments, as we have seen, were closely connected everywhere with whatever export sectors developed. Perhaps nowhere as inextricably, however, as in Russia. One of the chief factors of the great economic boom developing in Czarist Russia at the end of the century was the steady development of foreign trade, which grew at an annual 3.8% between 1860 and 1910, a rate faster than the

European average. The vastness of the country, the size of her population, the variety of her mineral wealth all led Russia, economically underdeveloped though she was, to play a great part in European trade. At that time, Russia was the world's major grain exporter, providing 25-33% of the world's wheat imports. Her iron, ore and cotton exports were also considerable, and by the turn of the century the oil discovered in the Caucasus was among her export products.[43]

Russia's export structure was typical of that of all underdeveloped countries. In spite of changing ratios, the overall predominance of agricultural products, ores and industrial raw materials in Russian exports as a whole hardly varied. There can be no doubt that these exports and the incentives they gave to change were major factors in initiating development. For the growth of the volume of foreign trade — particularly initially — was not only an incentive to capitalization, but also a means to it. The receipts from exports were what covered the growing volume of industrial imports, as well as the state's contribution to the building of the infrastructure. The surplus export receipts even helped counter the severe balance of payments deficit, and contributed to payments of both interest and principal on the extraordinarily large amount of foreign loans flowing into the country. Russia's foreign trade, thus, cannot be regarded as an economic factor passively adapting to the needs of the industrialized nations; a means of relegating the country to the role of a raw material supplier, as befitted underdeveloped areas in the international division of labour. It would be a grave misunderstanding of complex economic processes not to notice that it was economic, social and political factors which made exports the motor of a dynamically growing economy, and enabled the difficult process of Russia's economic transformation

[43] P.A. Hromow, Ekonomika Rosszii perioda promiislennogo kapitalizma, Moskva 1963, pp. 191 ff.; and P. Bairoch, *op. cit.*, p. 254.

to get going — though only with a great deal of foreign help. It was Russia's political importance on the international scene, her size, the unique wealth of her economic resources, the conditions expecially suiting her for industrialization — her great iron and coal deposits — and her vast, though underdeveloped domestic market which enabled her to find the area of international specialization that activated her economy as a whole, rather than limiting growth to the development of her export branches at some rate dictated by foreign markets. Export receipts, favourable price trends and exchange rates permitted income from the export sector to be pumped into other spheres of the economy, with a consequent expansion of the domestic market.

The building of a railway network made the development of the domestic market part and parcel of the country's entry into international trade. At the time of the greatest industrial boom, 47% of all foreign investments went, apart from the export sectors, into the development of an autonomous economy, and into the building of technologically advanced heavy industries. We can hardly begin to make sense of all this if we do not bear in mind that Russia, unlike the other socio-economically backward countries we have been looking at, was a great power with quite different potential. The very size of the empire, its powerful bureaucracy and military strength were all factors which served to promote the country's economic take-off. There can be no doubt that after the turn of the century, Russia — backward as she was, predominantly agricultural, and dependent on foreign capital — had nevertheless got started on the road to transforming the structure of her economy, and had entered the age of industrialization.

The Southern European, Hungarian and Russian developmental processes examined so far, although they differ from country to country, nevertheless reveal two characteristic and essentially dissimilar patterns. In the Balkans and on the Iberian Peninsula — although there was a substantial increase in foreign trade, and foreign investments did introduce new elements into the economy — there was no real economic growth, nor any transformation of the economy. Contact with the countries of the developed centre, therefore, tended to reinforce the domestic socio-economic conditions and to retard and distort the development of these countries.

In the cases of Italy, Hungary, and Russia — despite all the real quantitative and qualitative differences in their development — foreign trade not only gave a boost to the economy, but there was also sufficient volume of foreign capital investment to accelerate economic development to such a degree that the independent internal forces of change could also become effective. Consequently, these economies began to be transformed through industrialization, though the rate at which this took place, the internal structure, and the level that was reached varied greatly from country to country. This pattern, too, shows elements of adaptation and subordination — differences in development between these regions and the centre, and the cycles of boom and recession made this inevitable — but there was no lack of internal development either, no exclusion from all the advantages of development, nor the unmitigated burden of its disadvantages.

The Scandinavian countries provide the third type of developmental pattern for the European periphery. In spite of their more developed social structure, bourgeois institutions, and somewhat more favourable economic position, these countries in the middle of the XIXth century must also be regarded as peripheral. That they were so is indicated by the lower level of development and of national income characteristic of the periphery, by their typically pre-industrial economic structure, as well by the complementary nature of their economic ties with the countries of the centre. Sweden's exports of 40-45,000 tons of iron a year, and her participation in world trade since the XVIIIth century [44]

[44] K. SAMUELSON, *From Great Power to Welfare State. 300 Years of Swedish Social Development*, London, 1968.

does not contradict this since we are not merely concerned with the volume of Sweden's exports, but also of their medium term and immediate economic and social consequences. But while the technological improvements of the industrial revolution and the advent of coke smelters diminished the role of Swedish iron on the world market, trade and manufacturing, merchants and craftsmen played a much more important part in the Scandinavian economy and in Scandinavian society than they did in other countries of the periphery.

It was after 1850 that British and Western European investments, and the new policy of free trade first stimulated Sweden to become a large-scale exporter of timber. Within 20 years, Sweden's timber exports rose from 0.4 million cubic meters to 1.8 million cubic meters. In the 1850s, 34% of all Swedish exports were timber and wood products; by the 1860s, the ratio was 44%, and did not change much to the turn of the century.[45]

Soon, however, there were incentives for development. To the extent that the Swedish iron industry was able to adapt the new British technology, and to the extent that British and Western European iron producers could not meet the ever growing iron needs of transport and industry, Sweden could still hope to become an exporter of iron again. In fact, iron became the second major stimulus to Sweden's economic growth. By the 1880s, iron products were accounting for 16% of all Swedish exports. Nevertheless, the export structure of the early 1880s was very close to the traditional pattern. 24% of all exports consisted of foodstuffs, 43% of raw materials, and 33% of industrial products. But it would be a grave mistake to forget that a foreign trade based on the export of iron and timber had the potential to greatly accelerate Sweden's capital accumulation,

---

[45] F. FRIDLIZIUZ, " Sweden's Export 1850-1960 ", *Economy and History*, 1963; and E.F. SÖDERLUND, " Short Term Economic Fluctuation and the Swedish Timber Industry ", *The Journal of Economic History*, 1953. M. FLINN, " British Steel and Spanish Ore ", *The Economic History Review* 1958, 8.

and consequently, the overall transformation of her economic structure. In this case, an increase in the volume of exports was bound to have more than just those consequences customary for a peripheral economy. While exports of wine or tropical fruits in whatever quantity could hardly lead to economic transformation, (the qualitative repercussions of quantitative growth being negligible in this case), and while wheat exports under optimal circumstances might initiate spin-off effects, the export of timber and iron had longterm and immediate effects on industry that were much quicker to appear. For one thing, the export even of unprocessed timber and iron ore gave rise to industrial activity which required a considerable labour force. For another, the very process of extraction — especially in a country with industrial traditions such as Sweden — was quick to lead to the development of primary processing. Given the favourable domestic preconditions, the booming world market soon initiated a process of industrial development in which timber processing, in combination with the nascent chemical industry, created the paper and cellulose industries that were to become major export sectors. Cellulose exports rose from an average of 7300 tons per annum between 1876 and 1880 to an annual 90,000 tons between 1891 and 1896; by 1900-1915, it was an annual 800,000 tons, and comprised three-quarters of all the cellulose produced. By 1911-1913, timber accounted for only 26 % of all exports, paper and cellulose having risen to 17.6 %.

Judging from the export structure, then, by the beginning of the XXth century Sweden had entered the ranks of the developed industrial nations. While in the more rudimentary Italian, Hungarian, and even Russian foreign trade structure it was the growing imports of machinery that provided the measure of responsiveness to the pull of the centre and of a certain degree of economic change, Sweden had already got much farther ahead, its specialised dairy and electrical machinery industries were also producing enough for large-scale exports. On the eve of

World War I, the machinery industry was producing 10% of Sweden's exports.[46]

Denmark's development was even more typical of that of a complementary economy, her export structure in the mid XIXth century being much more traditional, and remaining largely so until World War I. Denmark had no substantial store of mineral wealth, and no industrial traditions comparable to Sweden's. Danish export activity received its first impetus from her neighbours' needs for foodstuffs, since especially from the 1840s on there was no Danish food product that could not be sold on the neighbouring markets. It was largely in response to export demand that Denmark's agricultural production rose by 80% during the period under discussion. At this time, agricultural products comprised 80% of Denmark's exports, cereals, meat and milk products being the chief items.

When American grain flooded Europe's markets, however, Denmark cut back her grain production so far that she soon became an importer of cereals. She quickly and radically switched instead to livestock production. Initially, live pigs sold on the German market were the mainstay of this industry; later, when Germany withdrew behind a tariff wall, Denmark developed an impressive meat and food processing industry, a process in which inventions like the mechanical creamer played an essential part.

By the turn of the century, more than 90% of her exports consisted of agricultural and food products. Butter accounted for 47% of Denmark's exports, bacon and ham for 26%, and eggs for 8%. This extraordinarily intensive export activity, which was explicitly aimed at the English market (50-60% of it ended up on English breakfast tables) guaranteed Denmark

[46] Sweden had no need of protective tariffs for this; practically throughout the period she followed a free trade policy, hence primarily of complementary development, to transform her economic structure. For the data see L. JÖRBERG, in *Fontana Economic History of Europe*, vol. 4, The Nordic Countries, as well as K.G. HILDEBRAND, " Les traits caractéristiques de l'industrialisation des pays scandinaves ", in *L'industrialisation en Europe*, Lyon, 1973.

favourable terms of trade throughout the period, and brought her very considerable export receipts. Per capita national income doubled during this critical period, indicating a significant expansion of the domestic market. The national income grew 3.3 % *per annum* — thanks mostly to the 4.5 % annual export growth — and made possible the beginnings of the process of import-substituting industrialization.[47]

This industrialization, however, exclusively served the domestic market. Even just before the War, industrial products accounted for only 6-9 % of all Danish exports. Most of the consumer products Denmark needed, however, were produced domestically. Investment goods, on the other hand, continued to be imported. Denmark was a relatively backward agricultural country when she set out on the road to industrialization, but by the eve of World War I she had achieved a high national income purely through exporting processed agricultural products.

What we have here is a very special case of adaptation to the capitalist world economy, facilitated by Denmark's particular internal conditions and social and educational structure. Yet, Denmark's case also illustrates that — given adaptability to the fluctuations of the business cycle, export elasticity, a suitably high demand elasticity, and the development of spin-off effects — the relationship of the centre to the countries of the periphery in the mid-XIXth century was far from a closed, static system. The trend which emerged as dominant from the conflicting incentives and limitations to development that made up the centre's influence on the periphery was a function also of forces endogeneous to the particular peripheral country. Thus, for instance, the very effectiveness of the external factors, of the influence of the booming international market, was influenced by the flexibility with which a peripheral country exploited the opportunities its natural resources and geographic position presented.

---

[47] S.A. HANSEN, *Økonomisk vaekst in Denmark*, Copenhagen, 1974.

Denmark was a country particularly sensitive to foreign trade. Her exports provided a quarter of the national income. And while in her foreign trade relations she was still functioning as a more developed peripheral country, on the domestic scene there were already trends towards urbanization and industrialization which predominated. Here, then, foreign trade did not forestall but rather tended to accelerate the transformation of the economic structure.

We find much the same situation in the case of Norway, though with a more modest overall development. Norway's exports were all mainly related to lumbering and fishing; her greatest export receipts, however, came from her predominant role as carrier in international trade.[48]

The Scandinavian countries undoubtedly seemed to confirm the classical thesis that the indirect effects of a growth in demand included changes in old attitudes, the removal of the social obstacles to development, and the overcoming of the bottlenecks impeding economic progress.[49] The impetus international demand gave the economy through foreign trade was also closely related to foreign investment activity. Although to varying degrees, foreign capital, as we have seen, played an important part in all the Scandinavian economies, without, however, distorting their development, or being the main factor in it — it took on the complementary role of providing a necessary initial push. In the long run, foreign investments had a positive overall effect on the distribution of investments, on the economic structure, on foreign trade relations, and on the balance of payments.

Scandinavia's economic development began when these countries adapted to the needs of the export market. Here, too, it was complementary economies that first developed. In the case of the Scandinavian countries, however, the subordination of the

[48] J. Bjerke, *Trends in the Norwegian Economy 1865-1960*, Oslo, 1966.
[49] A. Cairncross, *Factors in Economic Development*, London, 1962, Chapter 13.

'second comers' to the centre's economic needs did not become a fixed pattern. In fact it was rather the forces making for independence, and for a rapid and multilateral development and transformation of the economic structure that came to predominate. Foreign influence worked, in the final analysis, not to keep these countries within an exploited periphery, but rather to integrate them with the countries of the developed centre.

What has been said so far should show that the relationships binding the European centre to the countries of the European periphery in the XIXth century contained elements — and left scope for forms of potential development — that were highly contradictory in nature. It would be a great mistake to think of these relationships as being comparable to those of the developed countries of the XXth century and the Third World. The industrial revolution and the world economic system it gave rise to had a major role in inducing development in all backward regions of Europe. There can be no question, however, of the fact that the incentives to development were by no means of equal force; and at the same time an area's responsiveness to the stimuli received differed even more radically.

The first group of factors determining this responsiveness was the relative scarcity or abundance of a region's natural resources, and the relative facility with which its geographic position and topographic configuration permitted the building up of the transport system necessary for entering the world market. The ability to adopt foreign technology, the existence of export opportunities, the strength of the traditions of international trade, and the flexibility with which the economy could adapt to new demands were other factors determining a region's response to the pull of the centre.

Above and beyond this, and operating in the context of the factors already mentioned, there were also domestic factors which determined the possibility, degree, and kind of the response to the external stimuli: the level of development or degree of backwardness of the economy; the country's social structure, educational system, ideology and value system; its international political status; and the government policy directing the independent state. The moment of impact of the external stimuli was by no means a harmonious encounter of exogeneous and endogeneous forces; what we see was much more like a struggle, but it also implied interdependence. The reaction of the various regions of the European periphery to the challenge of the Industrial Revolution can be placed — to use H. Leibenstein's terminology — along a continuum ranging from " zero sum " (in the case of economies incapable of change, or capable only of becoming dependent colonial economies) to " positive sum " (in the case of economies where change resulted in the growth of the productive sectors and in the growth of the national income).[50] But the challenge of industrialization to the European periphery was not only earlier, but also quite different from the challenge to Africa and Asia. In Europe, development was not curbed and distorted through direct military and political control; there were opportunities for domestic decision-making, though only within the limits set by " structural force "; [51] this was of course, given further weight when necessary by appropriate political pressure exerted by the developed centre.

Nevertheless, the conclusions we can draw from the history of the European periphery permit no facile generalizations. Even in the worst case, we cannot, as we have seen, conclude that foreign capital activity led only to the stagnation of real wages, to the drawing off of a large slice of the national income, and to the conservation of domestic savings and of a backward do-

[50] H. LEIBENSTEIN, *Economic Backwardness and Economic Growth*, New York, 1957, pp. 188-189.

[51] It is John Galtung who uses the expression " structural force " (in *Kritische Friedenforschung*) D. Senghaas, ed., (Frankfurt 1971) to describe the economic relations of the developed and less developed nations. Our adoption of the term by no means implies total agreement with Galtung's theory.

mestic market. On the other hand, in the cases of both the Balkans and the Iberian peninsula we do find a situation more typical of colonial nations,[52] namely, that foreign demand promoted investments primarily in the export sectors, the money for the investments coming — due to lack of domestic accumulation — mainly from imported capital.

Nor can there be any doubt that export specialization — to the extent that it developed — had as a concomitant a strong dependence on foreign capital, a trend which was not counterbalanced by any opposite trend to limit foreign economic controls.[53]

Even in cases where foreign influence had more positive results (for instance, in Italy, Hungary, or Russia) or where foreign influence was unambiguously positive in its effects (for instance, in Scandinavia), it was not simply a matter of a country enjoying a comparative advantage in international trade. True, participating in international trade did promote specialization; it was an incentive to the more economical use of the factors of production, and tended to help maximize incomes. We cannot, however, ignore the consequences of such unequal partnerships. It remained the case of a dominant and of an accommodating economy, even where, as in the case of England and Denmark, it was precisely this accommodation that permitted the latter's rapid development in that particular phase of her history. However, the deterioration of the terms of trade at this time was not, as we have seen, inevitable for the countries of the periphery; in Scandinavia, there was a definite improvement, so that none of the area's income was drained off in this way either.

The stimulus the developed West gave to change certainly cannot be said to have led to balanced, overall economic deve-

[52] See TAMÁS SZENTES, *op. cit.* (1976), p. 145. Szentes gives a summary of G. Meier and R. Baldwin's views.

[53] See *idem*, pp. 146-147.

lopment in a number of the peripheral countries. It is just as true, however, that in other countries, notably in Scandinavia, the pull of the West proved an impetus whose positive effects were felt throughout the economy. As G. Myrdal noted, " The higher the level of economic development that a country has already attained, the stronger the spread effects will usually be. For a high average level of development is accompanied by improved transportation and communications, higher levels of education, and a more dynamic communion of ideas and value — all of which tends to strengthen the forces for the centrifugal spread of economic expansion or to remove the obstacles for its operation." [54]

Myrdal thus sees the spin-off effect as a function of development, but something that can also be promoted through state intervention. There can be no doubt that important as the nature and potential of foreign trade, and the volume and use of foreign capital imports were for the development of European periphery it was in government policy that we must seek the clue to a country's ability to react to the challenge of the industrialized West. Economic growth in the form of differentiation, substitution and absolute gain [55] could become a cumulative process only where there existed the internal structure, the institutions, and the adaptibility adequate to channel these in the direction that best served the country's interests. This is why it was that success as much as failure in any given country was the result of the encounter, struggle, and interaction between domestic forces and the forces of world capitalism.

---

[54] G. Myrdal, *Economic Theory and Underdeveloped Regions*, London, 1957, p. 34.

[55] Anna Gelei, *Növekedési trendek a gazdaságban*, (Economic growth trends), Budapest, 1971, pp. 24 ff.

ARTICLE 2

BANKING IN THE EARLY STAGES OF INDUSTRIALIZATION: CONCLUSION

by Rondo Cameron

It is not the purpose of this book to derive, on the basis of historical experience, a model of the "one best" banking system for all times and places, or even for all countries in the early stages of industrialization. On the contrary, our study shows the wide range of variability of successful systems; the system that "works" or is "best" in one place and time may not work successfully in another.[1] In every case the financial system is related to and must grow out of the distinctive legal, social, and political traditions and the objective economic conditions of its specific environment. The Japanese case is especially revealing: Japan deliberately attempted to imitate first the American banking system, then, disillusioned with that, the English banking system. What actually resulted was a system that in many respects was uniquely Japanese, consonant with other Japanese institutions and traditions. On the other hand, it is true that nations in the early stages of industrialization face many similar problems. Although the solutions to those problems may differ according to circumstances, there are certain common features in all successful solutions that deserve further study as possible keys to an understanding of the process of development. Moreover, some solutions are more successful than others.

In the historical cases reviewed in this book the banking systems of Scotland and Japan appear to have represented the best adaptations of structure and function to the growth requirements of their particular economies. Those of Belgium, Germany, England (the last two *in spite of* official pol-

[1] H. C. Wallich, "Comparison of Monetary Systems," 2: "The behavior of money is determined by the kind of economy in which it circulates. Monetary phenomena are not the same in a highly developed and in a significantly underdeveloped economy. They differ also depending upon the relative importance, within any type of economy, of the balance of payments, the public sector, and the private sector." Much the same is true of banks, the most important money-creating institutions.

icy), and Russia (after about 1890) qualify in one way or another as "moderately successful." Only the French banking system gets definitely unfavorable marks, although even it had some points in its favor.

Admittedly, such judgments are impressionistic and partly subjective. Lacking generalized, unambiguous criteria by which to measure the performance of banking systems, they cannot be otherwise. The basic criterion employed above is the contribution the banking system made to industrialization and over-all development, but the evidence is neither complete, precise, nor unambiguous. Moreover, the banking system is by no means the only, or necessarily the most important, element in national economic development, and, unfortunately or not, other things do not remain equal. It is conceivable that a nation with a theoretically perfect banking system (if such existed) might have a lower rate of growth than one with a less highly developed banking system but more favorable conditions among the other elements of growth. Such unruliness and apparent paradoxes are unavoidable when one moves from the realm of pure theory to those of history and policy.

In this chapter we shall first summarize and synthesize the preceding case studies in order to make clear the objective evidence (such as it is) for the judgments above. In so doing we shall attempt to distill from the evidence some generalizations of more nearly universal applicability concerning the role of banks in the early stages of industrialization, answering insofar as possible the questions raised in Chapter I. Finally we shall essay a few tentative remarks on the applicability of those generalizations to contemporary underveloped economies.

FINANCIAL STRUCTURE AND FUNCTION

Both theoretical reasoning and the historical evidence suggest that the banking system can play a positive, "growth-inducing" role as well as responding passively to the demand for financial services. According to Professor Hicks,

> The beginning of a process of expansion (pretty much, I suppose, what Rostow means by "take-off") might occur because of real factors (inventions and the like) raising the real (prospective) rate of profit. But it might also occur because of financial improvements, diminishing the size of our "gap" [between the rate of return on real investment and the return received by holders of idle balances]; thereby permitting access to funds, for improvements which could have been made earlier, if the necessary funds had been forthcoming. It is not savings only that

> are required, but a channel of communications between potential savings and potential real investment.[2]

Moreover, the potential role of the financial system is greater in cases of derivative or imitative industrialization (i.e. in every case since England's original industrial revolution) because of the relative ease with which financial innovations can be introduced and the relative amenability of the banking system to sensible policy measures. For this reason the banking system should be of particular interest to policy makers in developing nations. It also helps to explain the paradox of Professor Gerschenkron's "advantages of backwardness."

Whether in fact the banking system actually plays the growth-inducing role that is theoretically possible depends upon its structure and behavioral characteristics. To review briefly the theoretical considerations in Chapter I from a slightly different standpoint, the financial requirements for industrialization can be conveniently summarized under three major headings: (1) the accumulation of capital; (2) the mobilization of capital; and (3) the efficient utilization of capital. (This classification can, of course, be applied to any economy, not only to those undergoing industrialization; but the precise nature of the problem will differ for underdeveloped and highly industrialized economies.) In addition, a fourth requirement is closely related: the necessity for an extensive monetization of the economy to alter the flow of resources from production directly to consumption to the more roundabout but productive channels of specialization, exchange, and markets. Financial institutions are especially concerned with the second of these requirements, and if they fail to function satisfactorily in that respect they can effectively prevent any significant degree of industrialization. This appears to have been the fate (not to the exclusion of other causes of backwardness) of portions of the Austro-Hungarian Empire and the countries of Mediterranean Europe in the nineteenth century, as well as of Russia before 1890 and (to some extent) of France and Germany before 1850. The role of financial institutions is not restricted to the mobilization of capital, however. They can facilitate its accumulation by stimulating saving and by serving as vehicles for the introduction of capital from abroad, as well as by "forcing" society to save by putting newly created money in the hands of entrepreneurs. Responsibility for the efficient utilization of capital lies primarily with nonfinancial entrepreneurs, but financial institutions exercise a large measure of discretion in the selection of investment alternatives and—potentially, at least—in the follow-up supervision of the productive processes in which the capital is employed. Finally, as the chief

[2] Hicks, *Capital and Growth*, 290n. Hicks also states that "I am prepared to believe that this last point has considerable historical significance."

dispensers of money in recent times, banks have played a major role in the monetization of the economy.

To repeat, the effectiveness with which an actual financial system discharges its functions depends in large measure on its structural characteristics. The following summary of the historical cases is intended to highlight structural features which are especially favorable or unfavorable.

The Legal Status of Banking

The legal status of banking as an industry varied considerably in the countries constituting our sample, but in every case banks were subjected to greater restriction and regulation than other forms of business enterprise.

Bankers had the greatest liberty in Scotland. There were two well-established chartered banks there before the onset of industrialization, and eventually Scotland obtained five chartered banks before the middle of the nineteenth century. Their larger size may have been related to their special legal status, but other than that it is difficult to see what privileges their charters conferred. Apart from the restrictions on small notes (less than £1) that prevailed from 1765 to 1797 and again after 1829, no legal obstacle prevented private individuals, partnerships, or unregistered joint-stock companies (without limited liability) from setting themselves up as bankers before 1845. It is clear that many did so, adding a degree of competition not to be found in any other country. (A few American states for short periods in the first half of the nineteenth century present the only comparable examples.) Although some private bankers did not issue their own notes, most did. Freedom of issue was evidently a greater inducement to enter banking than the privilege of incorporation, for no new banks were set up in Scotland after 1845, when incorporation remained a simple matter of registry but new note issues were prohibited.

English banking benefited from freedom of entry, but in other respects was severely hampered by restrictive legislation. Until 1826 the privileges of the Bank of England effectively prevented any but very small partnerships (fewer than seven persons) from engaging in banking operations. Freedom of note issue enticed many entrepreneurs into banking, giving England one of the highest banking densities in the world, but the restrictions on size doomed the English system to instability and periodic breakdowns. After 1826 the authorization of joint-stock banks eliminated the artificial limitation on size, but concurrent developments, which culminated in the Bank Act of 1844, imposed new restrictions on note issues and freedom of entry without remedying the instability of the system.

France, as the country most encumbered by restrictive legislation on banking, stood at the opposite extreme from Scotland. Until 1863 the privilege of incorporation was reserved to a very few institutions specially chartered and regulated by the state. Even more harmful to the development of banking, the right of note issue was similarly restricted until 1848, when it was conferred on a single monopolistic institution. Moreover, France did not grant legal recognition to checks until 1865. As a result, and in spite of vigorous efforts at financial innovation to surmount the obstacles imposed by legislation, the French banking system exhibited—with a few notable exceptions—very little of the vigor associated with free competition.

Belgium was unique in the combination of competition and monopoly imposed by legislation. Until near the end of the period of industrialization covered in this volume, the privilege of incorporation for banks was reserved to institutions specially chartered by the state. On the other hand, the Belgian government was somewhat more liberal in granting charters of incorporation than the French. (Private banks existed, but they were relatively unimportant.) Nevertheless, Belgium had only two banks that gained national significance before 1850. The resulting cutthroat competition led the government to confer a monopoly of issue on a single state-controlled institution in 1851. Thereafter the Belgium banking system evolved a variety of highly specialized institutions with relatively little competition among them.

Germany presented a kaleidoscopic spectacle. In all states the right of note issue—if it existed at all—was restricted to institutions chartered and usually controlled by the governments. Joint-stock banks were similarly restricted until 1848—even longer in Prussia, with minor exceptions. Private banks existed in considerable numbers, but without the right of note issue or legal recognition of checks they were obliged to develop novel financial techniques which in subsequent years made Germany a leader in industrial finance through the banking system.

The legal status of banking in Russia shows the continued primacy of the state both as regulator and direct participant in the economic system. The government itself assumed the responsibility for the issue of paper money until 1897, when it gave a monopoly of issue to the state owned and operated State Bank. At the same time, the government made a deliberate effort to enlist private enterprise in filling out the financial structure—subject, however, to continued paternalistic supervision. The poor results of the government's encouragement in the early years appear to have been due primarily to the backwardness and sluggishness of the private business community—itself in part a product of a long tradition of paternalism—requiring the authorities to assume a more direct role than they apparently intended in order to achieve their goals. Nevertheless, the omnipresent influence of the state and the formidable bureaucratic processes involved in chartering and operating ordinary commercial banks should not be overlooked as factors in the slow development of the financial system. When, in the 1890's and afterward, joint-stock commercial banks began to play a prominent role in the economy, they did so frequently with the assistance, both financial and technical, of foreign banks—again with government encouragement.

Japan exhibited a number of similarities to Russia in this respect, with the significant difference that private entrepreneurs in Japan responded far more vigorously to their opportunities to establish privately owned banking companies. Whether the pecuniary incentives were higher in the case of Japan, or whether other factors, such as some special quality of the Japanese entrepreneurial mentality, were more important is difficult to test; but the rapid growth of national banks after the liberalization of charter provisions in 1876 argues for the former. Whatever the motivation, the decision to charter a number of banks of issue using government bonds as partial backing for the issues proved to be a stroke of genius, even though it was subsequently revoked. Apart from its favorable political effects (including the value it imparted to government bonds), it undoubtedly hastened the formation of banks and accustomed the public to using bank money, thereby preparing the way for ready acceptance of the Bank of Japan's convertible notes as well as the deposit money of ordinary commercial banks. This instance fully verified Bagehot's prescription for habituating the population to the use of bank money.[3]

This brief survey of the legal status of banking seems to indicate that, for a country in which the banking system is but little developed, the right or privilege of note issue is one of the most effective means both of eliciting a rapid growth in the number of banks and of habituating the public to the utilization of financial intermediaries. To that extent, for reasons already indicated, it is (or was) favorable to development. Historically, free incorporation seems to have been less of an incentive to potential bank entrepreneurs than freedom of issue, but the limitation of corporate charters certainly discouraged the formation of new banking enterprises where such limitations were in effect. Other restrictions associated with the limitation of corporate charters tended to make the banking system weaker and more unstable. The prohibition or lack of clear legal recognition of financial assets other than specie and banknotes inhibited the growth of currency substitutes, thereby hampering the ability of the banking system to gain control of the idle financial resources of the society. On the other hand, where legislation or accepted custom permitted the use of deposits, bills of exchange, and similar assets as means of payment, the banking system was able to secure control of a larger fraction of the society's resources and could therefore exert greater leverage in the process of development.

[3] See above, p. 127.

Quantitative Aspects of Financial Structure

1. *Density.* The "density" of financial institutions is significant primarily in terms of their accessibility, but it also has an important bearing on the degree of competition in banking. Utilization of bank facilities, whether by borrowers or lenders (depositors), is costly in terms of time and convenience (witness the popularity of "drive-in banking" and "bank-by-mail" schemes in the United States), in addition to being costly in terms of the interest paid by borrowers, and the gap between cost and benefit is considerably less than infinity. Other things being equal, the larger the proportion of the population that has access to bank facilities (and the larger the share of financial assets passing through the hands of bankers), the more effectively the banking system can discharge its functions.

Density can be measured in various ways: number of offices per unit of area or per number of inhabitants, value of assets per number of persons or as a percentage of income payments, etc.[4] Any single measure of density is by itself inadequate, however. The number of offices per unit of area or inhabitants says little about the volume of bank resources available to service the area or its inhabitants. The ratio of bank assets to area or population does not tell us whether or not the assets are in fact available to the entire area and its population. Ideally one would like to be able to measure regional as well as national variations in density. Unfortunately, for the historical cases in hand we do not even have statistics to make all of the international comparisons we should like. Moreover, there are serious conceptual difficulties involved in the international comparisons because of differences in currencies and fluctuations of exchange rates and price levels. The difficulties are especially serious in comparisons of two or more countries in different chronological periods. In spite of these difficulties, it may nevertheless be worthwhile to try to distinguish significant variations, bearing in mind the inherent limitations of the concept and its measures.

The simplest measure of banking density is one which relates the number of institutions or offices to area. If distances between centers of population were great and communications facilities poor there would be a special advantage in a high banking density, but in fact those are precisely the circumstances in which density is likely to be lowest. Thus the ratio of bank offices to territory is not likely to be very significant for purposes of economic analysis. At the ends of the respective periods of interest in this volume Japan had the highest geographical density of bank offices, 26.4 per thousand square miles of territory. England and Wales also had a relatively high density—almost 19. The figure for Scotland was 12.8; it is probable

[4] Cf. Goldsmith, *Financial Intermediaries*, 5.

that Belgium was in the same range, Germany somewhat lower. France in 1870 had only 2.2 offices per thousand square miles. Because of Russia's vast extent of thinly populated territory, its figure is so small as to be almost entirely meaningless. (For some modern comparisons, for what they are worth, the United States in recent years has had about seven offices per thousand square miles, England and Wales about 170.)

These figures clearly suggest that geograpical density of banks is related to the geographical density of population. Accordingly, a somewhat more significant measure of density relates bank offices to total population. The actual index used here is constructed as follows:

$$\frac{\text{no. of bank offices} \times 10{,}000}{\text{total population}}$$

We arbitrarily define a ratio over 1.0—that is, more than one office per 10,000 inhabitants—as "high." A ratio between 0.5 and 1.0 is "moderate," whereas one below 0.5 is "low." A ratio of less than 0.1 is "very low." Unless otherwise indicated, the index refers to commercial banks only, excluding saving banks, mortgage banks, etc. Banks of issue are included if they did a significant amount of business with the general public, but not if they functioned principally or exclusively as central banks.

For England in 1750 the index was about 0.0625. Of course, when the total number of banks is small the index has very little significance. Most of the banks, in fact, were located in London; there were not more than a dozen in the rest of the country. By 1785 the density had more than tripled, raising the index to about 0.21. At the beginning of the nineteenth century every county but one in England and Wales had at least one bank, and the index stood at 0.48. By this time the inclusion or exclusion of London's banks and population made very little difference in the index, indicating a fairly widespread network. The index remained almost stable until the crisis of 1825–26, then rose rapidly again with the growth of joint-stock banks to a maximum for the entire period of about 0.77 at the end of the 1830's. (For modern comparisons, the banking density in England and Wales in 1960 was 2.2. In the United States in 1957 it was 1.27; it had reached a maximum of 2.94 in 1920).

Scotland shows both a relatively high initial density and a high rate of increase. Already in 1750 its index stood at approximately 0.1. It rose rapidly to about 0.3 in the 1770's, and to 0.58 in 1800. By 1845, when its index stood at 1.4, Scotland had almost twice the banking density of England and, overall, probably the highest for any country at a comparable stage of development. In the latter part of the nineteenth century the Scottish index rose to about 3.3.

France, on the other hand, shows one of the lowest banking densities of any developed country. In 1800 the index was less than 0.025. As late as 1840 it was still less than 0.1, or "very low," and had risen only to 0.12 in 1870. Even in the twentieth century, in spite of the spread of great banking chains, France had an unusually low density. In 1960 the index stood at about 0.67, lower than England in 1840. (The 1960 index figure does not take into account the several *banques d'affaires* or the numerous branches of the Bank of France, all of which performed some commercial banking functions.)

For the other countries, except Japan, we have insufficient data for even this simple measure. For Belgium in the early 1830's the index was approximately 0.22. For subsequent years the lack of information on private banks and branches of joint-stock banks makes it impossible for us to compute an index, but it is probable that if we could it would register 0.5 or more. The Prussian figures show surprisingly little variation during the period in question, falling slightly from 0.3 in 1820–21 to 0.27 in 1849, then rising to 0.34 in 1861. It is likely that an all-German index would be higher than this, especially in the 1860's. After 1870, of course, with the formation of joint-stock banks and the spread of branch banking (the latter less pronounced than in Britain or France, however [5]), the density increased substantially. Russia in 1914 had some 900 joint-stock bank offices, about 1,100 mutual credit societies, and perhaps 500 municipal banks, giving an index of approximately 0.17. If one includes branches of the State Bank, which also conducted some commercial bank operations, the index is slightly higher.

Japan experienced an extremely rapid increase in banking density, especially if one includes the officially registered quasi-banks. In 1876 the index was approximately 0.01; by 1884 it had risen to 0.32. It remained relatively stable in that range until 1895, when it again rose rapidly with the spread of branch banking. In 1901 the index reached a maximum of 0.77, then fell off slightly to 0.75 in 1914. Over-all, the Japanese index rose more than twice as rapidly as the English; in less than 30 years it reached the same point that the English had taken almost 100 years to reach.

Some measure of the geographical distribution of bank offices with respect to population would be more relevant than either a strict area or population ratio in discussing the adequacy of banking facilities. Only for England do we have reasonably precise data. (See map, Chart II.2, p. 28.) These indicate that, at the beginning of the nineteenth century, England was already well served by banks, in the sense of convenience of location. By 1841 not only were there more banks, they were also more uniformly distributed through the country with respect to both area and population. In fact, the areas with the smallest population densities had the highest ratios

[5] Reisser, *The German Great Banks*, 685–8, 695.

of banks to people. In Scotland, too, although the figures are less precise, there is no doubt that the country was well covered. Naturally the largest concentration of bank offices was in the heavily populated industrial belt between the firths of Forth and Clyde; but by the 1820's every town or village of as many as 1,000 inhabitants—and some smaller—had at least one office, even in the Highlands. Towns of 5,000 to 10,000 typically had three or four offices. In 1836, when the North of Scotland Banking Company was established in Aberdeen, then a city of 50,000, the city already had two local banks, each with several branches in the immediate area, and four branches of Edinburgh banks.[6] More extreme examples could be cited.

In Belgium in 1833, in addition to the Société Générale and its five branches, there were 84 private bankers. These provided reasonably good coverage of the main commercial and industrial centers, though without more information on the size and nature of the operations of the banks it is difficult to form a judgment about their effectiveness. In subsequent years there is no doubt that coverage increased in both depth and breadth; in the second and third quarters of the century private bankers flourished most, and there was also a growth in the numbers and branches of joint-stock banks; but precise figures are not available.

France suffered from financial overcentralization as well as political overcentralization. The large number of private bankers in Paris had relatively few counterparts in the provinces. Most major commercial centers had one or two, sometimes more, full-time bankers; but until after midcentury the majority of financial transactions in the provinces was probably handled by merchants or tax officials who were part-time bankers at best. So important a city as the port of Le Havre depended principally on agents of Paris bankers for financing its commerce.[7] Several important industrial centers, such as St. Étienne and Mulhouse, apparently had no banks at all until after 1850.

The largest concentrations of private bankers in Germany were naturally to be found in the larger commercial and administrative centers, but owing to the greater importance of these centers in Germany than in France the result was a more comprehensive coverage of the country. By means of correspondent relationships among the bankers, including those with part-time bankers in the smaller centers, each region of the country maintained financial contact with the others directly, instead of by way of a single major financial center as in France. The dearth of financial institutions in Russia outside the two major cities of St. Petersburg and Moscow, and the efforts of the government to counteract it, have already been discussed;[8]

[6] Keith, *North of Scotland Bank*, 15.

[7] Gille, *Banque et crédit*, 64.

[8] See above, p. 186.

nothing further need be added. Taking into account the quasi-banks and similar small credit institutions, the Japanese banking system was apparently relatively well dispersed in accordance with the density of population.

2. *Relative size of the banking sector.* The ratio of bank assets to national income might be regarded as simply another measure of banking density. The importance of this ratio is considerably greater than that of other measures of density, however, in that it also measures the size of the banking sector relative to the whole economy. (Ideally, perhaps, the best measure of size would be the ratio of assets in banking to total assets—financial and nonfinancial.[9] In the present state of our knowledge, however, the computation of such ratios is out of the question.) In a sense, the relative size of the banking sector, although not an infallible indicator, is the best practical measure of its efficiency in the performance of one of its two essential functions: securing control of disposable funds for investment.

Table IX.1 presents in summary fashion some of the relevant statistics that have been worked up in earlier chapters regarding the relative size of the banking sector in the different countries, plus a section on the United States to lend additional perspective. The specific figures and ratios are subject to all the reservations and qualifications mentioned previously. Despite the inevitable inaccuracies and crudeness, the table reveals some hitherto unsuspected patterns of indubitable significance.

Not much weight can be attached to international comparisons of bank resources per capita, except perhaps for comparison of Scotland with England and Wales and of Belgium with France, because of the problems of price fluctuations and exchange rate variations. (Price indices appropriate for the deflation of these series to permit valid comparisons do not exist.) The growth of bank resources per capita in Scotland was very much more rapid than in England. Although starting from farther behind, by 1800 Scottish banks controlled a larger fraction of national wealth in relation to population than did the English, and that in spite of substantially lower average incomes in Scotland. By 1845 the ratio was about two and one half times that of England. A similar result emerges in the comparison of France and Belgium. The two banking systems were on a par with respect to resources per inhabitant in the 1830's, but Belgium soon pulled ahead. By 1860 the Belgian banks commanded almost twice the resources per capita as those of France, and in the 1870's more than two and a half times as much. It is tempting to push the comparisons further, but more can be learned from studying the ratio of bank resources to national income.

Column 5 of Table IX.1 indicates a surprisingly wide range in the ratio

[9] See Goldsmith, *Financial Intermediaries;* Kuznets, *Capital in the American Economy,* esp. chap. 6; Gurley and Shaw, "The Growth of Debt and Money in the United States."

TABLE IX.1

COMPARATIVE FINANCIAL RATIOS

(1) Year	(2) Total Bank Resources	(3) Resources per Capita	(4) National Income	(5) (2) ÷ (4) × 100
		A. England and Wales		
	(£ million)	(£)	(£ million)	(per cent)
1775	20.5	2.7	135.0	15.2
1800	54.8	6.0	196.7	27.9
1825	80.0	6.2	270.0	29.6
1844	139.0	8.4	403.8	34.4
		B. Scotland		
1750	0.6	0.5	[a]	6.0
1800	13.25	8.3		41–69[a]
1845	60.0	21.9		89.6
1865	77.2	24.2		80.0

[a]See note on sources.

(1) Year	(2) Total Bank Resources	(3) Resources per Capita	(4) National Income	(5) (2) ÷ (4) × 100
		C. France		
	(million francs)	(francs)	(million francs)	(per cent)
1800	102.	4.	7,400(1801)	1.5
1830	576.	18.	9,600	6.0
1840	948.	28.		
1850	1,236.	35.	13,800(1847)	9.1
1860	2,250.	62.	19,300(1859)	11.6
1870	3,744.	101.	24,000	15.6
		D. Belgium[b]		
1825	81.	21.		
1830	63.5	16.	1,000–1,200	5.–6.
1835	102.7	25.		
1845	331.	78.		
1850	225	51.	1,600–1,700	13.–14.
1860	584	115.		
1875	1,323	249.	3,200–3,300	40.–42.

[b]Joint-stock banks only.

(1) Year	(2) Total Bank Resources	(3) Resources per Capita	(4) National Income	(5) (2) ÷ (4) × 100
		E. Russia		
	(million rubles)	(rubles)	(million rubles)	
1914	7,203 (comm. banks only)	41.	11,805	61.
1914	15,244 (all fin. inst.)			129.

TABLE IX.1 (cont.)

COMPARATIVE FINANCIAL RATIOS

(1) Year	(2) Total Bank Resources	(3) Resources per Capita	(4) National Income	(5) (2) ÷ (4) × 100
		F. Prussia		
	(million thalers)		(million thalers)	
1855	270		1,530	17.6
1865	610		1,960	31.1
		G. United States		
	($ million)		($ billion)	
1871	2,003	49.	GNP 6.71 (1869–73)	29.8
1884	4,221	76.	11.3 (1882–86)	37.5
1894	7,291	103.	13.6 (1892–96	53.5
1899	10,679	104.50	17.3 (1897–1901)	62.0
1900	10,011	103.10	NNP 15.8 (1897–1901)	63.3
1922	47,267	429.00	67.9	69.8
1929	65,621	540.00	95.8	68.5
1939	64,997	495.00	83.3	78.0
1952	188,603	1,219.00	323.0	58.5
		H. Japan		
	(million yen)	(yen)	(million yen)	
1878–82	107.5	2.9	667	16.1
1883–87	161.6	4.2	607	26.6
1888–92	191.3	4.8	809	23.6
1893–97	355.4	8.5	1,208	29.4
1898–1902	802.7	18.3	1,978	40.6
1903–7	1,328.6	28.5	2,522	52.7
1908–12	2,026.2	41.2	3,366	60.2

SOURCES FOR TABLE IX.1

A. Table II.1.

B. Table III.1; national income estimated as on p. 94, with the range in the ratio for 1800 depending upon whether one takes a high, low, or medium estimate of relation of Scottish to English per capita income.

C. Tables IV.1 and IV.2.

D. Table V.1. National income estimated on the assumption that per capita incomes in Belgium and France were approximately equal.

E. Table VII.10.

of bank resources to national income. The countries with the highest ratios at the end of the period investigated (and the highest growth rates within it) were also the countries that experienced the most rapid spurts of industrialization. Scotland is outstanding both for the very high ratio it achieved and for the rapidity with which it grew. Japan and Russia, both latecomers among the industrial nations, achieved a relatively high ratio in a very short time. (Lack of national income data for Russia inhibits the comparison, but one can infer a high rate of growth in the ratio from the fact that assets of the commercial banking system grew from less than 1 billion rubles in 1875 to more than 7 billion in 1913.) Although the Belgian ratio stood at less than 50 per cent in 1875, it should be remembered that the Belgian figures for bank assets refer to joint-stock banks only. It is quite likely that if private banks could be included the ratio would be in the vicinity of 60 per cent. In any case, the Belgian ratio is remarkable for the rapidity of its rate of increase, especially from 1850 to 1875. Given the fact that in each of these cases the main outlines of the banking system and the principal banking institutions had been established prior to the onset of rapid industrialization, the pattern strongly suggests a major role for the banking systems as determinants of that rapid industrialization.

The French ratio is remarkable for its very slight development even by 1870, leading one to suspect the figures on which it is based. If they are not wildly inaccurate, however, the weak and tardy development of the banking structure which they depict may very well hold the key to understanding the peculiar pattern of economic development in France. That is to say, the monetary and banking policies of the French government, which repressed and retarded the free development of the banking system, must bear a large share of the responsibility for the relatively slow progress of industrialization.

Comparable data for Germany are unfortunately not available. The Prussian data, such as they are, indicate that Prussia achieved a financial capability by 1855 that France had not yet achieved in 1870, and by 1865 was as highly developed financially as England in the 1830's or the United States in the 1870's. This suggests that private bankers in Prussia before

F. Tables VI.4 and VI.6; national income from W. Hoffmann, *Deutsche Volkseinkommen*, p. 186. Figures for bank resources do not include equity capital or liabilities of the *Landschaften*. Inclusion of the latter would raise the ratios to 29 per cent in 1855 and 45 per cent in 1865.

G. 1871–1899: U.S. Bureau of the Census, *Historical Statistics of the United States, Colonial Times to 1957* (Washington, D.C., 1960), Series A-2, F-1, and X-21 ("all banks—total assets or liabilities"). 1900–1952: net national product as above, series F-6; bank resources from Raymond Goldsmith, *Financial Intermediaries*, Table A-2, "operating commercial banks—total assets."

H. Tables VIII.5 and VIII.6, Ohkawa, *Growth Rate*, and Bank of Japan, *Historical Statistics*.

1855 were unusually successful in circumventing the restrictions placed on the monetary and banking systems by official policy, but it is also significant that the rapid growth in the ratio between 1855 and 1865 corresponds in timing with both the liberalization of Prussian banking and monetary policy and with Prussia's first spurt of rapid industrialization. In terms of the general relationship between banking and industrial growth, it is also significant that the period after 1870, when Germany achieved its highest industrial growth rates of the century, was also a period of virtually complete freedom (except for note issue) for German banks.

England might in some respects be considered the "normal" case, with the others constituting variations or extremes. Although the English series stops in 1844, other data suggest that the ratio continued to rise gradually during the century, probably reaching 50 or 60 per cent at the beginning of the twentieth century. In the 1920's the ratio of the assets of London clearing banks to United Kingdom national income was approximately 50 per cent, and England and Wales accounted for approximately 85 per cent of United Kingdom income. The United States, which we have not considered in previous chapters for the reasons indicated in Chapter I, might also be considered a "normal" case. Projecting the United States series backward requires somewhat more daring and imagination, but it is not inconceivable that the ratio for the United States in 1800 was less than 5 per cent, and that it attained perhaps 15 or 20 per cent in the 1830's. It seems likely that in the cases of England and the United States the banking systems were essentially passive or permissive, responding fully to the demand for financial services but not inducing growth directly. (This generalization should be accepted very tentatively for the United States pending further study. Certainly there were periods in American history—the 1830's and possibly the period from about 1885 to 1929—when the banks played a more positive, active role.) If so, this may help explain why English and American banking theorists generally deprecate the contribution of banks to industrial growth.

These findings are broadly consistent with those of Goldsmith in his study of financial structure in developed economies, which deals primarily with the twentieth century.[10] Goldsmith chose to relate financial assets to national wealth rather than income, and included savings banks along with commercial and central banks. The results of his computations are presented in Table IX.2. If the proportion of national wealth to national income is on the order of four or five to one, then the ratios are closely comparable to what might result from a projection of the series in Table IX.1 They show that France closed the gap somewhat between 1870 and 1912—as one would

[10] Goldsmith, "Financial Structure and Economic Growth in Advanced Countries," in Abramovitz (ed.), *Capital Formation and Economic Growth*.

TABLE IX.2

BANK ASSETS AS PER CENT OF NATIONAL WEALTH

Country	Date	Bank Assets as Per Cent of National Wealth
United States	1912	16
Germany	1913	14
United Kingdom	1913	12
France	1913	10

Source: Goldsmith, "Financial Structure and Economic Growth in Advanced Countries," in Abramovitz (ed.), *Capital Formation and Economic Growth*, Table 9, p. 151.

expect from the widespread development of branch networks by the large Paris joint-stock banks—but still lagged behind the industrial leaders. Goldsmith's data, like those of Table IX.1, also suggest a fairly close correspondence in rank order between the rates of growth of the banking system and the rate of industrial growth.

The empirical relations indicated in Table IX.1 prompt a number of questions. What are the determinants of the ratios of column 5 and their rates of growth? Is there an "optimal" growth path that the ratio should follow in the course of development? Is there a practical ceiling on it, or can it continue to rise indefinitely? Do the ratios reveal—or conceal—any causal relationships between financial development and industrial growth?

One obvious determinant of the relative size of the banking sector is the extent to which bank liabilities fulfill the role of money. The contrast is striking in the cases of Scotland and France, and explains much of their deviation from the mean. The same factor helps to explain the rapid growth of English banking before 1800, Belgian after 1850, and Japanese after 1875. The use of bank liabilities as money in these cases (except for Belgium) is, in turn, related to the question of freedom of note issue. In almost every case the most rapid expansion of the banking system occurred during the phase when banknotes constituted the most important liabilities of the banks. (Even Belgium is not exceptional in this respect.) In most countries deposits subsequently assumed greater importance, but it is also worthy of note that demand deposits were most widely used in countries that had previously become accustomed to the note issues of competing independent banks. The public of countries with monopolistic issuing agencies evidently regarded banknotes as a form of fiat money; in any event, they did not develop the "banking habit" to the same degree as did the public in countries in which relative freedom of banking prevailed. The consequence of this may be observed in France and Germany, where the general public still makes relatively infrequent use of checks.

The "latecomers" to industrialization (at least in the cases considered here) typically not only demonstrated a higher rate of industrial growth, but also a more rapid growth of the banking sector. This is consistent with the hypothesis of the "advantages of backwardness" of Professor Gerschenkron.[11] Just as underdeveloped nations have the option of adopting the most modern technological devices, they have the option of adopting the latest and/or most efficient techniques of banking—and at considerably less real cost. This was obviously important in the cases of Japan and Russia, as well as in Germany in the 1850's and 1860's. (Though they are outside the range of this study, the Scandinavian countries also benefited from the banking experience of other countries, especially that of France and of Germany.)[12]

Another factor of considerable importance in the cases of Russia and Japan was the role of government. Government initiative and desire to industrialize have frequently been cited as contributory factors in the rapid industrialization of those two countries. It is important to recognize, therefore, that government influence was exercised in large part through the banking system. In addition to such measures as governmental presure on the central banks to lend liberally in periods of financial strain of crisis, government deposits accounted for a substantial part of the total resources of the banks in countries in which private resources were extremely limited. Unlike the governments of some contemporary underdeveloped countries, who use the money-creating facilities of the banking sector as an alternative to taxation, the governments of Russia and Japan transferred resources obtained by taxation to the banking system. In effect, they used a combination of the tax power of the state and the specialized talents of bankers to shift resources from consumption to production.

The question of an "optimal" growth path is difficult to answer. If the banking system expands its loans and liabilities too rapidly it may generate inflationary pressures, which, especially if the economy is relatively open to foreign competition or is tied to an international monetary standard, may cause balance of payments problems that will result in setbacks to the effort to industrialize. At one time or another all the countries in our sample experienced such strains as a result of too rapid expansion on the part of banks: for example, Scotland in the 1760's leading up to the crisis

[11] *Economic Backwardness in Historical Perspective*, esp. pp. 11–16.

[12] This was brought out by Professor Hildebrand at the Bellagio conference mentioned in the Preface. See also Cameron, *France and Europe*, 129, 164–5, 490–91.

of 1772; England in 1825 and 1836; France in 1837; Belgium in 1838; Germany in the 1850's and again in 1872–73; Japan in 1881–82; Russia in 1900–1901. (Not all financial crises resulted from such expansion; essentially noneconomic events were responsible for the crises of 1797 and 1848, among others. Still others were propagated internationally, and had little if anything to do with bank expansion in certain of the affected countries.) Russia was persistently hobbled in its effort to industrialize by balance of payments problems; Japan solved similar problems by *de facto* depreciation. Thus it is possible for the banking system to expand too rapidly. On the other hand, it can certainly expand too slowly. The "optimal" rate of expansion would seem to be that which is consistent with the full employment of resources, reasonable price stability, and the ability of the economy to absorb and utilize technical innovations; unfortunately, such criteria cannot be specified in advance or generalized to cover all situations.

It is tempting, nevertheless, to conceive of a theoretically optimal long-term path of bank expansion. It would probably resemble a logistic curve, rising rapidly at first owing to the monetization of the subsistence sectors of the economy, the substitution of bank liabilities for commodity money, and the existence of banks as the sole or only significant institutions capable of supplying external finance for industry and commerce. Subsequently the rate of expansion would drop off (relative to the economy as a whole, at least) as business firms become able to supply more of their own financial needs by retained earnings, and as other financial institutions arise to compete with the banks for the investable funds of surplus spending units. The growth of non-bank financial intermediaries might eventually reduce the relative share of banks in total assets; both Goldsmith and Gurley and Shaw have noted this fact in highly developed economies, especially in the United States.[13]

Chart IX.1 shows how actual performance might compare with the hypothetical optimum. The data come from column 5 of Table IX.1, but the time scale is a composite one. This introduces an element of arbitrariness to match the crudeness and incompleteness of the data; if different "time-zeros" were chosen the chart would convey a slightly different impression (e.g. if t_0 for the U.S. were 1789 or 1865 instead of 1800), but the individual growth rates are not affected. As it is, the following "time-zeros" have been selected:

England	1750
Scotland	1750
France	1800

[13] Goldsmith, "Financial Structure and Economic Growth in Advanced Countries," in Abramovitz, *Capital Formation and Economic Growth;* Gurley and Shaw, "Financial Aspects of Economic Development," and other writings (see List of Works Cited)

Belgium	1800
Prussia	1800
United States	1800
Russia	1860
Japan	1870

The "curves" for Japan and Scotland come closest to the hypothetical optimum. (Selecting the median estimate for per capita income of Scotland in 1800 appears to give the line a constant slope; a higher or lower estimate would show more of a curve, but a logistic curve can easily be fitted to the three fixed points. The ratio for 1865, not shown in the graph, is actually lower than that for 1845, probably indicating a leveling of the curve in the years before 1845.) This accords with other evidence on the efficacy of the Scottish and Japanese banking systems, and thus supports the notion of a theoretical optimum. The data for Belgium and the United States, as well as that for Prussia and Russia, could also be fitted to a logistic curve, but the first two in particular seem to have suffered a long delay before beginning their financial "take-offs." The appearance could be the result of the arbitrariness of the "time-zeros." On the other hand, it is likely that the United States, at least, had already completed one cycle, with a sharp rise in the 1830's—the heyday of wildcat banking—and a leveling off before or during the Civil War. A similar, though more muted, phenomenon appears to have occurred in England: there was a sharp rise in the ratio associated with the growth of country banking; a leveling off after the Napoleonic Wars, perhaps indicating a saturation point for banknote issues; and a renewed rise with the advent of deposit banking on a large scale. The French data give only the barest hint of following a logistic pattern, and that at the very end of the period under consideration. The evidence presented in Chapter IV, of the desire of the French business community for increased banking facilities in the 1820's, 1830's, and 1840's, indicates that the French curve, too, might have followed the theoretical optimum but for the artificial restrictions imposed upon the banking system.

3. *Size and concentration of bank power.* The issue here is the size of the individual bank unit, and the degree of concentration of assets within the banking sector. Is there an optimal size or an optimal degree of concentration or dispersion? The question is obviously related to that of branch vs. unit banking, and to the degree of competition within the banking sector, both to be discussed below. Here we indicate merely the trends in the experience of the countries studied. Since in most cases our estimates of total resources in banking were based on estimates of the average size of the estimated number of banks, there is little point in breaking the aggregates down to obtain the original estimates. The best we can do with this

CHART IX.1

BANK ASSETS AS PER CENT OF NATIONAL INCOME

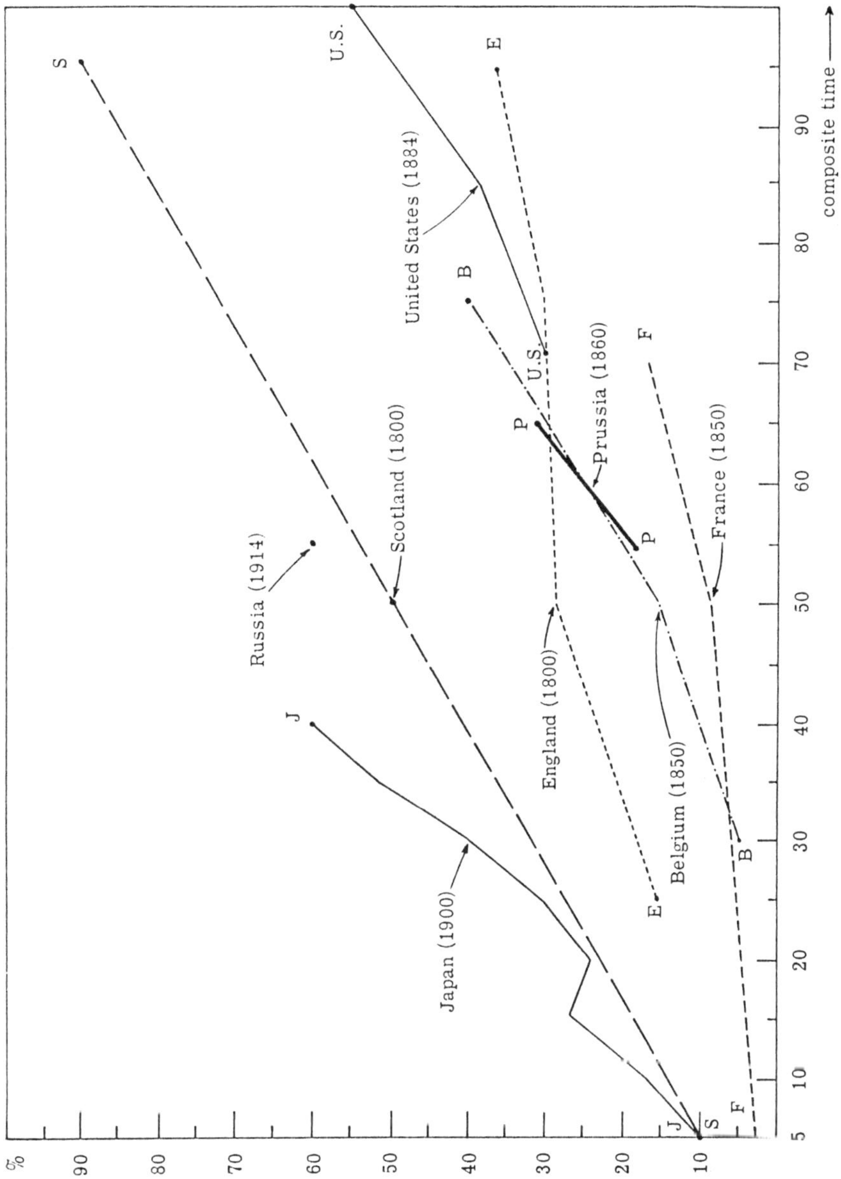

essentially quantitative question is to deal in nonquantitative generalities, with a few percentages thrown in where they are available.

At one time or another all of these countries had one or more very large banks with special privileges—most of which eventually became central banks—and numerous very small banks. The relative importance of the privileged banks tended to diminish over time, but it always remained great. Typically their assets constituted between one-fourth and one-third of the total assets of the banking system; proportions as high as one-half and even higher are not unknown. Usually but not always the large size of the largest bank was related to a monopoly or near-monopoly of note issue. In 1865 the Bank of Prussia controlled 43.6 per cent of the assets of all joint-stock banks in North Germany. The Russian State Bank owned about 30 per cent of the resources of the commercial credit system of Russia until the 1890's. Until after 1905 the Bank of Japan owned between one-fourth and one-third of the assets of the total financial system, and by itself was one-third to one-half the size of the whole commercial banking system.

Nonprivileged banks usually went through two distinct phases with respect to size and concentration. The first phase was that of increasing average size but decreasing concentration. At first the number of private banks (including privately owned joint-stock banks, if any) would be small. The formation of new banks of course reduced the degree of concentration, but average size generally increased as existing banks grew by the reinvestment of profits and new banks entered with larger capitals than earlier banks had. The second phase saw continued growth in average size along with increasing concentration. In most cases the increase in concentration set in at or very near the end of the periods treated in this volume, sometimes (in Germany, for example) shortly afterward. One might almost date the end of the "early stages of industrialization" by reference to increasing concentration in the banking sector. Usually the movement toward concentration was associated with the growth of joint-stock banks and the establishment of branch banking. Joint-stock banks were sometimes created by the transformation or merger of existing private banks. In other cases they frequently absorbed older private banks into their growing chains. The tendency was for private banks to disappear altogether, but a few hardy sports still survive in every country except Scotland and Soviet Russia. After devouring most of the private banks the giant joint-stock banks turned on one another: the ultimate development in each country saw a relatively small number of huge banks, serving the entire nation, engaged in oligopolistic competition. But that development occurred long after the passing of the early stages of industrialization.

Branch Banking vs. Unit Banking

Both the branch banking system and the unit banking system have advantages and disadvantages. At first glance the advantages of branch banking seem greater, as do the disadvantages of unit banking.

Other things being equal, branch chains have larger resources, which enable them to withstand sudden emergencies more easily. They can also diversify their risks to a greater extent, and thus they are less susceptible to emergencies. Branch banks can also shift resources geographically with greater ease and less cost; normally this should result in a smoother functioning national capital market, with a greater tendency toward equalization of rates of return on capital in different regions and industries. Great chain banks can also benefit from economies of scale by employing specialists in various branches of finance such as the call money market, the bond market, etc. Under certain circumstances, depending on governing legislation, they can also work with smaller cash reserves in proportion to total assets.

Offsetting these advantages, large chain banks can suffer from the diseconomies of scale involved in the multiplication of supervisory officers and other staff positions characteristic of large-scale bureaucratic organizations. How important these diseconomies will be depends in part on the geographic spread of the branch network and the technology of communications. The general tendency of the progress of communications has been to reduce their significance. Prior to the advent of the railway and electric telegraph only Scotland had developed branch banking to any significant extent. More serious is the loss of intimate personal contact and detailed knowledge, both within the bank organization and between the bank and the public. The Scottish pioneers of branch banking solved this problem by selecting their "agents" (not branch managers) from the outstanding businessmen of the communities in which they located branches, giving them a large measure of decentralized decision-making authority, and at the same time involving them in the financial responsibility for their operations. Subsequently both the Scottish banks and other chain banks adopted more bureaucratic forms of organization.

Another potential danger of branch banking is the reduction of competition that may result if the banking system falls into the hands of a few oligopolistic chains through the elimination of independent local banks. A related danger, one that showed itself in France after 1870 in particular, is that the large banks may use their network of branches merely as a means of draining the hinterland of surplus funds, and then use those funds in low risk but relatively unproductive employments in the metropolis or abroad, without seeking to develop local resources. How serious a problem this was in other nations is difficult to ascertain with the evidence available. No doubt it tends to be more serious in nations such as France where there is a high degree of concentration of both political and financial power in a single city.

The disadvantages of unit banking are virtually the mirror image of the advantages of chain banking: relatively limited resources (this does not rule out the possibility of a few large unit banks in major cities, or of joint-stock banks of the *mobilier* type that operate on a national scale from a single location); difficulties in the way of diversification of risks and of forming a unified, efficient nationl market; and inability to benefit from economies of scale. The major compensating advantage of independent local banks is that they are more likely to be attuned to the needs and potentialities of local communities and interested in developing local resources. The contrast between the English country banks and the local Scottish banks, on the one hand, and the French and Russian joint-stock banks of the late nineteenth century, on the other, is striking both in methods of operations and results. For that matter, though the English and Scottish systems after mid-century lie outside the scope of this study, it would be interesting to compare the performance of the country banks with that of their latter-day successors, the branches of the English "Big Five." It was one of the last surviving private banks that, in 1910, enabled the founder of Morris Motors to get his start (see above, p. 57). One wonders if he would have been so fortunate in dealing with a branch manager of one of the Big Five.

In considering the relative utility of branch and unit banking, one must take into account the historical trend, which seemingly favored the former. The earliest banks in all countries engaged initially in unit banking. Widespread branch banking had to await the adoption of the joint-stock form of organization for banking; after that branch banking grew rapidly at the expense of unit banks, except for special purpose banks or where branch banking was restrained by special legislation, as in the United States. This record would seem to indicate the overwhelming superiority of branch banking, at least from the viewpoint of stockholders. Yet one wonders if there may not have been a divergence between private and social benefits in that case.

No account has been taken in this discussion of branch systems of central banks or banks with monopoly privileges in the matter of note issue. The historical record is mostly negative on this point. Neither the Bank of England nor the Bank of France, the two leading examples, established branches until forced to by competition or governmental pressure. The Prussian Bank, the Banque Nationale de Belgique, and the Bank of Japan were all latecomers to their respective scenes, and were all creatures of the government, as was the State Bank of Russia. The evidence, such as it is, seems to indicate that insofar as such banks act as lenders of last resort and monopolistic note issuers—if there is to be monopoly—they should be endowed with a network of branches.

Competition in Banking

The question of competition in banking is intimately related to the legal status of banking, bank density, size and concentration, and the relative merits of unit and branch banking. What follows, therefore, is in the nature of a summary of summaries.

Majority opinion, even among groups otherwise favorable to competition in economic life, seems to have regarded the banking system as an area to be preserved from competition or, at least, protected from "excessive" competition. Virtually every country at one time or another has imposed restrictions on banking, with the effect—intentional or not—of limiting competition among banks. In view of this well-nigh universal current of opinion it should be stated in the most unequivocal terms that there is no historical justification whatever for believing that competition in banking is, in general, detrimental to the interests of society as a whole. On the contrary, insofar as the criterion for judging bank performance is the contribution of banks to growth, the best results have been achieved when competition was freest and most unfettered. Once again the examples of Scotland and Japan (and of some American states during the period of free banking) spring to mind. Even Germany in the era of the "great banks," when competition was free if not perfect, was no exception. This suggests that the theoretically optimal growth path of the ratio of bank assets to national income is also the "natural" growth path that would result from free competition. Competition in banking is also, in general, favorable to financial innovation, although in certain circumstances the artificial restrictions placed on banking may likewise serve as a spur to financial innovation if the demand for financial services is strong.

Competition in banking is related to the question of competition in industry. In general the two flourish—and decline—together. Whether this phenomenon is a joint by-product of other circumstances, or whether it results from the decline or restriction of competition among banks, is a matter worthy of further research. It is a striking coincidence, in any case, that industrial structure—competitive, oligopolistic, or monopolistic—tends to mirror financial structure.

FINANCIAL INNOVATION

A financial innovation is any new financial technique or institution intended to facilitate the exchange of goods and services, the settlement of debts, or the provision of producers of goods and services with the resources they require. In the broad sense, banks and even money itself were at one time financial innovations, although far antedating the period of concern in this volume. Just as industry and agriculture progress by means of new techniques—innovation—so the introduction of new financial techniques is the mark of progress in the financial sector. Not uncommonly financial innovation has been the prerequisite of successful technical innovation.

Theoretically the success of an innovation can be measured by reduction in cost or the increased productivity of labor and other resources it permits. The same is true in principle of financial innovations, but measurement of the productivity of new techniques of finance may be extremely difficult because the benefits may be external to the firm introducing the innovation. Some financial innovations, such as the introduction of clearing houses or the use of business machines for record keeping, are not substantively different from technical innovation in industry, and the measurement of their productivity can be easily handled by conventional methods. Other, more important types of financial innovation, such as the introduction of banknotes and the practice of making payments by check in place of currency, are essentially social innovations. Benefits to the firms introducing them accrue not in the form of cost reductions for services performed, but in the long-run expansion of their business. The reduction in cost is widely dispersed throughout society as a form of social saving.

Financial innovations by banks may be classified under three headings: (1) those relating to a bank's ability to attract resources (i.e. those affecting its liabilities); (2) those relating to its use of resources (i.e. its assets); and (3) innovations in the form of financial institutions. The first is of special interest, in that certain bank liabilities performed the functions of money. The eighteenth and nineteenth centuries were prolific in financial as in technical innovations. It is therefore a matter of some interest to know who made the innovations, when, why, under what conditions, and what effects they had.

In the middle of the eighteenth century the dominant form of money was commodity money, principally gold, silver, and baser metals. Banknotes existed, but they were of limited use in a few localities only. Payment by assignment in bank (checks, *giro*) had been practiced for several centuries, but also on a very restricted basis. Bills of exchange may also have been used as a means of current payment, although their characteristic use was for distant payments at deferred dates.

The outstanding feature of the monetary history of the last two centuries, the nineteenth in particular, has been the substitution of various forms of bank-created money for commodity money.[14] The first modern banknotes were issued by the Bank of Stockholm, forerunner of the Swedish Riksbank, in 1656. The experiment was not successful and they were soon withdrawn. Soon afterward goldsmith bankers in London began to issue deposit receipts payable on demand, which, in effect, were banknotes. The Bank of England issued banknotes from the beginning of its operations, and John Law's Banque Générale in France issued them in superabundance. But it was the Scottish banks, soon followed by English country banks, that generalized banknotes in practice. Their major refinement was the introduction of notes of relatively small denominations, £1 and even less. The motivation in both cases seems to have been the scarcity of specie, especially coins of small denomination. This was brought about in England by the inefficiency of the Mint, but it was aggravated in the case of Scotland by its unfavorable balance of payments with England. From Britain the use of banknotes spread to the Continent, though haltingly at first. Eventually banknotes became the most important component in the money supply of several countries. Unlike Britain, most Continental countries did not tolerate the private issue of money claims, or did so only under strict government control. The result was that the issue of paper money came to be centralized in the hands of specially chartered and regulated institutions, as indeed it did in England after 1844.

Bank deposits first existed in medieval Italy in the form of *giro* accounts, which required that both payor and payee have accounts with the same bank. They were not widely used in the sixteenth and seventeenth centuries except in Amsterdam and two or three other major commercial cities. The goldsmith bankers of London also accepted deposits payable on demand, but such payments were usually made directly over the counter to the drawee or his agent rather than being settled by transfers on the bankers' books. The London Clearing House, dating from 1773, indicates that checks were becoming increasingly common, at least among the more important merchants and bankers of the metropolis. Probably not until the 1820's or 1830's, however, with the growth of non note-issuing joint-stock banks, did deposits subject to check come to occupy an important place in the stock of money or current means of payment in England. By that time they were also widely used in Scotland. Scottish banks originally operated almost entirely on the basis of note issues, but with the strong demand for capital

[14] Cf. Triffin, *International Monetary Standard*.

that arose in the 1780's they made vigorous efforts to attract deposits by means of interest differentials for demand, or current account, deposits and time deposits. The Scottish banks also democratized deposit banking, as they had note issues, by accepting deposits for as little as £10.

The use of checks in Britain was facilitated by legislation and judicial decisions dating from the seventeenth century which provided for summary diligence against drawers of bad checks. Curiously, no similar legal provisions appear to have existed on the Continent. German bankers made fairly extensive use of *giro* credit and current account overdrafts, but demand deposits subject to check made very slow headway on the Continent in spite of the British example. This was no doubt partly due to the lack of legal definition until the 1860's, but since checking remained of secondary importance even afterward it may have reflected the preference of bankers for liabilities over which they had greater control, such as acceptance credits and time deposits. Russian banks followed Continental practice, whereas Japan adopted and adapted the British and American systems.

One other attempted innovation affecting bank liabilities failed, apparently due to official hostility, although lack of market demand may also have been a factor. The Société Général de Belgique raised additional funds in the 1830's by issuing its own debentures. In later years some German, Russian, and Japanese banks, especially special purpose banks, did likewise Jacques Laffitte in the 1830's and the founders of the Crédit Mobilier in the 1850's attempted to carry the idea one step further by issuing short-term interest-bearing bonds as a means of circumventing the Bank of France's monopoly of note issue. Laffitte had moderate success, but the crisis of 1848 put an end to his bank. The Crédit Mobilier made only one issue, then bowed to official pressure against them and made no others. Interestingly, the idea has been picked up again in modified form by large New York and London banks in the twentieth century.

The gradual assumption by bankers of the function of money supply not only made the stock of money more elastic and expansive, but also facilitated economic growth by economizing on the use of specie, a form of money with much higher real cost. It is significant that France, the country that made a fetish of the use of specie, also suffered the lowest rate of growth. The most decisive advantage of the use of bank money, however, was that it gave to bankers, the specialists and presumed experts in the business of lending to deficit units, a large leverage over real resources which they could transfer directly to enterprising entrepreneurs.

Innovations with respect to the use of bank assets mostly took the form of novel ways of granting credit. Loans on mortgages and promissory notes were already standard in the eighteenth century, but the former had the disadvantage of tying up bank assets for long periods of time, whereas the latter frequently lacked the guarantees looked for by cautious bankers. Moreover, until after the French Revolution usury laws effectively hindered the use of promissory notes on the Continent. As a result self-liquidating bills of exchange became the favored instrument of credit in orthodox circles. Bills of exchange had the additional advantage that they could be used as means of current payment. In areas where banknotes were not permitted, or where, as a result of sad experience, the public preferred not to use them, businessmen resorted to the use of bills of exchange as means of payment on a large scale. The effective use of bills in this manner required the existence of banks as ultimate sources of credit by means of discount.

Not all areas nor all entrepreneurs could offer the kinds of security required for bills of exchange. Frequently the need for fixed capital in modest amounts, or even working capital with a turnover period longer than the customary 90 days, took precedence over the need for short-term credit for the movement of goods. In such cases entrepreneurs and their bankers had to devise new methods of finance. Where banks depended upon note issues for their profits—principally in Scotland and England (and also in Japan under its national banking system)—current account advances or "cash credits" sufficed. Elsewhere—in Germany, for example—more subtle and elaborate devices were needed, and banks came to use current account overdrafts and eventually bankers' acceptances. Out of these practices developed a highly elaborate system of *giro* accounts with bankers performing the function of a clearing house for payments between its client firms.

Institutional innovation took the form of the creation of new institutions to perform new functions, and of adaptation of existing types of institutions to suit special circumstances. Among the former were savings banks, mortgage banks, and mutual credit societies. The most important from the standpoint of industrial finance was the joint-stock investment-promotional bank, of which the outstanding examples were the Société Général de Belgique and the Crédit Mobilier of France. Outstanding examples of institutional adaptation were the amalgam of banks of the *mobilier* type with the practices of the German private bankers in the great German Kreditbanken, which performed "mixed" banking functions, and the adaptation by the Japanese of American, British, and Continental models to suit the peculiar needs and circumstances of the Japanese economy.

The significance of this record for contemporary underdeveloped economies is that they should not depend upon slavish imitation of the financial techniques and institutions borrowed from their wealthier neighbors, but should not hesitate to adapt them to their own particular circumstances and, where possible or necessary, to invent new devices to solve their own special problems.

PAST AND PRESENT

One of the clearest lessons to be derived from this study of history is that no single model of a banking system is appropriate for all economies. It is not to be expected, therefore, that the historical experience of the industrialized nations of Europe or elsewhere will be directly applicable in countries that differ widely in culture, resources, historical traditions, and many other respects. In any case it would be the height of folly to attempt to prescribe banking or monetary policy or reforms for a given country on the basis of either history or theory without a detailed and intimate knowledge of that country's economy and society. We make no such prescription. It does not follow, however, as we have already argued, that nothing at all can be learned from history.

In order to ascertain whether and to what extent the experience of the past is valid in the world of the present it is necessary to specify the major similarities and differences in the respective circumstances relevant to our problem. Apart from the obvious differences in culture, tradition, level of technology and the like, there are a number of fundamental differences relating specifically to financial, monetary, and banking matters. Unlike the people of Europe at the beginning of the nineteenth century, today there is virtually no sizable social group that is not familiar with banknotes or other forms of paper money, and that does not have at least a rudimentary consciousness of the existence of banks and their functions. Although many underdeveloped economies are as yet incompletely monetized (that is one of the measures of their backwardness), the scope for industrial capital formation by means of rapid monetization through the banking system is greatly reduced.

With respect to monetization, today's underdeveloped countries are considerably more "advanced" than were their counterparts of 150 or 200 years ago. In other respects, however—proportion of labor force in agriculture, standards of literacy and education, industrial and commercial traditions, etc—many of them are considerably more backward. This indicates that not only will extensive—and expensive—programs of agrarian reform, educational development, and such be required to increase productivity and habituate the populace to modern techniques and procedures, but also that novel financial institutions and practices may be necessary to activate the latent entrepreneurial talents and productive energies of hitherto tradition-minded peoples. In that connection it should be pointed out that, in most contemporary underdeveloped economies, banking practice and policy, although not exactly "orthodox" (central banks in particular), follow far too slavishly the canons of their more advanced neighbors, where conditions and traditions are quite different.

In one important respect contemporary underdeveloped countries have liberated themselves from bondage to an outmoded and expensive fetish. No longer do gold and silver figure prominently—or at all—in the monetary circulation of any nation. As everywhere, however, liberty has its price. Perhaps the outstanding merit of a metallic monetary standard for nations that were committed to one was that it kept governments as well as bankers and businessmen within reasonable bounds when computing expected income and expenditure. Today's underdeveloped nations no longer have to bear the burden of a large stock of unproductive capital in their monetary systems, but all too often they must pay an even higher price in the form of an improvident, unproductive, and corrupt government bureaucracy and a derangement of the monetary system.

Central banks under state control with legal monopolies of note issue are probably here to stay. This seems to be an unavoidable but not necessarily felicitous conclusion. Historically local banks of issue have played an important role not only in habituating the populace to the use of financial instruments and institutions, but also in the development of small-scale local industry and agriculture. Conceivably this can be done by means of deposit money, but in relatively primitive conditions of production and marketing, where laborers and entrepreneurs alike have but little sophistication in the use of modern financial instruments, there is much to be said for local banks of issue, using, perhaps, a standard form of banknote such as the national banknotes of the United States and Japan. It would, in any case, be highly instructive if such experiments could be carried out in countries like Brazil and Nigeria, where there are still some traditions of regional autonomy.

It is frequently argued that the shortage of capital and of entrepreneurs in contemporary underdeveloped economies are reasons why the state must play a larger direct role in economic development than was characteristic of the now-industrialized nations in their respective early stages of industrialization. Such arguments are unsound historically. It is by now abundantly clear that the rate of savings and capital formation among the industrializing nations of the eighteenth and early nineteenth centuries was much less than had previously been supposed. Many if not most of today's developing nations have higher rates of capital formation. If the results are not wholly satisfactory, it may be due to the efficiency—or inefficiency—with which the capital is used.

As for the shortage of entrepreneurs, that argument can only be tested after the fact. There was no reason to believe in 1750 that the Scots would soon people the world—the English-speaking world at least—with a race of hardy entrepreneurs (witness the fate of the inglorious Darien expedition at the end of the seventeenth century), or to believe in 1870 that Japan would soon produce some of the world's foremost industrial and financial entrepreneurs. (As late as 1930 most Westerners still regarded the Japanese as "quaint.")

Arguments for a larger direct entrepreneurial role for the state also overlook the fact that decisions and actions must still be made by men. The kinds of men to be found in the service of the state—and, more importantly, the kinds of incentives that impel them—are not necessarily those most conducive to innovation and economic growth. There is, in fact, a considerable danger—and much contemporary evidence—that such efforts will duplicate the mistakes of Colbert and the cameralists: a sobering example of Santayana's warning. There is little doubt that government officials today are much more "growth-minded" than their counterparts in the past, but this is not necessarily a valid reason for substituting their energies and decisions for those of private entrepreneurs. While there will obviously be exceptional cases, in general the state would do better to devote its own limited resources to creating the conditions in which private entrepreneurs and private capital can flourish. This hortatory suggestion is intended specifically to include private entrepreneurs in banking. It should be evident from the historical record that the banking system is a strategic location with a large leverage for economic development. Therefore, insofar as there is an actual shortage of capable entrepreneurs, the argument for attracting them into the banking system gains force.

Another advantage in allowing greater latitude for the private banking system is that it decentralizes decision-making and encourages initiative on a broad front. It is a common observation that state-sponsored development projects tend toward the grandiose and monumental, and are frequently of doubtful economic value, while small-scale industry and agriculture are often neglected. The most successful historical cases of industrialization, however, show the prominence if not the predominance of small-scale enterprise in the earliest stages of industrialization. If small enterprises are successful, it is because they have met the test of the market, and they will inevitably grow larger. On the other hand, if they fail the losses to society are much less than in the case of the failure of a large government-sponsored enterprise, which for political reasons may not be allowed to appear to have failed. Of the countries in our sample, Russia is the one in which the government played the largest role, both directly and indirectly; and although it had some credits to its record, over-all its achievements were more spectacular than solid. The comparison of the precise role of government in Russia and Japan, two countries with much in common as far as their conditions of backwardness were concerned, is extremely instructive.

In the final analysis, today's developing nations must work out their own problems with such assistance as they can obtain from their wealthier and technically more advanced neighbors. Conditions are different now from what they were in the past, and they will be different in the future from what they are in the present. But present and future grow gradually and ineluctably from the past. A knowledge of history on the part of the leaders of today is the best guarantee that the world will benefit from the mistakes of the past as well as from its triumphs.

ARTICLE 3

PATTERNS OF DEVELOPMENT IN NINETEENTH CENTURY EUROPE*

By N. F. R. CRAFTS

I

It is now quite a long time since Kuznets summarized the state of empirical knowledge on economic growth in the long run (1966) (1971). Since then there have been several developments in the literature which taken together permit a further assessment of major aspects of long run growth experience. First, the work of Chenery–Syrquin (1975) on patterns of development since World War II provides interesting results with which nineteenth century experience can be compared. Second, there have been improvements and extensions in basic data relating to national income in nineteenth century Europe (Bairoch, 1976; Crafts (1983b). Third, the work of Kravis *et al.* (1978) on the measurement of biases in official exchange rate comparisons of national income in the cross-section of 1970 facilitates the measurement of differences in nineteenth and twentieth century development of patterns.

Whilst Chenery and Syrquin felt able to extract a considerable number of generalizations concerning recent patterns of development, the tendency among economic historians has increasingly been to react unfavourably to grand generalizations such as those of Rostow (1960), and even to those of Gerschenkron (1962) which sought to distinguish between the experience of "backward" countries and earlier European developers. Thus we find, Ashworth (1977, p. 140) arguing that the evidence adduced to support the 'alleged universalities' is inadequate, we find Trebilock (1981, ch. 6) although finding merit in Gerschenkron's ideas nevertheless suggesting that much of Europe fits very uneasily into the scheme of things, whilst O'Brien and Keyder (1978, p. 18) in their controversial book on comparative economic growth in France and Britain complain that "Economic theory lends no support to assumptions, present all too often in writings on French backwardness, that there is one definable and optimal path to higher per capita incomes and still less to the implicit notion that this path can be identified with British industrialization as it proceeded from 1780 to 1914," and criticise earlier writers for "a normative assumption ... that the 'English Way' constitutes not merely initial but best or normal practice." In other words nineteenth century European economic development is increasingly coming to be regarded as hard to describe in terms of generally recognizable patterns and, presumably, unpromising ground for Chenery–Syrquin type generalizations.

* I have received useful comments from P. A. David, C. K. Harley, B. R. Mitchell, P. K. O'Brien, M. F. Thomas and J. G. Williamson and participants in seminars at the Universities of Berkeley, British Columbia, Chicago, Virginia, Stanford and Wisconsin. An anonymous referee's suggestions significantly improved an earlier version. Mistakes are mine alone.

For British economic history, in particular, there is a further angle to be developed. There has been a prominent tendency in recent commentaries on British economic performance to suggest that "failure" is the result of very long-standing features of the British economy and society. Although some of the more foolish and superficial versions of this story have been subjected to sharp and deserved criticism by economic historians (Floud and McCloskey, 1981, ch. 5), the argument developed by Kaldor (1966) that Britain's early move away from agriculture made subsequent growth harder to achieve continues to attract examination and interest (McCombie (1981), Chatterji and Wickens (1982)). In describing nineteenth century European economic development patterns, it seems worth examining the extent to which Britain's experience does appear to differ from any "normal" patterns which can be inferred from the evidence.

The rest of the paper is organized as follows. In Section II the Chenery–Syrquin methodology is applied to nineteenth century Europe and a picture is developed of both the "average" patterns and dispersion about the average in nineteenth century accumulation, resource allocation and sectoral productivity, and demography and a comparison is made with more recent experience. In Section III the position of Britain within the nineteenth century European context is examined, although the specific comparison with France made by O'Brien and Keyder is reviewed separately in Crafts (1984). In Section IV the implications of the results of Section II for the grand theories of Rostow and Gerschenkron and further research in European economic history are briefly considered.

II

Recent work has led to revised estimates of real national product for nineteenth century Europe. These estimates are reported in Table 1 whilst full sources and methods are discussed in Crafts (1983b). In Table 1 national product per head is measured in terms of 1970 dollars and can be compared with estimates of national product per head for developing countries in 1970 as developed by Kravis *et al.* (1978). A selection of Kravis *et al.*'s estimates is given in Table 2.

Obviously the index number problems involved in measuring nineteenth century output are very substantial. The intention is to measure real output in 1970 prices and thus to investigate the income levels of past generations seen through the eyes of the present generation. Kuznets (1966) attempted the same kind of comparison. In practice, what we have are only approximations to this ideal. For countries with good data, we have measures which are based on chain-linked calculations working backwards from Kravis *et al.*'s figures for income in 1970 in 1970 dollars. For countries with less good data, as indicated in the footnotes to Table 1, estimates can be obtained using a procedure similar to that of Beckerman and Bacon (1966). In all, the

TABLE 1
Estimates of Gross National Product per Capita in Nineteenth Century Europe[a]
(in 1970 U.S. Dollars)

	1700	1760	1800	1820	1830	1840	1850	1860	1870	1880	1890	1900	1910
Great Britain[b]	333	399	427		498	567	660	804	904	979	1130	1269	1302
Belgium							534	637	738	832	932	1013	1110
Denmark				358	382	402	489	497	563	617	708	850	1050
Switzerland									589	715	750	895	992
Germany							418	481	579	602	729	868	958
Netherlands							481	522	591	707	768	840	952
France					343	392	432	474	567	602	668	784	883
Austria								373	466	556	664	740	802
Sweden								292	351	419	469	597	763
Norway								420	441	486	548	605	706
Hungary								297	345	435	503	604	684
Finland								300	390	407	458	529	561
Portugal											483	488	550
Italy								451	467	466	466	502	548
Spain								382	391	410	464	554	547
Greece									312	350	380	393	455
Russia								236	252	253	276	342	398

[a] Estimates for Netherlands, Switzerland, Austria, Hungary, Portugal, Spain and Greece have been obtained using a modification of the procedure developed by Beckerman and Bacon (1966). The discussion in Crafts (1983b) gives full details of the method and describes tests of its ability to predict accurately. It should be noted that the estimates shown here for Hungary are probably too high and that for these seven countries in particular it would be unwise to use the above estimates to calculate decadal growth rates of income.
[b] England and Wales prior to 1800; United Kingdom 1860 to 1910.

situation is probably no worse than that which confronted Chenery–Syrquin and it seems worthwhile to proceed on the basis of what is available.

An important point to note is that the outcome of recent research suggests that nineteenth century European countries were at least in the same ballpark in levels of national income as recent developing countries, a point on which Kuznets was troubled (1971, p. 28). A further point of importance is that the estimates for Britain (and England and Wales) prior to 1830 have

TABLE 2
GDP Per Capita in 1970 (in 1970 U.S. Dollars)

Mali	145	Paraguay	621
Bangaladesh	198	El Salvador	696
Afghanistan	244	Peru	810
Zaire	284	Iraq	840
Uganda	341	Malaysia	860
Kenya	394	Brazil	1143
Pakistan	451	Mexico	1292
Sri Lanka	474	Portugal	1347
Phillippines	494	Chile	1531
Egypt	532	Hong Kong	1347

Source: derived from Kravis *et al.* (1978), pp. 232–236, based on Column (5) of their Table 4.

TABLE 3

Chenery–Syrquin Type Regressions for Nineteenth Century Europe[a,b] (t-statistics in parentheses)

Dependent variable	*Constant*	ln *Y*	ln *N*	*FR*	*BR*	*RUSS*	$\bar{R}^2$	*SEE*	*N*	*Mean of dependent variable*
CBR	80.370	−8.031	1.849	−11.512	0.307	5.665	0.780	2.729	114	33.328
	(16.423)	(−10.214)	(7.120)	(−10.874)	(0.343)	(3.821)				
CBR	87.787	−9.198	1.565				0.436	4.366	114	33.328
	(12.379)	(−8.210)	(4.896)							
CDR	71.846	−8.575	2.581	−4.273	−0.992	−0.896	0.590	3.520	114	23.113
	(11.384)	(−8.457)	(7.707)	(−3.130)	(−0.859)	(−0.469)				
CDR	71.981	−8.534	2.189				0.563	3.631	114	23.113
	(12.208)	(−9.161)	(8.236)							
AGLAB	238.655	−30.186	2.808	−9.444	−26.724	−17.482	0.794	7.290	81	46.860
	(15.033)	(−12.128)	(3.304)	(−2.962)	(−10.993)	(−2.216)				
AGLAB	259.237	−33.067	−0.169				0.479	11.589	81	46.860
	(10.588)	(−8.668)	(−0.145)							
AGY	205.486	−26.585	0.132	−6.439	−8.681	−7.740	0.641	8.058	73	35.178
	(11.332)	(−9.062)	(0.138)	(−1.875)	(−3.055)	(−1.520)				
AGY	204.338	−26.250	−1.445				0.602	8.480	73	35.178
	(12.605)	(−10.261)	(−2.041)							
MANY	−48.036	11.083	1.270	11.216	1.175	−11.849	0.815	3.713	73	25.588
	(−5.749)	(8.199)	(2.867)	(7.089)	(0.897)	(−5.050)				
MANY	−71.682	15.031	0.995				0.482	6.220	73	25.588
	(−6.028)	(8.010)	(1.916)							
SCHOOL	−1.592	0.308	0.004	0.097	−0.249	−0.123	0.624	0.093	83	0.369
	(−7.009)	(8.743)	(0.393)	(2.347)	(−5.650)	(−2.165)				
SCHOOL	−1.229	0.256	−0.016				0.383	0.119	83	0.369
	(−5.096)	(6.917)	(−1.453)							
INVT	−25.160	6.145	0.244	−2.486	−7.392	−2.558	0.428	3.135	48	12.794
	(−2.549)	(3.904)	(0.580)	(−1.493)	(−5.577)	(−1.033)				
INVT	−7.383	3.352	−0.474				0.066	4.007	48	12.794
	(−0.720)	(2.087)	(−1.127)							
CONSN	123.336	−6.503	−1.246	0.429	3.034	3.906	0.415	3.202	48	79.429
	(12.233)	(−4.045)	(−2.902)	(0.252)	(2.241)	(1.544)				
CONSN	121.745	−6.314	−0.749				0.386	3.352	48	79.429
	(14.186)	(−4.698)	(−2.127)							
INFL	9.740	−1.200	−0.875	−0.952	−2.602	3.338	0.515	1.942	48	−0.660
	(1.593)	(−1.231)	(−3.363)	(−0.923)	(−3.171)	(2.176)				
INFL	23.911	−3.522	−0.786				0.367	2.218	48	−0.660
	(4.212)	(−3.962)	(−3.376)							
GOVT	16.933	−1.542	−0.069	1.022	1.683	1.628	0.123	2.265	48	7.454
	(2.374)	(−1.356)	(−0.226)	(0.849)	(1.757)	(0.910)				
GOVT	14.511	−1.188	0.205				0.948	2.281	48	7.454
	(2.485)	(−1.299)	(0.854)							

been considerably amended by recent research as compared with the well known earlier work by Deane and Cole (1962). The rationale behind the changed picture is given in Crafts (1983a) where it is argued that growth between 1760 and 1830 was considerably slower than was thought until recently.

The objective of Tables 3 and 4 is quite modest, namely to offer a way to present the "stylised facts" of nineteenth century European development and to contrast them with the "stylised facts" of economic development in the 1950's and 1960's. Later on this preliminary picture will be refined somewhat as information is presented on dispersion about this average behaviour and about differences between earlier and later developers in nineteenth century Europe.

Estimation of equations to describe nineteenth century European development is based on as wide a range of countries as can be achieved, whilst remaining within the bounds of reasonably reliable national product estimates. The coverage is wide enough to permit a more thorough investigation of European economic development than has previously been possible but on the whole Eastern European countries have had to be omitted and their absence does limit the generality of the findings. No attempt has been made to include extra-European evidence, in particular because the author's knowledge of source material is not sufficient.

The functional form of the equations used to examine nineteenth century European development is slightly different from that chosen by Chenery–Syrquin (1975) but can nevertheless be thought of as coming from the same stable. In the equations reported in Table 3, there are no terms in $(\ln Y)^2$, $(\ln N)^2$, F or time dummies. F represented inflow of foreign capital and is omitted for data reasons in order to retain wide coverage. The $(\ln Y)^2$ and

---

[a] Variables are defined as follows: CBR is the crude birth rate, births per 1000 population. CDR is the crude death rate, deaths per 1000 population. AGLAB is the percentage of the labour force in agriculture and extractive industry. AGY is income originating in agriculture and extractive industry as a percentage of national product in current prices. MANY is income originating in manufacturing and construction as a percentage of national product in current prices. SCHOOL is the fraction of the population aged 5–19 enrolled in primary or secondary schools. INVT is the percentage of gross national expenditure devoted to gross investment including stocks in current prices. CONSN is the percentage of gross national expenditure devoted to private consumption in current prices. INFL is the deficit on current account of the balance of payments as a percentage of gross national product in current prices. GOVT is the percentage of gross national expenditure devoted to government expenditure of current goods and services in current prices. $\ln N$ is the log of population size in millions, based on boundaries at the date in question. $\ln Y$ is the log of per capita GNP in 1970 U.S. dollars. FR is a dummy variable taking the value 1 for France and 0 for other observations. BR is a dummy variable taking the value 1 for Britain and 0 for other observations. RUSS is a dummy variable taking the value 1 for Russia and 0 for other observations.

[b] It should be noted that an alternative procedure is available to estimate the equations. It is possible to estimate the equations using throughout the values of Y estimated by the Beckerman-Bacon method as described in Crafts (1983b). None of the independent variables used in that forecasting equation are dependent variables in Table 3. Using that "two-stage" procedure makes little difference to the results reported here. That this is so should be comforting to readers who are uncomfortable at the idea of $1970 since this indicates that very similar results can be obtained using what is in effect a "standardised income score."

$(\ln N)^2$ terms were generally unimportant statistically and much more impact on the fit of the equations is achieved by use of country dummy variables. One of these is Russia, by far the largest country in population and, perhaps, following a distinctive "East-European" development pattern; another is for Britain which in all years is the richest country, in some years by a significant amount. The British dummy variable thus partly absorbs the role of $(\ln Y)^2$ for Chenery but, of course, it also picks up differences between Britain and the rest of the sample, which will be of particular interest in Section III. Experiments with time dummies proved unsuccessful in general in the sense that they were rarely statistically significant.

Equations thus were estimated for all available observations for each country and for each end-decade point pooled. The availability of national product data is seen in Table 1. Population data is available in all cases, but information on other variables is usually only available for a subset of observations. Data were taken from Mitchell (1975) and the national accounts sources listed in Crafts (1983b). In fitting the equations, no attempt was made to look for more complicated expressions which might fit better. Rather the research design was parsimonious and the intention was to investigate how well a simple equation along Chenery–Syrquin lines could describe the European historical data. In the demographic equations, it is well known that French experience was "idiosyncratic" and given both this and the intense interest that there has been in French/British comparisons the reported equations also include a French dummy variable in the case of all dependent variables.

Before discussing these results it will also be helpful to have available the evidence of Table 4. This presents pictures of structural change during transition based on the equations of Table 3 and Chenery-Syrquin's evidence. Chenery–Syrquin's equations have been transformed into 1970 U.S. dollars real measures for national income on the basis of Kravis *et al*'s findings for the relationship between nominal and real national incomes by country in 1970.

The transitions for the nineteenth century countries are reported for the equations including the three country dummy variables and can be said then to apply to the average of the other countries of Table 1. The Chenery–Syrquin picture is for all countries in their group, rather than any of the subgroups which they analyse and in each case the estimates reported are for a country of 10 million population.

The first point to note from Table 3 is that it seems as legitimate to talk about "uniformity" of development in the nineteenth century European case as in the 1950's and 1960's. As we shall see, this "uniformity" also conceals various interesting features. The regressions in Table 3 are not strictly comparable to Chenery–Syrquin; the former have country dummies and not time dummies, they do not have $(\ln Y)^2$ and $(\ln N)^2$ terms and they deal with a much narrower range of income experience. Nevertheless, as measured by $\bar{R}^2$ they suggest that it is about equally easy to find a "simple

TABLE 4
Patterns of Development in Nineteenth Century Europe Compared With the World in 1950–70
(Income Level in 1970 U.S. dollars; simulations for countries of 10 mn. people)

Forecast Values at	\$300	\$400	\$550	\$700	\$900
(a) 19*th Century Europe*[b]					
CBR	38.8	36.5	34.0	32.0	30.0
CDR	28.9	26.4	23.7	21.6	19.5
AGLAB	72.9	64.3	54.6	47.4	39.8
AGY	54.2	46.5	38.0	31.6	24.9
MANY	18.1	21.3	24.8	27.5	30.3
SCHOOL	0.174	0.262	0.360	0.435	0.512
INVT	10.5	12.2	14.2	15.7	17.2
CONSN	83.4	81.5	79.4	77.9	76.2
INFL[a]	0.9	0.5	0.1	−0.1	−0.4
GOVT[a]	8.0	7.5	7.0	6.7	6.3
(b) 1950–70 *World*[b]					
CBR	44.8	42.5	38.8	35.8	32.6
CDR	19.0	17.0	14.1	12.5	10.9
AGLAB	66.7	62.7	57.3	52.4	46.6
AGY	46.3	41.3	34.6	29.7	24.9
MANY	14.5	17.0	20.5	23.3	26.2
SCHOOL	0.354	0.429	0.522	0.592	0.663
INVT	15.4	16.7	18.3	19.5	20.8
CONSN	73.0	70.9	69.1	67.7	66.1
INFL	2.4	2.1	1.7	1.4	1.1
GOVT	13.4	13.6	13.5	13.4	13.5

[a] Coefficient on ln Y in the regression of Table 3 insignificant.
[b] "19th century Europe" comprises the countries listed in Table 1; 1950–70 includes 101 countries which are listed in Chenery–Syrquin (1975, pp. 12–13).

explanation" for the behaviour of important aggregates in development. In almost all cases the ability to predict is similar to that of Chenery-Syrquin; they too had similarly less success in accounting for government consumption and personal consumption. It is also striking to find that such simple equations as those using as independent variables only ln Y and ln N have such high $\bar{R}^2$ as those recorded for death rates and sectoral shares of output. Even in the case of birth rates, where generalizations based on the Demographic Transition have been widely discarded (Coale, 1973), there is quite a strong statistical association to be found.

Something which economic historians have not paid attention to but which figures in several of the regressions and tended to be found also by Chenery and Syrquin is the role of population size. Recall that Russia with a population size unlike the other European countries has been assigned a dummy variable in the regressions. The log of population shows up as significant, nevertheless, in the cases of the birth rate, the death rate, the percentage of the labour force in agriculture, the percentage of output in manufacturing, the percentage of national expenditure which is consumption and the current account deficit as a percentage of gross national product. Chenery and Syrquin argued that Ln *N* might capture the effects of economies of scale or transport costs (1975, p. 17) and, presumably, therefore might be related to sources of comparative advantage. The rationale for the differences between small and large countries showing in equations such as those of Table 3 may perhaps be placed on the agenda for future work by economic historians.

Table 4 offers a comparison of the "stylised facts" of transition in nineteenth century Europe from being a $300 to being a $900 country with those of the same transition in 1950–70. Many of the results are not unexpected. The crude death rate is in each case apparently slowly reduced by rising incomes but is obviously at a much lower level in 1950–70 throughout transition. These findings are very consistent with the arguments of Preston that life expectancy is somewhat increased by rising incomes but that the twentieth century has seen a very substantial decline in mortality at any age given income level (1980, pp. 306–7). *Ceteris paribus*, population pressure was much lower in nineteenth century Europe. It is also not surprising to find accumulation of both physical and human capital at a faster pace in the twentieth century. Perhaps a little more interesting is that net foreign investment is the case for the Europeans by $700 of the transition. This might be taken to indicate a considerable difference in position *vis-a-vis* international capital markets for nineteenth century European and twentieth century developers.

Another notable contrast between twentieth century development and that of nineteenth century Europe is to be found in terms of sectoral allocation of labour and particularly in relative labour productivities by sector. In both cases the share of income originating in the primary sector declines, although nineteenth century countries at all income levels had higher proportions of income originating in both agriculture/mining and in manufacturing/construction. At levels of income above $400, however, nineteenth century European developers had smaller proportions of the labour force in the primary sector. As a result the "sectoral productivity gap" ((=100−AGY)/(100−AGLAB)×(AGLAB/AGY)) between the primary sector and the rest of the economy was much smaller and had narrowed substantially by $550 whereas for the twentieth century countries the gap is growing throughout. Thus Table 5 shows a gap of 2.32 at $300 and 2.63 at $900 for the twentieth century developers compared with 2.27 at $300 and 1.99 at $900 for the nineteenth century Europeans. A number of hypotheses might be advanced to explain this phenomenon which once again might be part of a useful research agenda.

Having illustrated the proposition that nineteenth century European economic development can be arranged in terms of stylized facts of uniformity, it is now time to explore the diversity of that experience. In doing so, it must be borne in mind that Eastern Europe is sadly underrepresented in what follows. Table 6 presents values for the variables considered in Table 4

TABLE 5

The Sectoral Productivity Gap[a]

Forecast values at	$300	$400	$550	$700	$900
(a) 19th Century Europe	2.27	2.07	1.96	1.95	1.99
(b) 1950–70 World	2.32	2.39	2.54	2.61	2.63

[a] This is defined as noted in the text as (100 − AGY)/(100 − AGLAB) × AGLAB/AGY, which is labour productivity in non-agriculture divided by labour productivity in agriculture.

TABLE 6

Nineteenth Century European Economies at $400, $550 and $900[a]

	CBR	CDR	AGLAB	AGY	MANY	SCHOOL	INVT	CONSN	INFL	GOVT
(a) $400										
England and Wales 1760	33.9	28.7	49.6	37.5	20.0	na	5.7	74.4	na	12.7
Denmark 1840	30.4	21.0	na	52.9	20.0	na	na	na	na	na
Germany 1850	37.2	25.6	54.6	46.0	20.4	na	12.9	82.6	−0.7	5.2
France 1840	27.9	23.7	51.9	33.3	39.7	na	11.8	na	na	na
Austria 1860	38.2	26.8	na	na	na	0.290	na	na	na	na
Sweden 1880	29.4	18.1	61.2	36.3	22.0	na	10.4	85.7	1.8	5.8
Norway 1860	33.3	17.2	na	na	na	na	na	na	na	na
Hungary 1880	42.8	37.8	64.0	na	na	0.307	na	na	na	na
Spain 1880	35.5	30.1	70.5	na	na	0.372	na	na	na	na
Finland 1880	36.5	23.9	71.2	58.0	17.4	0.060	na	na	na	na
Greece 1900	33.6	20.3	na	na	na	0.287	na	na	na	na
Russia 1910	43.1	27.4	na	42.3	14.6	0.151	12.1	79.7	2.3	1.5
(b) $550										
Great Britain 1840	35.9	22.2	25.0	24.9	31.5	na	10.5	80.4	−1.2	7.9
Belgium 1850	30.0	21.2	48.9	27.0	24.0	0.385	na	na	na	na
Denmark 1870	30.3	19.0	47.8	29.0	20.0	na	12.0	82.0	0.1	6.1
Germany 1870	38.5	27.4	50.0	39.9	29.7	na	15.2	81.0	−0.7	4.3
Netherlands 1860	31.6	24.8	37.4	na	na	0.406	na	na	na	na
France 1870	25.9	28.4	49.3	33.5	36.0	0.476	12.5	78.4	−2.1	7.0
Austria 1880	37.5	29.7	55.6	na	na	0.374	na	na	na	na
Sweden 1900	27.0	16.8	53.5	27.2	30.1	0.486	12.0	84.2	2.2	5.9
Norway 1890	30.4	18.0	49.6	27.2	22.5	0.435	16.8	80.8	2.4	4.8
Hungary 1900	40.7	32.5	64.0	na	na	0.377	na	na	na	na
Finland 1910	31.7	17.4	69.2	47.0	25.3	0.212	na	na	na	na
Portugal 1910	31.7	19.2	57.4	na	na	na	na	na	na	na
Italy 1910	33.3	19.9	55.4	38.2	23.9	0.326	16.4	74.3	0.2	9.5
Spain 1910	32.7	23.1	56.3	na	na	0.366	na	na	na	na
(c) $900										
Great Britain 1870	35.2	24.2	15.3	18.8	33.5	0.168	8.5	80.5	−6.2	4.8
Belgium 1890	29.0	20.8	32.1	11.0	30.0	0.342	na	na	na	na
Denmark 1900	29.7	16.8	46.6	29.9	26.4	0.503	20.8	75.7	2.8	6.3
Germany 1870	38.5	27.4	50.0	39.9	29.7	na	15.2	81.0	−0.7	4.3
Netherlands 1910	28.6	13.6	28.4	na	na	0.502	na	na	na	na
Switzerland 1900	28.6	19.3	34.6	na	na	0.541	na	na	na	na
France 1910	21.3	21.9	41.8	27.6	38.3	0.577	13.6	75.3	−3.1	8.0

[a] All measurements are of per capita national product at 1970 U.S. dollars.

Sources: Mitchell (1975) and for national income data the sources listed in Crafts (1983b).

where available for countries at *c.* $400, *c.* $550 and *c.* $900 in pre World War I Europe. The experience at $550 is perhaps the most informative. At such a level of income structural transition associated with development is well under way. The information available to us also permits consideration of at least some aspects of development in countries frequently considered as "failures" in pre-World War I development, for example, Portugal and Spain. Examining economic characteristics at a given income level rather than "during take-off" or "at the time of the great spurt" is also an interesting variant on the discussions usually made by European economic historians.

Table 6 gives us a somewhat different perspective on nineteenth century Europe. At all income levels it becomes obvious that there was a fairly wide range of demographic experience, both in terms of vital rates and natural increase. France's fertility, of course, stands out as exceptional. Perhaps most interesting, however, is to look at the sectoral allocation of resources, and, especially the behaviour of the percentage of the labour force in agriculture. In this regard the latecomers appear different from early Western European industrializers. At $550 there is a strong tendency for countries later in reaching $550 to have a greater share of the labour force in agriculture at that income level. The rank correlation is 0.83. Britain already, even prior to the abolition of Corn Laws, appears to be something of an outlier.

At $900 we can, of course, only observe the early, "successful" developers. When we do so, within this group there appear again to be interesting differences in the allocation of resources, reflecting presumably both comparative advantage and trade policies. By 1870, Britain, not surprisingly, had specialized away from agriculture to a considerable degree, even though "her sectoral productivity gap" had been eliminated by 1840.

The point that there were differences between early comers and latecomers in European development can be strengthened by reference to Table 7. This table gives Chenery-Syrquin type results for the later developing European economies; the sample of observations is based on those countries which had income levels below $500 in 1880. It is apparent that the "stylised facts" of development for this group of countries are that they would have more labour in the agricultural sector at any income level over $400 *and* that the "sectoral productivity gap" would be larger. In fact, the "sectoral productivity gap" rises steeply for the later developing European countries; its forecast value based on Table 7 is 2.08 at $300, 2.29 at $400, 2.64 at $550 and 3.02 at $700. These gaps are *larger* at $550 and $700 than those reported in Table 5 for the 1950–70 countries.

Problems of low agricultural productivity growth in the "backward" European economies are, of course, a prominent part of Gerschenkron's approach to their development experience. Nevertheless, this feature should also be seen in the context of international trade relations. Taken out of the context of the success of the early developers, agricultural problems in what

TABLE 7

The Transition in Late Developing European Countries[a]

Forecast values at	$300	$400	$550	$700
AGLAB	69.3	64.4	59.0	54.9
AGY	52.0	44.1	35.3	28.7
MANY	12.8	17.8	23.3	27.4

[a] Defined as countries with less than $500 income in 1880.
Source: Derived from the following regressions

$$AGLAB = 161.524 - 17.004 \ln Y + 2.070 \ln N \qquad \bar{R}^2 = 0.244 \qquad N = 28$$
$$(3.994)(-2.648) \qquad (1.449) \qquad SEE = 7.637$$
$$AGY = 211.523 - 27.545 \ln Y - 1.046 \ln N \qquad \bar{R}^2 = 0.359 \qquad N = 29$$
$$(4.982)(-4.083) \qquad (-0.817) \qquad SEE = 9.658$$
$$MANY = -83.148 + 17.300 \ln Y - 1.192 \ln N \qquad \bar{R}^2 = 0.741 \qquad N = 29$$
$$(-5.241) \quad (6.864) \qquad (-2.492) \qquad SEE = 3.609$$

TABLE 8

Export Experience of European Countries prior to World War I

(a) Percentage of Exports by Value in Manufactures[a] at $550

Great Britain	1840	90.5	Hungary	1900	36.6
Denmark	1870	3.8	Sweden	1900	24.4
France	1870	53.2	Spain	1910	24.1
Germany	1870	52.5	Italy	1910	28.8
Norway	1890	8.8	Portugal	1910	7.7

(b) Percentage of Exports by Value in Manufactures in 1910

Great Britain	76.1	Netherlands	20.9	Portugal	7.7
Belgium	37.0	France	59.2	Italy	28.8
Denmark	5.1	Austria-Hungary	64.0	Spain	24.1
Switzerland	75.5	Sweden	24.1	Greece	0.0
Germany	74.3	Norway	10.9	Russia	3.8

(c) Rates of Real Export Growth of Later Developing European Countries[b]

	1860–80	1880–90	1890–1910
Hungary	na	1.6	3.6
Greece	2.1	6.7	2.9
Russia	4.9	3.5	3.6
Sweden	3.3	3.7	3.2
Norway	6.6	1.8	3.8
Finland	7.2	1.6	5.5
Italy	4.4	−2.0	4.2
Spain	2.0	2.9	0.9
Portugal	3.2	0.5	2.1

[a] "Manufactures" are not easily defined in terms of nineteenth century trade statistics and there is considerable scope of error here. The data were taken from British Parliamentary Papers (1913) and Kuznets (1967).

[b] Taken from Berend and Ranki (1980, p. 546).

turned out to be late developing countries might seem to suggest comparative advantage in manufacturing and development skewed to a strong manufacturing performance, which is indeed suggested in the Gerschenkron approach.

Table 8 indicates, however, that, in general, later developing countries did not establish exports mainly in manufacturing. There is a tendency, particularly if Denmark is excluded, for later arrivals at the $550 income level to have a lower share of their exports as manufactures.

Table 1 indicated that there was widespread growth in Europe prior to World War 1. Table 8 shows that this growth was combined with a considerable growth of exports among the later developers. It also suggests a European Economy with countries specialized in different sectors but it does not suggest that a $550 country in 1910 could be expected to have comparative advantage in manufacturing as Britain, France and Germany had at the same income level.

Thus, whilst as a first approximation nineteenth century European economic development can be described fairly well in terms of simple Chenery-Syrquin equations, it seems unlikely that this first approximation is fully adequate to deal with the richness of experience that economic historians confront. In particular, not only is there quite considerable diversity about the average experience but there seem to be tendencies towards a different kind of structural change in the later developing countries.

Comparisons between nineteenth and twentieth century development and also between early and late comers in nineteenth century development reveal differences in the observed pattern of structural change. Even among the early comers of the nineteenth century specialization for export was in different goods. That this was so should not be surprising for it is consistent with standard predictions from the theory of international trade. The evidence of this section underlines the validity of Bhagwati's comment on the use of the Chenery–Syrquin methodology for prediction as opposed to description: "a country's trade pattern and volume and its production pattern are . . . the result of *interaction* between the country's own endowments and demands and the rest of the world's endowments and demands . . . Chenery's regressions, if used for prediction, would imply that one could read off the changes in the country's trade and production without reference to what had happened in the rest of the world, *whereas* a trade theorist would consider this to be an incomplete . . . procedure" (1977, p. 498)

III

We have already seen some differences between British experience in the nineteenth century and that of other European countries. This section seeks to explore these differences in rather more detail. Table 9 sketches out the transition experienced by the British economy between 1700 and 1910.

TABLE 9
Britain's Development Transition Compared with Norms for a Country of the Same Population

Income level	333	399	427	567	904	1130	1302
Year	1700	1760	1800	1840	1870	1890	1910
CBR Actual	33.1	33.9	37.7	35.9	35.2	30.2	25.1
European Norm	36.8	35.8	36.1	34.8	31.8	30.4	29.6
CDR Actual	26.5	28.7	27.1	22.2	22.9	19.5	13.5
European Norm	26.3	25.4	26.0	25.0	21.9	20.6	19.9
AGLAB Actual	57.1	49.6	39.9	25.0	20.0	16.3	15.1
European Norm	68.0	63.2	62.4	55.5	42.4	36.3	32.6
AGY Actual	37.4	37.5	36.1	24.9	18.8	13.4	10.3
European Norm	51.3	46.5	44.8	37.3	25.0	19.1	15.3
MANY Actual	20.0	20.0	19.8	31.5	33.5	33.6	31.8
European Norm	18.5	20.7	22.1	25.9	31.6	34.3	36.2
SCHOOL Actual	na	na	na	na	0.168	0.385	0.542
European Norm	na	na	na	na	0.514	0.582	0.626
INVT Actual	4.0	6.0	7.9	10.5	8.5	7.3	7.0
European Norm	10.9	12.1	12.6	14.5	17.5	18.9	19.8
CONSN Actual	91.2	74.4	76.8	80.4	80.5	81.6	73.8
European Norm	83.5	82.0	81.0	78.5	75.0	73.3	72.1
INFL Actual	na	na	0.6	−1.2	−6.2	−5.2	−11.0
European Norm[a]	na	na	0.4	−0.4	−1.3	−1.8	−2.1
GOVT Actual	4.8	12.7	15.3	7.9	4.8	5.9	8.2
European Norm[a]	7.9	7.6	7.5	7.0	6.2	5.9	5.6

[a] Coefficients on ln Y insignificant in regression in Table 3.

[b] All variables apart from income level and national accounts data are measured for Great Britain from 1800 on and for England and Wales for 1700 and 1760. National accounts data are for United Kingdom for 1870, 1890 and 1910, Great Britain for 1800 and 1840 and England and Wales for 1700 and 1760.

Source: "European norm" based on the equation with country dummies in Table 3 for each variable with values for a non-dummy country of the population of Britain in the given year. Sources for British data as in Tables 1 and 3 together with Crafts (1983a) and sources listed there.

This transition is compared with the European norm for a country of the same population that Britain had in the given year. Britain's population rose substantially during the nineteenth century, of course, but it should be noted that the European norms which would come from retaining the Chenery-Syrquin approach of simulating values for a country of constant population are not in general very different from those calculated in Table 9.

Interestingly enough Britain only appears similar to the norm of the European "stylised facts" in two aspects of Table 9. Demographic behaviour in Britain seems generally to be rather similar to elsewhere and government consumption, whilst low by twentieth century standards, looks similar to nineteenth century European experience. (1760 and 1800 were, of course, episodes of expensive war.)

Some of the idiosyncrasies of the British experience which show up in Table 9 are relatively well known. Britain's exceptional commitment to foreign investment and very low home investment by comparative standards are quite clear. Also, it emerges that the second half of the nineteenth century was one in which personal consumption was a relatively high share of total expenditure. The difference of the presentation in Table 9 is that the comparisons are normalized for income level as opposed to being made by country in a given chronological period. Similarly, Britain's lower attention to the accumulation of human capital appears in the school enrollment figures.[1] Whether this late nineteenth century experience was a "failure" or reflects different opportunities available to the pioneer industrializer will, no doubt, continue to be debated (McCloskey (1979); Kennedy (1982)).

Other distinctive features are becoming well known but deserve comment. It is especially noteworthy how long it took for Britain to move from being a $400 to a $550 country, namely between 75 and 80 years. This has become clear recently (Harley (1982); Crafts (1983a)). Indeed, this may well be the key to understanding the peculiarly bitter debate about living standards in Britain during industrialization as compared with discussions relating to other countries. It seems likely that real wage rates grew at a similar rate to national income per head (Crafts, 1983c). In terms of European comparisons it is worth noting that the time for going from $400 to $550 where this can be measured varied from about 40 years in the case of Norway to a little less than 20 for Sweden and averages only about 30 percent of the British time. The relatively slow British growth probably owes something to the disruptions of war, but it should also be noted that overall productivity growth was slow and that rapid productivity growth was confined to a few sectors, at least prior to 1830. The possibility of imitating technological change was, of course, low for Britain at this time and Britain was not pervasively successful at innovation (Crafts, 1985, ch. 4).

The feature of Britain's development transition which requires most comment, particularly in view of the recent economic history literature, relates to the structure of production and sectoral productivity. The first important point to note is that the share of the labour force in agriculture and mining was much lower in Britain than the European norm *throughout* the transition.[2] Note that even in 1840, before the abolition of the Corn Laws, the share of the labour force in agriculture and mining was only *half* the European norm for the same income level. The most striking feature of the British industrial Revolution is the very rapid and early change in the share of the labour force in the primary sector which took place whilst income per head grew only slowly.

[1] The school enrollment figure for 1870 is undoubtedly misleadingly low.

[2] Measurement of sectoral shares in the labour force as far back as the 18th century is in some ways a dubious proposition, see Lindert (1980). The estimate given here is based on Lindert and Williamson (1983) and comparing their data for 1760 with those for 1801 and Deane and Cole's (1962) figure for agriculture's share of the labour force in that year. Although the precise figure may be arguable, there is no question but that England and Wales belongs with France and Germany rather than Spain and Finland. For a fuller discussion see Crafts (1985, ch. 2).

This, in one way, is very consistent both with a long standing theme in British economic history and also with recent research on agriculture's contribution to the economy prior to the classic Industrial Revolution period. It is a commonplace of the literature that agrarian tenure relations were transformed in Britain much sooner than elsewhere on the continent and that migration out of agriculture was thus more likely (Kemp (1969); Cohen (1975)). More recent research has established that the agricultural sector experienced significant growth prior to 1760 (Crafts (1983a)), that there were substantial improvements in yields (Overton (1979)) and that the agricultural sector released labour significantly during the years before the "Industrial Revolution proper" (Crafts (1980)). Rapid productivity growth in agriculture was also characteristic of the first half of the nineteenth century (Crafts, 1985, ch. 4).

The early release of labour by agriculture has other implications, however. The second important point to note is that, in the sense that we used the term earlier, the "sectoral productivity gap" had been eliminated in Britain by 1840. Britain appears thus as a striking exception to the regularities of sectoral productivity differences which feature so prominently in Kuznet's account of modern economic growth (1971; ch. 5); this point has not received the attention devoted to the agricultural revolution. It should be noted that a "sectoral productivity gap" re-emerges after 1870, when productivity growth in agriculture and mining was much less impressive than in manufacturing (Matthews *et al.*, 1982 pp. 228–9). Even so, this gap was still much lower than that shown by the European Norm; the figure for 1910 is 1.55 compared with a European Norm of 2.68.[3]

There seem to be further implications of this point. First, if one wishes to look for structural problems affecting the overall capacity for growth in the manner of Kaldor (1966) or Cornwall (1977) and further wishes to look for the differences arising in the elasticity of labour supply from agriculture, then pre World War I Britain should be a period of some interest. Certainly the difference between the British experience and the industrialization based on Lewis-like dualism which Mokyr describes for Belgium (1976) is great.

Second, relatively low levels of productivity of labour in manufacturing at the *same level of income* are a long-standing feature of the British economy. Difficulties of data exist here[4] but the following seems to be the case. The ratio of the percentage of manufacturing output in total output to the percentage of the male labour force in manufacturing (MLM)[5] was 0.67 in

[3] The figures in Table 9 include property income from abroad which by 1910 was about 7% of national income; if these flows of income are excluded from the calculation the productivity gap in 1910 was 1.42.

[4] In particular, despite the noble efforts of Bairoch (1968), there appear to be serious inconsistencies of practice between countries in the definition of manufacturing workers and accuracy of the census description. See below fn. 6, especially.

[5] Despite difficulties of definition it seemed interesting to run a Chenery–Syrquin regression for MLM on the basis of Bairoch's data (1968). "Manufacturing" was defined as "manufacturing" and "construction" but excluding extractive industry which has earlier been taken as part

1840 compared with a Chenery-Syrquin type estimated norm of 1.00, in 1870 it was 0.68 (norm = 0.83), in 1890 0.66 (0.78) and 1910 0.59 (0.76).

Similarly, calculations of manufacturing output per industrial worker in 1910 suggest that at the income level Britain had in 1850, Germany (in 1890) had labour productivity at least 25% higher and France in 1890 perhaps 40% higher.[6] By 1910, of course, in terms of absolute labour productivity at a point in time in Britain was at a lower level than Belgium, Germany and Sweden.[7]

Third, the above results indicate that to put French industry into comparative perspective, as O'Brien and Keyder (1978) wished to do, it is likely to be informative to widen the comparisons beyond Britain alone. Readers of O'Brien and Keyder (1978) would do well to take heed of their point that there were different paths to maturity for nineteenth century Europe (1978, p. 18) but to do so, in particular, by noting that in very important respects, Britain's was a very idiosyncratic progress. The point that there were different paths seems robust, whereas there are undoubtedly difficulties of the specific productivity comparisons they made[8] which have apparently led some prematurely to dismiss their work. (Kindleberger, 1979).

Indeed, an obvious but nonetheless important point to emerge in this section is that writers looking to develop a general model of economic development would be ill-advised to attempt generalization from the pre-World War I British experience.

In searching for reasons for Britain's different experience, it would seem important that future research considers Britain in the context of her comparative advantage, especially given Britain's commitment to free trade after 1846 and the well known high share of exports relative to GNP. In that

of the primary sector. The equation which resulted was

$$\underset{}{MLM} = -128.351 + \underset{(-7.831)}{}\; 23.938 \ln Y + 0.836 \ln N + 0.862\,Fr + 13.035\,BR + 0.054\,Russ$$

$$\quad (-7.831)\quad (9.397)\qquad (1.061)\qquad (0.269)\qquad (4.698)\qquad (0.008)$$

$\bar{R}^2 = 0.729$, SEE = 6.308, $N = 72$.

The exclusion of property income from abroad would only modify the detail and not the substance of the argument.

[6] Calculations were made using growth of industrial output from Hoffmann (1965) for Germany and Markovitch (1965) for France and working back from the value of industrial output in £ for 1910, converting at official exchange rates. Workers were taken from Bairoch (1968). The French figures have subsequently been severely criticised by Carre *et al.* (1972) and the German income level for 1890 was about 10% higher than Britain in 1850, hence the guarded statements in the text. The calculations of productivity in £ of 1910 turned out to be for Germany: 1910: £106.8, 1900: £86.6 and 1890: £68.9, and for France £84.1 in 1910 (£80.7 using Carre *et al*'s labour force figures), £80.1 in 1900 (£69.6) and £76.0 in 1890. Britain in 1850 was £47.0 (in 1910£) and in 1910 was £90.4.

[7] Calculations were made as in fn. 6; the figures for Belgium are £102.5 and for Sweden £112.3. Other countries may also have had higher labour productivity (e.g., Switzerland, Denmark) but were not included because of deficient data.

[8] In fact French measurement of the labour force in industry seems extremely problematic and it seems quite probable that O'Brien and Keyder's comparison of industrial labour productivity in Britain and France places the French in a misleadingly "superior" position, see Crafts (1984).

sense, discussion of sectoral productivity should take account not only the traditional agricultural revolution points reviewed earlier but also the unusually high share of manufactures in her exports that Britain achieved as the pioneer. Moreover, in considering factor proportions, Pollard's recent emphasis on the role of natural resources (1981) may be helpful. In particular, Britain's coal *consumption* was extraordinarily high; the figure of 3.82 tons/head of 1870 was not equalled elsewhere in Europe before World War I, (by which date Britain had reached 5.04 tons/head). A plausible picture of Britain's comparative advantage at this time is that it was based on exports of horse-power intensive products exchanged for imports of human-capital intensive products.[9]

IV

Earlier sections have suggested that Bhagwati's reactions to the Chenery-Syrquin generalizations about development are well-taken for nineteenth century Europe, namely that there are strong signs of differences of transition both between the pioneer industrializer, Great Britain, and the other countries and between later developers and early developers. This seems to be so, even though regressions of a Chenery-Syrquin type are a good first approximation to Europe's experience.

This finding, obviously, tends to argue that general models of development of the Rostovian kind are not particularly illuminating in looking at European development. Moreover; recent research has strengthened claims that Britain did not experience a Rostovian "take-off" (Crafts 1983a) and the present paper's version of the "normal transition" given in Table 4 certainly does not show the dramatic rises in the investment to income ratio once thought to be the hallmark of "take-off".

More intriguing, however, is where the evidence in this paper leaves the Gerschenkron "backwardness" approach. Evidently, the approach used in this paper also indicates that development was rather different for latecomers but this was described in the context of some elements of latecomers' international position, whereas the best known tests of Gerschenkron's hypothesis were conducted solely in terms of the internal attributes of the countries considered (Barsby (1969)). Moreover, these tests were based on a small group of countries, whereas it is clear that growth before World War I was a very widespread phenomenon in Europe.

Barsby stated Gerschenkron's hypotheses on European economic growth as follows: "the greater a country's relative backwardness on the eve of its [initial period of industrialization] (1) the more rapid was the subsequent rate of manufacturing growth, (2) the greater was the stress on the bigness of

[9] Coal consumption was calculated using data on production and trade from Mitchell (1975); by 1910 Belgium consumed 3.53 tons/head, Germany 3.34 and France 1.45. In 1870 their respective consumption had been 2.02, 0.92 and 0.64 tons/head respectively. Regressions to support this proposition are presented in Crafts and Thomas (1983).

the size of plant and enterprise, (3) the greater was the stress on producers' goods as opposed to consumers' goods, (4) the less rapid was the increase in the level of consumption, (5) the greater was the role played by special institutional factors designed to speed industrialization, and (6) the less the agricultural sector contributed to economic growth as measured by the rate of increase in agricultural labor productivity" (1969, p. 449). Behind this set of arguments lay a vision that backward countries had different factor endowments (especially entrepreneurial human capital), savings capacities and institutional arrangements in agriculture and that the existence of relatively advanced countries pressured their development. On the basis of rank correlations ordering countries in terms of several measures of backwardness, Barsby concluded that (1) and (3) were consistent with the data but (6) was not.

The data given in Table 6 can also be related to Gerschenkron propositions, ranking the countries by variable in terms of its value at the \$550 income level and by time in terms of rank order of reaching \$550 with Britain = 1, etc.[10] Among the results obtained are an inverse correlation of 0.61 between time of reaching \$550 (DATE) and INVT ($n = 7$), an inverse correlation of 0.54 between DATE and CONSN ($n = 7$), an inverse correlation of 0.49 between SCHOOL and DATE ($n = 15$), AGLAB and DATE were positively correlated at 0.83 ($n = 14$), whereas the sectoral productivity gap and DATE were correlated at 0.45 ($n = 9$). Additionally there is virtually no correlation between growth of coal consumption or industrial output growth and DATE.

\$550 is surely a good point at which to consider "initial industrialisation" and is also an income level at which a relatively wide range of information is available. The results of the above correlations whilst admittedly a very crude test, do not seem to offer such strong support for the Gerschenkron hypotheses as Barsby, who examined only 6 countries and used somewhat different ways to order countries, though he had.[11] In particular, the correlations for consumption, growth of coal consumption and industrial output growth must be regarded as rather weak results by comparison with those obtained by Barsby. On the other hand, the correlation for the sectoral productivity gap, given the evidence of Table 7, is perhaps rather more favourable than Barsby would have anticipated. The key point is probably the size of sample; Ashworth (1977) seems to be right that there is a danger that generalizations much used by European economic historians have been based on unduly small samples.

The Gerschenkron story about latecomers seems to be in need of some restatement, even though it *is* correct to argue that their development differs

[10] Where possible Switzerland, assumed to have arrived at \$550 simultaneously with the Netherlands was also introduced.

[11] It is perhaps useful to point out that Barsby's approach included Russia while this approach does not; however, Gregory (1974) (1980) has strongly questioned the applicability of Gerschenkron's hypotheses to even this, apparently its strongest case support.

from that of the first European industrializers. It is a major requirement of further research in comparative European economic history to find a better summary statement about latecomers as compared with early industrializers. The work in this paper can do no more than provide some clues but there do seem to be some pointers as follows.

First, it is instructive to consider the use of coal by early industrializers. Excluding Denmark, the rank correlation between DATE and the order of countries in coal consumed/worker in manufacturing[12] at $550 is 0.84 ($N = 13$). Early industrializers seem to have often had the ability (factor endowments?) to achieve a substantial use of coal early on. Coal was expensive to transport in nineteenth century Europe, especially prior to the railway, and was located in concentrated fashion (Pollard (1973)); coal inputs were obviously very important to nineteenth century technology.

Second, as the example of Denmark illustrates, with the right circumstances at home it was possible even among the first rank of developers to achieve high income levels on the basis of exploiting a different basis of comparative advantage from that used by the Western European "core" countries.

Third, later developers appear to have reached similar income levels in 1910, in each case high by eighteenth century standard, whilst following different economic policies and having rather different economic structures. The obvious contrast is perhaps between Italy, which looks reasonably close to the standard Gerschenkron story, and Finland which at face value looks more like a second generation Denmark.

Fourth, recall the evidence of Table 8. This suggested that whilst European late developers were indeed in an important way a part of an international trading economy, it did not appear that they were able to establish comparative advantage in manufacturing exports to anything like the same extent that their predecessors had at the same income levels.

Fifth, later developers therefore faced a "choice"—whether to develop passively with comparative advantage or to attempt to protect and stimulate industrialization. The policy response depended, no doubt, on a mixture of internal political forces and external political circumstances. There were, of course, similar choices to be made in the 1950s and 1960s and the more refined versions of Chenery-Syrquin reflect this very clearly, whilst our sample is too small to obtain different transition regressions by policy orientation.

Thus, it appears that the elements which will be required to restate adequately a Gerschenkron-like story, in addition to an adequate description of the development transition, will be a closer examination of revealed comparative advantage, a method to "predict" political response to the circumstances in which late European developers found themselves and the

[12] Figures are from Mitchell (1975) and Bairoch (1968; the denominator is approximated by multiplying population aged 15–64 by the fraction of male workers employed in manufacturing. The highest coal consumption/worker was ranked 1.

recognition that the Gerschenkron stereotype adopted by Barsby is only likely to apply, at best, to a subset of the later developers. However, it whould be clear that this represents a constructive programme for research rather than the pessimistic scepticism which threatens European economic history.[13]

University College, Oxford

REFERENCES

ASHWORTH, W. (1977), "Typologies and Evidence: Has Nineteenth Century Europe a Guide to Economic Growth? "*Economic History Review* 30, 140–158.

BAIROCH, P. *et al.* (1968), *The Working Population and its Structure*. Brussels.

BAIROCH, P. (1976), "Europe's Gross National Product: 1870–1975, "*Journal of European Economic History* 5, 273–340.

BARSBY, S. L. (1969), "Economic Backwardness and the Characteristics of Development," *Journal of Economic History* 29, 449–472.

BECKERMAN, W. and BACON, R. (1966), "International Comparisons of Income Levels: A Suggested New Measure." *Economic Journal* 76, 519–536.

BEREND, I. T., and RANKI, G. (1980), "Foreign Trade and the Industrialization of the European Periphery in the XIXth Century," *Journal of European Economic History* 9, 539–584.

BHAGWATI, J. (1977), "Comment," pp. 496–8 in Ohlin, B. *et al.* (eds.), *The International Allocation of Economic Activity*. New York: Holmes and Meier.

British Parliamentary Papers (1913), *Statistical Abstract of Principal Foreign Countries*. Sessional Papers, vol. LXXIV.

CHATTERJI, M. and WICKENS, M. (1982), "Productivity, Factor Transfers and Economic Growth in the UK." *Economica* 49, 21–38.

CHENERY, H. B., and Syrquin, M. (1975), *Patterns of Development, 1950–1970*. Oxford: Oxford University Press.

COALE, A. J. (1973), "The Demographic Transition Reconsidered." *International Population Conference* Vol. 1. Liege: IUSSP, 53–72.

COHEN, J. S., and WEITZMAN, M. L. (1975), "Enclosures and Depopulation: A Marxian Analysis." *In European Peasants and their Markets* ed. by W. N. Parker and E. L. Jones. Princeton, N.J.: Princeton University Press, pp. 161–176.

CORNWALL, J. (1977), *Modern Capitalism: Its Growth and Transformation*. New York: St. Martin's Press.

CRAFTS, N. F. R. (1980), "Income Elasticities of Demand and the Release of Labour by Agriculture During the British Industrial Revolution." *Journal of European Economic History* 9, 159–168.

CRAFTS, N. F. R. (1983a), "British Economic Growth, 1700–1831: A Review of the Evidence." *Economic History Review* 36, 177–199.

CRAFTS, N. F. R. (1983b), "Gross National Product in Europe 1870–1910: Some New Estimates." *Explorations in Economic History*, 20, 387–401.

CRAFTS, N. F. R. (1984), "Economic Growth in Britain and France, 1830–1910: A Review of the Evidence." *Journal of Economic History*, 44, 49–67.

CRAFTS, N. F. R. (1983d), "English Workers' Real Wages During the Industrial Revolution: A Second Look." Mimeo, Stanford, California.

CRAFTS, N. F. R. (1985), *British Economic Growth during the Industrial Revolution*. Oxford: Oxford University Press.

[13] For a much more sceptical position see Trebilcock (1981; ch. 1, 6 and 7).

CRAFTS, N. F. R. and THOMAS, M. F. (1983), "Comparative Advantage in Britain's Foreign Trade in 1910 and 1935." Mimeo, Stanford, California.

DEANE, P. and COLE, W. A. (1962), *British Economic Growth*, 1688–1959. Cambridge: Cambridge University Press.

FLOUD, R. C., and MCCLOSKEY, D. N. (1981), *The Economic History of Britain Since* 1700. Vol. 2. Cambridge: Cambridge University Press.

GERSCHENKRON, A. (1962), *Economic Backwardness in Historical Perspective*. Cambridge, Mass.: Bellknap.

HARLEY, C. K. (1982), "British Industrialization before 1841: Evidence of Slower Growth During the Industrial Revolution." *Journal of Economic History* 42, 267–289.

HOFFMANN, W. G. (1965), *Das Wachstum der deutschen Wirtschaft seit der Mitte des* 19 *Jahrhunderts*. Berlin: Springer.

KALDOR, N. (1966), *Causes of the Slow Rate of Economic Growth of the United Kingdom*. Cambridge: Cambridge University Press.

KEMP, T. (1969), *Industrialization in Nineteenth Century Europe*. London: Longmans.

KENNEDY, W. P. (1982), "Economic Growth and Structural Change in the United Kingdom, 1870–1914." *Journal of Economic History* 42, 105–114.

KINDLEBERGER, C. P. (1979), "Review of O'Brien and Keyder." *Economic History Review* 32.

KRAVIS, I. B., (1978), "Real GDP Per Capita For More Than One Hundred Countries." *Economic Journal* 88, 215–242.

KUZNETS, S. S. (1966), *Modern Economic Growth: Rate, Structure and Spread*. Yale: Yale University Press.

KUZNETS, S. S. (1967), "Level and Structure of Foreign Trade: Long Term Trends." *Economic Development and Cultural Change* 15, 1–140.

KUZNETS, S. S. (1971, *Economic Growth of Nations*. Cambridge, Mass.: Bellknap Press.

LINDERT, P. H. (1980), "English Occupations, 1670–1811." *Journal of Economic History* 40, 685–712.

LINDERT, P. H., and WILLIAMSON, J. G. (1983), "English Workers' Living Standards during the Industrial Revolution: A New Look." *Economic History Review* 36, 1–25.

MARKOVITCH, T. (1965), "L'Industrie francaise de 1789 a 1964." In J. Marczewsi (ed.), *Cahiers de l I.S.E.A*; A.F. 4.

MATTHEWS, R. C. O., Feinstein, C. H., and Odling-Smee J. C. (1982), *British Economic Growth*, 1856–1973, Stanford, Stanford University Press.

MCCLOSKEY, D. N. (1979), "No It Did Not: A Reply to Crafts." *Economic History Review* 32, 538–541.

MCCOMBIE, J. S. (1981), "What Still Remains of Kaldor's Laws?" *Economic Journal* 91, 206–216.

MITCHELL, B. R. (1975), *Abstract of European Historical Statistics*. London: Macmillan.

MOKYR, J. (1976), *Industrialization in the Low Countries*, 1795–1850. New Haven: Yale University Press.

OVERTON, M. (1979), "Estimating Crop Yields From Probate Inventories: An Example From East Anglia, 1585–1735." *Journal of Economic History* 39, 363–378.

O'BRIEN, P. K., and KEYDER, C. (1978), *Economic Growth in Britain and France*, 1780–1914. London: Allen and Unwin.

POLLARD, S. (1973), "Industrialization and the European Economy." *Economic History Review* 26, 636–648.

POLLARD, S. (1981), *Peaceful Conquest*. Oxford: Oxford University Press.

PRESTON, S. (1980), "Causes and Consequences of Mortality Decline in Less Developed Countries during the Twentieth Century." In R. A. Easterlin (ed.). *Population and Economic Change in Developing Countries*. Chicago: University of Chicago Press.

ROSTOW, W. W. (1960), *The Stages of Economic Growth*. Cambridge: Cambridge University Press.

TREBILCOCK, C. (1981), *The Industrialization of the Continental Powers, 1780–1914*. London: Longmans.

François Crouzet

*Wars, Blockade, and Economic Change in Europe, 1792-1815**

THE wars which raged almost continuously from 1792 to 1815 and which are generally, but not quite properly, called in English the Napoleonic wars, are the longest period of warfare which Europe has known since the early eighteenth century, and as they took place at a crucial stage of economic development, when the Industrial Revolution had just taken off in England and when its preliminary stirrings were showing in various places of the Continent, their impact upon the growth of industry in Continental Europe was quite serious. Unlike the twentieth-century world wars, the Napoleonic wars were not marked by large-scale physical destruction; though the productive potential of some towns or districts suffered from military operations or civil disturbances, such destruction was quite limited in space and time. On the other hand, most European countries suffered during the wars from bouts of paper-money inflation, which had undoubtedly serious consequences —especially in France during the Revolution, when the working capital of many merchants and manufacturers was destroyed through the combination of a sharp rise in their costs, of price control under the *maximum*, and of payment of government orders in depreciated *assignats*. Also, the diversion of resources to military purposes and the heavy burden of taxation and exactions must not be underestimated.

However, the impact of the wars upon the long-run development of industry, which this paper will try to investigate, was felt mostly through the dislocations in international trade which were brought about by the twenty-year-long conflict between Britain and France and by the progressive involvement of all other European countries in this bitter struggle in which economic warfare played a prominent part. Those trade dislocations were at work during the whole war period and not only during the Continental blockade—which was, of course, the most dramatic episode of the wars from the economic point of view, but which was also only their high-water mark; in fact most of the developments which are often thought of as charac-

* Due to limitations of space, some footnotes to this article had to be omitted. These may be obtained on request from the writer.

teristic of the blockade years (1806-1813) had started much earlier, so that a longer view is needed to set the problem in its proper perspective.

Three main factors worked during the Napoleonic wars toward disturbing traditional trade relations: maritime blockade by the British, "self-blockade" of the Continent imposed by the French, and lastly the large-scale redrafting of Europe's political map. Each of them had its own pattern of consequences for the Continent's industries.

I

First the British, thanks to their superior sea power, imposed upon the seaborne trade of Continental countries which were at war with them a number of restrictions which caused the decline and eventually the almost complete interruption of their seaborne trade; this led to serious difficulties and even to a complete and lasting collapse of the large sector of Continental industry which was dependent upon overseas trade.

The eighteenth century can be truly called the Atlantic stage of European economic development. Foreign trade, and especially trade with the Americas, was the most dynamic sector of the whole economy (for instance, French colonial trade increased tenfold between 1716 and 1787), and furthermore the demand from overseas markets was stimulating the growth of a wide range of industries as well as increased specialization and division of labor. Owing to the superiority of sea transport over land transport, the eighteenth-century European economy was organized around a number of big seaports, the most prosperous being those with the largest share in the growing colonial trade, such as Bordeaux or Nantes; each of these had, not only its own industries, but also its industrial hinterland in the river basin of which it was the outlet. For instance, Bordeaux had shipbuilding yards, sugar refineries, distilleries, tobacco factories, and glassworks, while along the Garonne and its tributaries were to be found industries such as sail and rope making, foundries making guns for West Indiamen and boilers for sugar mills, manufactures of linens for slaves and woolens for planters, as well as cornmills producing fine flour for export to the West Indies. The seaboard provinces of France were undoubtedly the most industrialized in the eighteenth century, but the influence of the great seaports penetrated far into the interior; for instance,

Pierre Léon has shown how much the industries of a landlocked province such as Dauphiné were also interested in the West India trade. And if this "Americanization" of trade and industry was the most pronounced for countries which owned a colonial empire (such as Britain, France, Holland, and Spain), its influence extended also farther to the east, to countries which had no colonies but were able to send goods to America as reexports from the colonial powers, especially through Cadiz; so German linens, cutlery, and hardware reached the West Indian and Spanish-American markets. And the growing demand for colonial produce and for raw materials used for shipbuilding stimulated seaborne trade and production for export along the whole European seaboard. So the eighteenth-century European economy was outward-looking, ocean oriented, and its most dynamic trading and industrial centers were geared to seaborne trade and American markets.

This "Atlantic" sector was of course bound to suffer whenever relations with America were hampered. This had been experienced during the wars of the eighteenth century; but during the Napoleonic wars, the situation was much worse owing to the more complete mastery of the seas which was achieved by the Royal Navy and to the far longer duration of hostilities.

French colonial trade collapsed quite early in the war, owing mostly to the slave revolt in Santo Domingo which completely ruined "the jewel of the Caribbean"; three quarters of French colonial trade had been with Santo Domingo, and the sudden and complete destruction of such a valuable branch of trade dealt to several French ports and industries blows from which they did not recover. During the early years of the war, other maritime countries profited a good deal from the decline in French (and Dutch) trade and shipping, and ports such as Barcelona (up to 1796), Leghorn, Lisbon, Hamburg, Copenhagen, and later Trieste, enjoyed periods of great prosperity; but their trade booms all came to an abrupt end when these ports were occupied by the French or when their governments sided with France against England.[1] Still, up to 1807, most enemy ports were not put by the British under rigorous blockade and they could go on trading under neutral flags; though direct relations between enemy ports and their colonies were forbidden

[1] No significant industrial development resulted from such temporary trading booms; and if many merchants, in the Hanseatic towns or Trieste, for instance, made money through smuggling activities or speculation, they used such capital after the wars for resuming their traditional trade and not for industrial investment.

by the British government, neutral ships (mostly American) were allowed to carry on an indirect trade by calling at some United States port on their way. So war with England did not cut at once all relations between Continental countries and the Americas or East Indies, but this trade did in fact decrease a good deal (especially exports of manufactured goods, because American ships often went to England for their return cargoes, and so Continental manufacturers lost ground in their own colonies).[2] Eventually such trade came to a complete stop when the British government decided, partly as a retaliation to the Continental blockade but also in order to placate some pressure groups—the West India and shipping interests—to tighten its blockade. In November 1807, Orders in Council prohibited in fact the indirect trade which neutral ships had been carrying between enemy countries and their colonies; an earlier order of January 1807 had forbidden neutrals to trade between enemy ports, thus destroying most of the active international coasting trade, especially between northern and southern Europe. Moreover, the last French and Dutch colonies were soon to be conquered by the British; and when, in 1808, the Spanish colonies sided with the Spanish insurgents against Napoleon, they fell into the economic orbit of England and their markets were lost to their traditional Continental suppliers of manufactured goods.

So in the later years of the Napoleonic wars, from 1807 onward, Continental countries, nearly all of whom had by then sided willy-nilly against England, were subjected to a drastic sea blockade by the British; they were completely cut off from overseas markets and the most valuable part of their seaborne trade was destroyed. Some coasting trade was of course maintained, and the English permitted—even encouraged—relations with Britain herself; but though they were very active at some periods, they were no helpful substitute for Continental industries, because the Continent could export to England only foodstuffs and raw materials.[3]

It is well known that, as a consequence, the great seaports of the Continent, which had been the hubs of its economic life in the

2 In Barcelona, indirect trade with the colonies through neutrals was in 1806 only a third of the direct trade in 1804. In Nantes, ships entering the port and clearing out had in 1806 a total tonnage of 54,000 as against 80,000 in the peace year of 1802. Moreover, trade through neutral ships did not prevent unemployment of national shipping and depression of shipbuilding.

3 The only exception was German linens, large quantities of which were imported into Britain (and reexported) in 1809-10; American ships were also able to carry them from Tonningen or ports in the Baltic to Latin America.

eighteenth century, were completely crippled from 1807 onwards. Harbors were deserted, grass was growing in the streets,[4] and in large towns like Amsterdam, Bordeaux, and Marseilles, population did actually decrease. However, the collapse of industrial production in the ports and in their hinterland has not been as much noticed. It resulted from the loss of overseas markets and to a lesser degree from the difficulty in obtaining raw materials. In Marseilles, the value of industrial output fell from 50 million francs in 1789 to 12 in 1813.

Among the victims were, of course, shipbuilding and its ancillary industries, such as sail and rope making (Tonneins, east from Bordeaux, had 700 rope makers in 1783, 200 in 1801, none ten years later). Other industries to suffer were those processing imported colonial produce—especially sugar refining (Amsterdam had 80 refineries in 1796, 3 in 1813; Bordeaux 40 before the Revolution, 8 in 1809), also tobacco factories, and tanneries because of the shortage of hides which had been imported from South America. Also affected were industries which had been preparing foodstuffs for overseas markets or for provisioning ships, such as cornmills, distilleries, breweries (in Holland, and glassworks as a consequence), and those for salting, curing, drying, preserving of fish, vegetables, condiments (Marseilles had many such industries and their output fell very low). In Nantes and Hamburg, cotton printing was almost entirely wiped out. However, the most important casualty was the linen industry, which had been very widespread in several Continental countries in the eighteenth century, specially in western France, Flanders, Holland, and Germany. It was an export industry (in the 1780's, Bas-Maine had been exporting two thirds of its output) with its main markets overseas, in the West Indies and in Spanish America, to which large quantities of French and German linens were reexported through Cadix. During the Revolution, French exports to those markets were almost completely stopped, and in Normandy, Brittany, and Maine, output collapsed, falling by as much as two thirds in some districts; there was little recovery under the Empire, owing to the peninsular war; in 1810, Laval

[4] On March 26, 1808, the American consul at Bordeaux reported, "From the Baltic to the Archipelago nothing but despair and misery is to be seen. Grass is growing in the streets of this city. Its beautiful port is deserted except by two Marblehead fishing schooners and three or four empty vessels which still swing to the tide" (quoted by F. E. Melvin, *Napoleon's Navigation System: A Study of Trade Control during the Continental Blockade* [New York: Appleton, 1919], p. 48).

was making 18,000 pieces of linens, as against 36,000 in 1789. In Germany, the situation was not so bad, up to 1806-1807, because exports to America went on and may have increased thanks to the fall in French competition; but after that date German linens also lost their overseas markets, and the linen-making districts of Westphalia, Saxony, and Silesia underwent a serious and lasting crisis; exports of linens from Landshut in the Riesengebirge fell from 174,000 *schock* in 1804 to 29,000 in 1811.

This decline of the linen industry (which can be observed also in Holland, Switzerland, and Italy) was of course in accordance with long-term trends in economic development: it was an archaic industry which was doomed anyway and which had already suffered in the eighteenth century from competition by the fast-rising cotton industry. Still there is no doubt that its decline and decay were greatly accelerated by the loss of overseas markets which it suffered during the Napoleonic wars, and the resulting crisis was so sharp that it made impossible for most centers of the industry any successful effort to adapt or to turn to other activities. In such a domestic manufacture, organized according to the putting-out system and with little fixed capital, the merchant-manufacturers were the key men; once they had lost their working capital, the industry was left in a hopeless condition.

The same remarks apply to the various industries which in the ports and in their hinterlands had been making consumption goods for overseas markets. They were of course small-scale industries, with generally an archaic technology, and at that stage they were not very susceptible to technical improvement and fast expansion. However their decline was unfortunate, as it did decrease opportunities for industrial employment and for accumulation of capital which could have been shifted later to more productive industries.

Moreover the "Atlantic" sector of the European economy was not to recover when peace had been restored in 1815. There was of course a revival of traffic in the ports, but even where a fairly high level was attained, most of them had lost their position as international entrepôts and had become mere regional ports. As for their industries, they were relatively far less active. Such was the case, for instance, of Nantes, Bordeaux, Barcelona (P. Vilar writes that its career as a great colonial and international port was over), and Amsterdam, the "general trade" of which did not revive. This was due to a number of reasons, including having lost touch with international trade during a period of rapid change and losing therefore the necessary skills, information, and capital; but including also the long-term decline in the importance of the West Indies in the world economy, the ruin of Santo Domingo, and the fact that the Latin American markets were henceforth, and for nearly a century, monopolized by the British, while the United States market had little demand for most of the Continent's traditional exports. These developments can also be considered as inevitable, owing to the technical advance and therefore lower prices of British industry and to superior British commercial organization; in fact, before the war, British manufactures had been infiltrating the Spanish and even the French colonies, indicating that Continental industries had been dependent upon the maintenance of a high wall of protection under the old colonial system. Still, war conditions greatly helped the British to engross overseas markets and made it impossible for Continental industries to adapt fast enough to resist English competition and to regain a foothold in those markets after the peace.

Because of the permanent injury inflicted on many Continental industries by interruption of overseas trade, the war brought about a lasting deindustrialization or pastoralization of large areas (with, in some parts of France and Holland, a definite shift of capital from trade and industry towards agriculture). Seaboard districts failed generally to attract the new industries which developed during the wars, or when some new activities were attracted (like the wool industry in Denmark), they did not strike deep roots and declined after the peace—largely because those districts were outlying parts of a Continental empire at a time when land transportation was most expensive. Such was the case of nearly the whole French seaboard, especially the western and southwestern provinces, and also Languedoc, on the Mediterranean, which had in the eighteenth century an important woolen industry working mostly for the Levant markets; in 1809, a German traveler commented that this industry was unable to modernize, owing to stagnating production, lost markets, and reduced profits. These provinces have been ever since, and still are, much less industrialized than northern and eastern France, and to my mind their relative underdevelopment has one of its major causes in the collapse of their traditional industries during the Napoleonic wars (of course, one has still to explain why they failed to industrialize again after 1815, but this is another story).

In France, this crisis was partly offset by the rise of new industries, but this did not happen in some other countries, such as Spain. There, industry had revived during the eighteenth century, especially in Catalonia, largely to answer the growing colonial demand; it was prosperous up to 1796, stagnated during the next twelve years, and collapsed in 1808; the loss of the colonial empire—and the devastation during the struggle against Napoleon—destroyed a vital part in the economic mechanism and worked towards a lasting pastoralization of the country. In Portugal, there was some industrial progress up to 1806, mostly to supply the Brazilian market; but this was lost to the British in 1808, and in 1811, over one third of industrial establishments were described as in a state of decay or closed. In Holland, the wars did not promote any significant industrial growth, while not only the "port industries," but the manufacturing of woolens in Leyden, of silks, and of earthenware in Delft was completely ruined. In Norway, war with England did bring about a sharp decline in its only industries—timber sawing and iron making—and the ruin of the business upper-middle class; and recovery after 1815 was hampered because England had turned to Canada for her timber supply, while Norway, annexed by Sweden, lost the Danish market for her iron.

The dislocation and eventual interruption of the Continent's seaborne trade, owing to maritime war and British blockade, brought about undoubtedly a collapse of the "Atlantic sector" in the Continental economy, which had serious and lasting consequences.

II

There was however a second type of trade dislocation, the consequences of which were not so negative. This was the "self-blockade" by Continental countries against British trade and especially their attempts at preventing imports of British manufactured goods. This was of course the purpose of the Berlin Decree and the Continental blockade, but in France and her satellites, this policy was no innovation and had been enforced since the war had started. The self-blockade was never perfectly effective, and, except for short periods, many British goods were smuggled into the Continent; still, serious obstacles were erected against the import of such goods. Although the self-blockade was intended first and foremost as a means to force England into surrender, it acted practically, and from the point of view we are interested in, as a system of extreme protection, and its effect was to foster the growth of the industries which were the most seriously threatened by British competition.

The most important example in this respect is the cotton industry, and especially the spectacular rise of machine spinning upon a large scale. The cotton industry had grown quite fast during the eighteenth century in several parts of France, in Switzerland, and in Saxony and was well established there before the wars; but, though jennies had been introduced in France in the 1770's and in Germany in the 1780's and a handful of cotton mills had been built in France before 1793, hand spinning remained overwhelmingly predominant everywhere when the war started. From 1796 onward, however, machine spinning increased a great deal in France, and at the turn of the century cotton mills, using water frames or mules, and some of them driven by waterpower, were established in several other countries. Their simultaneous rise is very striking, as the first mills were built in Bohemia in 1797, in Switzerland (St. Gall) in 1800, in Saxony (Chemnitz), Austria, Belgium (Ghent), and Holland in 1801, in Alsace in 1802. This was an answer to English competition, which during the 1790's had met with few obstructions except in France, and which was threatening with extinction the hand-spinning industries of countries like Saxony and Switzerland; but shortage of labor and difficulties in getting British yarns during the war of the second coalition may have had also some influence. However, during the first years of the nineteenth century, machine spinning spread relatively slowly (in each center of the industry only a handful of mills were built), not only because of teething troubles in the use of the new machinery, but also because British competition was unabated and even stronger owing to a continuous fall in the price of English yarn.

In February 1806, imports of yarn were prohibited or surtaxed in France, and soon the Continental blockade made such imports much more difficult and expensive for most countries than formerly—though they were never completely stopped, especially in Switzerland. Moreover, the simultaneous prohibition of British and of Indian cotton goods greatly increased the demand for locally produced yarns, and in Alsace, for instance, cotton printers were induced to start spinning mills and weaving sheds. There is a definite connection between these prohibitions and the boom in machine spinning which started in 1806 and developed up to 1810, with only one interruption in 1808.

Though in France, owing to an already high protection, progress had been rather fast during the preceding years, there was a sharp upward swing, and many new, relatively large mills were established.

The output of machine-spun yarn seems to have doubled from 1806 to 1808 and doubled again at least in the next two years. Lille, Roubaix, and Tourcoing had 32,000 spindles in 1806, 114,000 in 1808, and 177,000 in 1810; towns like Ghent and Mulhouse, where cotton spinning had been almost unknown before 1800, show a fast increase of the same order. However, the 1806 turning point was still more pronounced in Saxony, which before that date had been exposed to British competition and which enjoyed suddenly an unhoped-for prosperity; the number of mule spindles increased from 13,000 in 1806 to 256,000 in 1813. A fast growth took place also in lower Austria and in Switzerland, where the building of new mills went on during the unprosperous years after 1810, owing to increased difficulties in getting English yarn. And on the left bank of the Rhine, where machine spinning had been quite unknown before 1806, it made dramatic progress in the Gladbach-Rheydt district. At the end of the Empire, Continental Europe had built up, within a dozen years, a spinning industry which had, in very rough figures, over one million and a half spindles, of which one million were in France, over 250,000 in Saxony, and 150,000 in Switzerland. (In 1811, Britain had 4,900,000 spindles.) This fast growth was undoubtedly a direct consequence of the Continental blockade, which of course did not create the industry but guaranteed its expansion. It brought about a concomitant decline in hand spinning, which by 1815 had almost entirely disappeared except in some isolated places. And it was accompanied by a good deal of technical progress, and especially by the substitution of the mule for the jenny and the water frame, which allowed the spinning of much finer counts of yarn. The mule was almost unknown in 1800 but became dominant after 1806, its triumph being helped by the crises of 1808 and 1811 which also increased concentration by wiping out the smaller firms and encouraging the creation of large, integrated concerns.

Technical progress took place also in other branches of the cotton industry: the flying shuttle was introduced in Switzerland in 1801, in Ghent shortly after, and in Mulhouse in 1805; cylinder printing started in Paris in 1797 and spread to Alsace and Switzerland around 1805, but was not taken up in Saxony. However, spinning was the only branch to undergo a complete technical revolution, and in others the innovations (including the Berthollet process for bleaching) did not come into general use. As a matter of fact, the sharp progress of machine spinning must not be taken as an index for the cotton industry as a whole.

In France, where the industry enjoyed a continuous protection against English competition, there was undoubtedly an important increase in cotton-goods output. The cotton industry was the only one which did not collapse during the Revolution and even managed to increase its output; though some centers were hurt, especially in printing, which had been working for overseas markets, some others developed fast during the Directory and the early 1800's, particularly Paris and Lille; around 1800, the French consumption of raw cotton was slightly larger than before 1789. Then the boom of the nineteenth century's first decade extended also to weaving and printing, especially in Ghent and around Lille and Mulhouse, and the raw-cotton consumption doubled at least within ten years; despite serious crises in 1808 and 1811, output remained generally at a high level at the end of the Empire.

In other countries, however, the story was rather different. In Saxony, English competition was increasingly dangerous between 1793 and 1806, hurting first the making of muslins, then the coarser goods. The industry was really saved by the Continental blockade, but it seems that overall cotton-cloth production did not rise much during the next few years, an increase in calicoes for printing being offset by a fall in muslins. In Switzerland and Berg, the situation was more unfavorable, because the cotton industries of both countries were heavily dependent upon exports, which were made difficult by the protectionist policy of Napoleon which closed the French and Italian markets. Though the Swiss managed to switch their exports to Germany, the industries of both countries suffered, especially in Berg (which had boomed during the French Revolution) and in the cotton printing and muslin making of Switzerland. However, the situation would probably have been worse if these industries had been exposed to the full blast of English competition, and one may say that the Continental blockade saved the Swiss as well as the Saxon cotton industry.

There is no doubt on the whole that the prohibition of British cotton yarns and goods which was instituted first in France and then in the countries which joined the Continental system gave to the Continental cotton industry a much needed protection, without which it would have been largely wiped out, as well as the opportunity of meeting the demand for cotton goods, which was expanding very fast owing to a shift in tastes and fashions.

On the other hand, one must stress that the balance sheet of the wars period was far from being entirely favorable to the cotton industry. First, the bitter and protracted struggle with England did certainly slow down the introduction of British machinery, production methods, and skilled workers (though it was in the thick of war that Lievin Bauwens smuggled out from England his first mules). Moreover, during the later part of the war period from 1808 onward, the cotton industry labored under a serious difficulty—the shortage and high prices of its raw materials. This was due partly to the Continental blockade but also to the American embargo and to a British Order in Council, of November 1807, which prohibited the carrying of raw cotton into enemy ports. However, except for a few months late in 1808, when many mills had to stand idle, there was never a real cotton famine and the shortage was alleviated by imports from the Levant overland and by cultivation in southern Italy, so that output could be increased again in 1810. Still, after 1807, manufacturers obtained their raw material at much higher prices than did their British competitors—at least twice and even four times higher; and the French were at a special disadvantage, because, owing to a more efficient customs system and to high duties upon raw cotton imposed in August 1810, prices were notably higher in France than in Germany, Austria, or Switzerland. This situation was one of the factors in the crisis which broke out in 1811 and hit the industry very hard, though output recovered the next year. There is no doubt that its expansion was hampered by these high prices of raw materials, and on the whole the cotton industry did not grow as fast on the Continent during the war period as in Britain. British consumption of raw cotton, which can be used as a rough index of output, did increase roughly fourfold between 1790 and 1810; we have no comparable figures for the Continent as a whole, but in France, which had the largest cotton industry, the consumption of raw cotton increased roughly threefold during the same period. Quantitatively, the Continent in fact lost ground to England, while in technology a twenty-year time lag persisted.

This fact has been stressed by various writers, such as Eli Heckscher, who have moreover pointed out that, when peace was restored in 1814 and 1815, the Continental cotton industries were hit very hard by British competition; in France, for instance, some of the largest manufacturers went bankrupt,[5] and prohibitions as

[5] In fact, the direct cause of this collapse was the sudden abolition in April 1814 of the high customs duties on cotton wool, which at a stroke depreciated manufacturers' stocks and brought sharply down yarn and cloth prices.

severe as during the Continental blockade had to be reimposed in order to save the industry. This has been seen as a proof of the blockade's futility, since it did not succeed in making the industry competitive with England, and so the Continental cotton industry has been denounced as a typical "hothouse" industry which developed under artificial conditions of wartime extreme protection and which constituted a serious misallocation of resources.

However this line of argument neglects two basic facts of the concrete economic situation in Europe during the late eighteenth and early nineteenth centuries. The first was that war was being waged and that England was using its mastery of the seas to dislocate the Continent's trade and to destroy some of its traditional industries. Though the Continental blockade was intended as offensive economic warfare, it was in fact a defensive reflex by the Continental economy, inasmuch as the war obliged it to labor under unfavorable conditions such as high prices of raw materials, a serious shortage of capital, and high rates of interest. It is most likely that if there had been no war, if relations with England had been maintained, if a moderate protection had been established, economic and technical progress on the Continent would have been faster. But after all, there was a war.

The second basic fact was the absolute ascendancy, the overwhelming superiority, which the British had achieved, so that no Continental industry, except silk, could be competitive with its English rivals—in fact, a full century was to elapse before foreigners were able to compete successfully with Lancashire mills except for some special lines of goods such as fine printed cottons from Alsace. By 1800, Continental Europe was threatened by pastoralization and the fate of India in the nineteenth century. In fact, countries like Spain, Portugal, and Sweden which fell into the economic orbit of England during the wars suffered a crisis or a collapse of their traditional industries without any compensating rise of new ones; and countries where the cotton industry did not develop during the wars, like Italy or Spain, were to miss the first bus of nineteenth-century industrialization. On the other hand, though the Continental blockade did not reduce much the time lag in techniques between England and the Continent and though the development of the cotton industry (and of some others) under protection was a misallocation of resources, it was a necessary one and the only way to introduce the Industrial Revolution. As a matter of fact, the cotton industry was the strategic sector at that stage. It was the only one likely to expand fast, the best suited to machine production, and it required little capital; its development and technical improvement were the best way to introduce the new technology and to train workers and managers in the skills of machine production and the factory system; also they were the only possible basis for a machine-building industry. And truly its hothouse development during the Napoleonic wars did just that: it gave to the Continent its first experience of large-scale, capitalist, integrated, mechanized industry; it created a new group of manufacturers with a progressive (and sometimes adventurous) outlook and a taste for technical advance and for expansion; moreover, jointly with technical change in the woolen industry, it gave birth to a machine-building industry in a few places, such as Paris, Liège, Verviers, Essen, Mulhouse, Zurich, and Chemnitz; it was of course much smaller than in England and not so advanced, but the quality of its mechanics and of its products improved and its rise was a necessary preliminary stage of industrialization.

On the whole, the cotton industry and especially machine spinning had struck rather strong roots in Continental soil during the Napoleonic wars, and so the foundations for further industrialization had been laid.[6] As a matter of fact, the beginnings of the Industrial Revolution on the Continent could be dated from the simultaneous establishment of machine spinning in various countries, around 1800, which has been pointed out earlier. As, moreover, it is not apparent to what other uses capital and labor resources could have been applied at this stage of economic development,[7] I can not think that the growth of the cotton industry was unfortunate from the point of view of economic growth. However, the "bad thing" was that it created vested interests which during the postwar crisis clamored for renewed protection and generally got it (especially in France), while in fact the industry could likely have developed more efficiently, with only a moderate amount of protection.

[6] During the postwar years, some centers of the cotton industry—especially in Germany and Austria—suffered serious difficulties under the impact of renewed British competition but most of them managed to survive and some went on growing.

[7] Improvement of agricultural techniques would have resulted in releasing labor which only industry could employ; moreover, home demand for foodstuffs was inelastic, export markets unstable, and the British market, which had a regular import demand, was largely closed right at the end of the wars by the new Corn Laws. Attempts at renovating archaic manufactures, especially the linen industry, would have led into a blind alley, as they were suffering a secular decline. Luxury industries, like silk, were not likely to expand fast and were unsuited to machine production.

As for the Continent's wool industries, the "self-blockade" had upon them an impact which was of the same kind as upon its cotton industry but not as powerful—because they were less in need of protection, because the Continent did not usually buy large quantities of English woolen goods, and because the possibilities of expansion were more limited. In France, the wool industry suffered badly during the Revolutionary period, as it was cut off by the war from its traditional markets in Germany, Spain, and the Levant; in 1795, production had fallen by two thirds as compared with 1789. But a general revival followed, and during the Napoleonic period the prerevolution level of output was exceeded, though some old centers of manufacture went on decaying. However, the most interesting development took place in territories which had been annexed to France and which benefited both from strong protection against the British and from the opening of the large French market, which they could take advantage of, thanks to decisive technical progress. This was especially true in Verviers, which at the turn of the century underwent within a few years (from 1799 to 1804) a true Industrial Revolution; the result was a sudden acceleration of growth, with output increasing at a mean average yearly rate of 6 per cent from 1800 to 1810, and a large increase in fixed capital and productivity. The woolen industry of Aachen also expanded a good deal, the value of its output increasing from 5 million and a half francs in 1786 to 11 million in 1811. But in many of the old French centers of the industry, technical change took place also upon a large scale; machine carding, spinning, and finishing became quite common. Protection through the self-blockade was, moreover, responsible for the taking off of a small woolen industry in Switzerland and in the Dutch district of Tilburg and for expansion of the Moravian wool manufactures. Elsewhere, there was little change.

However, two important industries were not much affected by the closing of the Continent to British competition. The first was the silk industry: far from needing any protection against the British, this was much larger and more efficient.[8] The other was the iron industry:

[8] However, different trade dislocations made their impact felt. During the Revolutionary period, the French silk industry underwent a serious crisis owing to a falling demand for luxury goods on the home market and to its being cut off from foreign markets; in 1801, its output was still about half that of 1789. At the same time, the silk manufacturers in Switzerland, Germany, and Austria were quite prosperous, owing to the elimination of French competition. However, the French silk industry recovered under the Empire and launched an export drive, which Napoleon tried to

most centers of production stagnated during most of the war period, the Liège and Düren districts being exceptions, and the overall increase in output of pig iron and iron seems to have been relatively small. As for technical progress, it was very limited—in France to the introduction of rolling mills. In various ironmaking districts (for instance in Berg during the blockade years), experiments were attempted to introduce the new English techniques of coke smelting, puddling, and crucible steelmaking, but nearly all of them failed and were given up. This stagnation may appear puzzling but can be easily explained: British exports of iron products to the Continent had been relatively small before the wars, and their interruption did not create a large vacuum to be filled up by local industries, while several centers of the iron industry (such as Berg) suffered from losing their overseas markets. As for the military demand for iron products, one must be careful not to overestimate it: it certainly helped to keep many ironworks active, but as it guaranteed to ironmasters a stable market for their output it acted, in fact, as a brake upon innovation. On the other hand, prohibition of English goods seems to have stimulated steelmaking and the secondary metal industries, such as the making of hardware, cutlery, tools, buttons, harness pieces, and tinplate articles, but their expansion was not sharp enough to take the primary iron industry out of the doldrums which discouraged technical progress and investment—though the disappearance of many small firms was a forerunner of the post-1815 development.

A complete balance sheet of wartime dislocations would include also a study of the chemical industries, in which the shortage of traditional imported raw materials such as barilla and potashes resulted in drastic technical change and especially in the introduction of the Leblanc process for soda making. Though these new industries expanded fast, they were still relatively small at the end of the war period; but their rise was significant for future growth, just as was that of the machine-building industry. On the other hand, the making of *ersatz* for colonial produce had no importance in the long run; as an example, the beet-sugar industry was wiped out after 1814, not to revive in Europe before the 1830's.

help as much as possible. This had unpleasant consequences in Switzerland and Germany, where the makers of ribbons in Basel and Barmen and of silks in Berlin were the main sufferers (but some centers of the industry remained prosperous). In Italy, there was little expansion of output.

III

To the trade dislocations wrought by the maritime and continental blockades must be added those which resulted from the large-scale redrafting of the political map of Europe during the wars, from the annexation to France of large areas, and from the drastic territorial changes which took place in Germany and Italy. These upheavals brought about frequent and important changes in the customs lines which crisscrossed the Continent and therefore, from the point of view of manufacturers, the closing of traditional markets and the opening of new ones.

The most important developments in this respect resulted from the annexation to France of Belgium and the left bank of the Rhine. Succeeding a period during which these districts had been battlefields and had suffered from French exactions, this had at first unfavorable consequences by interrupting traditional trade relations; but quite soon—roughly after 1800—the industries of Belgium and the Roer department expanded very fast and enjoyed a remarkable prosperity under the Imperial Régime. They shifted their sales from their former markets, mostly in Central Europe, toward the French market on which they concentrated. This integration with a huge protected and unified area of 30 million customers was the decisive influence which was responsible first and foremost for the fast expansion of the woolen industries of Verviers, Aachen, and Jülich, of the silk industry of Crefeld, of the metal industries of Liège and Düren, and for the rise of the cotton industry in Ghent and Gladbach. Its impact was especially strong on the left bank of the Rhine, where, from 1806 onwards, the exclusion from France of Berg manufactured goods brought about a massive immigration of manufacturers and workers from Berg. Industry struck strong roots in this district, which had been overwhelmingly agricultural in the eighteenth century; by 1811, the Roer department had 2,500 industrial establishments with 65,000 workers, and, according to H. Kisch, in a recent article, "French domination proved, in its long run effect, a true blessing." The annexation of Mulhouse in 1798 and the integration of Alsace to the French customs area (outside of which it had remained under the Old Régime) had similar effects for this province, which benefited from an immigration of Swiss capital and labor. This expansion of Belgian and Rhineland industries was helped by comparative advantages they enjoyed over those of *l'ancienne France:* they had suffered less from the economic blizzard which struck French industry during the Revolution; their labor costs were cheaper owing to the existence of a destitute urban proletariat; and the woolen manufacturers had the luck that William Cockerill settled in Belgium and supplied them with excellent machinery, which allowed Verviers to achieve a precocious, fast, and perfect Industrial Revolution, without much trial and error.

However, territorial changes had for other industrial centers far less happy consequences, because they were cut off from their traditional markets without enjoying the compensation which Belgium and the left bank of the Rhine found in France. This was aggravated by Napoleon's protectionist policy which closed the markets of France and Italy, not only to the British but to his own vassals and allies, while trying to persuade them to admit French goods. The continuous annexations to France of new territories which were at once closed to their neighbors made things steadily worse. Moreover, many states, especially in Germany, raised their tariff walls mostly for fiscal reasons. So the Continental market was far from unified and was, in fact, bristling with obstacles to trade.

The result was a general dislocation of trade relations and serious difficulties for countries which were exporting manufactured goods upon a large scale. The two main sufferers were Switzerland and Berg, which were both heavily dependent upon the French and Italian markets and which were badly hurt by the ultraprotectionist French tariff of 1806 and, a few months later, by their exclusion from the Kingdom of Italy. Berg exports fell one third from 1807 to 1810. However, the difficulties which the two countries suffered during the Continental blockade have been exaggerated; the growth of their industry was slowed down or stopped, they suffered from stagnation (especially the Swiss cotton, watchmaking, and jewelry industries and the ribbon making of Barmen), but no irreparable damage was done, there was no general deindustrialization except the disappearance of some Swiss centers of cotton printing. And one must not forget that Berg industries had grown fast up to 1806. As for Italy, which Napoleon made a *chasse gardée*, a preserve, an economic colony of France, surrounded by impassable custom walls except on the side of France, its industry survived nonetheless.[9]

[9] Grossly overdone are the lamentations and diatribes of B. de Cérenville against Napoleon (whom he charges with having deliberately tried to destroy Swiss industry), although the charges have been endorsed by E. F. Heckscher. See Cérenville, *Le Système continental et la Suisse, 1803-1813* (Lausanne: G. Bridel, 1906), pp. 17, 51,

On the other hand, recurrent and sudden changes in political frontiers and in customs regulations, as well as the violent and brutal methods which Napoleon used to enforce the blockade, did frighten and discourage businessmen, destroy security and confidence, and were baneful to entrepreneurship.

On the whole, there is no doubt that the manifold and serious dislocations which international trade suffered during the Napoleonic wars had an unfavorable and retardative effect upon the growth of Continental industry. In addition to causing the permanent decline of the "maritime" and of the linen industries, they slowed down the progress of many industries as a result of serious and protracted crises either during the French Revolution or in the later years of the Empire. During the 1790's, French industrial output collapsed and, though it recovered from 1796 onward, by 1800 it was at best 60 per cent of its pre-1789 level. The causes of this disaster were many, but the interruption of French foreign trade (not only through the British sea blockade, but also owing to the closing of Continental markets in enemy countries) was one of the most decisive. However, part of the damage was not irreparable; for instance, the wool and silk industries revived sharply under Napoleon, though some old wool centers had been mortally wounded. From 1802 to 1810, there was a powerful outburst of industrialization, but its first effect was only to make good earlier losses, and around 1810, the aggregate volume of industrial production was not much higher than in the 1780's (possibly an increase of 50 per cent).[10] As for other Continental countries, they had benefited from the French collapse during the 1790's, but (except for cotton spinning and for the industries of Belgium and the left bank of the Rhine) there is little evidence of fast growth at the time of the Continental blockade, and some important indus-

117, 132, 141, 148 ff., 170, 173 ff., 180, 210-20, 279-81, 321; and Heckscher, *The Continental System. An Economic Interpretation* (Oxford: Clarendon Press, 1922), pp. 307-10. On Berg, see Ch. Schmidt, *Le grand-duché de Berg (1806-1813). Etude sur la domination française en Allemagne sous Napoléon Ier* (Paris: Alcan, 1905), 326-38, 341-43, 354, 382-83, 386, 406-8, 417, 421.

10 This is a very rough guess. The *Statistique de la France publiée par le ministère de l'agriculture et du commerce. Industrie,* I (Paris, 1847), xv-xvii, gives the value of industrial production for 1812 (and for *l'ancienne France*) as double the 1788 figure; but this seems over optimistic. On the other hand, according to W. Hoffman's index, the volume of British industrial production did less than double between 1792 and 1815; Hoffman, *British Industry, 1700-1950* (Oxford: Basil Blackwell, 1955), table 54.

tries did certainly stagnate. Moreover, in 1810-1811, a sharp crisis, aggravated by various dislocations connected with the Continental blockade (overspeculation on colonial produce, enforcement of the Trianon decree), broke the momentum of the Napoleonic industrialization. The last years of the Empire were of stagnation, and many business firms disappeared in the final collapse of 1813-1814.

The balance sheet of the Napoleonic wars for the Continent as a whole might therefore be something like this: collapse of the "maritime" industries, decay of the linen industry, stagnation of the primary iron industry, modest growth of the wool, silk, and secondary-metal industries, relatively fast progress of the cotton industry, and a modest increase in overall industrial output. This was certainly smaller than the rise which had taken place meanwhile in Britain—though recent research has come to the conclusion that the Napoleonic wars had a retardative effect upon British industrial growth and investment in industry. So on the whole the impact of the wars was more unfavorable to the Continent than to England and made more serious its time lag. However, the rise of the cotton-spinning, machine-building, and chemical industries had laid foundations for future development which was also to be fostered by the reforms which French dominion had introduced into many parts of the Continent, such as freedom of enterprise and abolition of the manorial system and the guilds.

Moreover, some other important developments had taken place within the Continental economy. One was the shift, which was to be permanent, in the location of Continental industries. Many industries located in seaboard districts had declined or collapsed, and moreover such districts, with a few exceptions (mostly between the Seine and the Scheldt), did not benefit from the rise of new industries; even in Holland, industries developed inland, in Tilburg and Enschedé. On the other hand, industrial centers which expanded or maintained their levels of output were as a rule inland, within a large area which extended between the Seine and the Elbe rivers; and some of the most important developments (as well as much accumulation of capital, largely through smuggling activities) took place in the ex-*Lotharingia*, with the fast growth of industries in the *Nord* department of France, in Ghent, Verviers, Liège, around Aachen, and in Alsace. This shows how unfortunate it was for France to be deprived of her "natural frontiers" in 1814. The axis of the Continental economy had now moved from the Atlantic toward the Rhine,[11] and this was of course a forerunner of the growing concentration of European industry in northern and eastern France, Belgium, and northwestern Germany, which was to be so characteristic of the nineteenth and of the early twentieth centuries, and owing to which the so-called Golden Triangle (Paris-Hamburg-Milan) still possesses the largest part of the Common Market's industrial power. There were also some interesting developments along an axis situated farther to the east, in Saxony, Bohemia, Austria, and even Hungary.[12]

This shift of industry from the seaboard to the heartland of western Europe was moreover the sign of another significant change. In the eighteenth century, the European economy had been outward looking and geared to overseas markets; all important industries had been heavily dependent upon the export trade.[13] However, during the Napoleonic wars, and especially at the time of the Continental blockade, export industries had suffered the most and one could well say that the prosperity of each industrial center had been inversely proportional to its dependence upon international trade—as is well shown by the story of the linen-making areas, of Lyons, of Switzerland, and of Berg. Moreover, the British had engrossed overseas markets. So the Napoleonic wars made most European industries inward looking and geared first and foremost to national markets (a typical case was Catalonia which, after the collapse of her colonial trade, turned entirely toward the Spanish home market). This had been encouraged by the formation of larger economic units through the abolition of internal customs barriers. This development had run its full course in France, where the Revolution had created a national market which was relatively quite large, so that manufacturers to whom war and postwar conditions had made access to foreign markets difficult were encouraged to limit their ambitions within their own borders. In Central Europe, the situation

11 This was helped by the accumulation of capital, through smuggling, in the hands of merchants in Basel, Strasbourg, Frankfort, and Essen, and its subsequent use for investment in industrial ventures.

12 Industrial growth in these countries was stimulated by the shift toward the east of trade routes across Europe at the time of the blockade and by lower prices of raw materials. On Hungary, see György Mérei, "Ueber einige Fragen der Anfaenge der Kapitalistischen Gewerkentwicklung in Ungarn," *Etudes historiques*, II (Budapest: Académie Hongroise des Sciences, 1960), 753-59.

13 Because of obstacles to internal trade, high costs of land transport, and low incomes per capita, production upon a large scale could not be based on national or regional demand and had to rely upon export markets.

was different because *Kleinstaaterei* maintained many customs barriers; still, larger economic units had been created and there was a clear tendency of Berg and Saxony industrialists to increase their sales on the German markets, especially in the eastern parts of Germany. Later on, the creation of large national markets through the Zollverein and Italian unity was of course to strengthen this trend. With a few exceptions (like Switzerland or Alsace) and up to the worldwide economic expansion of Germany in the late nineteenth century, Continental industries were to work mostly for their own national markets and for those of the neighboring countries. This was in sharp contrast with England and her worldwide trading interests—but after all the inward-looking character of the Continental economy was a consequence of the decisive defeat which it had suffered in its struggle against British ascendancy. So the trade dislocations of the Napoleonic wars were responsible for some of the specific characters of Continental industry and for some of its differences with British industry—differences which they aggravated and which have persisted up to our own day. They are not unsignificant in order to understand some current problems of political economy.

University of Paris

Discussion

RICHARD TILLY: From the point of view of this meeting, the principal issue raised by the Crouzet paper is this: did the dislocations and attendant policies of the period, by leading to the substitution of Continental markets for British and transoceanic ones, accelerate or retard the economic development of Continental Europe? In general, it appears that the net effects of this substitution were negative, because of the previous extensive commitment of resources to British and transoceanic markets, the existence of intra-Continental tariff barriers, and the high cost of overland transportation relative to water transportation. François Crouzet mentions a number of industries and industrial centers that collapsed as the blockades became effective. To this one might add a number of agricultural areas (for example, the North German grain-producing regions) which experienced a serious decline after 1802 or 1803. At the same time, the weight of those sectors which did grow in this period was far from sufficient to compensate for the declining ones. Despite the retardation in Britain's industrial development registered in recent research—notably Crouzet's—the Continental industrial sector appears to have been still further behind Great Britain in 1815 than it had been in 1790.

This impression of the period is a relatively familiar one. Crouzet, however, has suggested that the process of adjustment to new markets had a significant positive side: the disruption of the Atlantic trading connections, in conjunction with import restrictions and military demands, led to a shift of resources into new lines. Crouzet argues that though this represented a misallocation of resources in short-run terms, it involved nevertheless an accumulation of manufacturing skills, techniques, capital, and locations which laid the foundations of subsequent industrial development. There are two rather Schumpeterian aspects to this argument: (1) the distinction between the initiation of change or innovation and the development in perceptible quantitative terms of results; and (2) its emphasis upon the difficulty of altering the circular flow pattern of resource allocation (in this case, that related to overseas markets) in the absence of "exogenous" assistance of the kind represented by the Napoleonic wars and blockades. The implication of Crouzet's argument is, I think, that certain important innovations, such as machine spinning, and certain accompanying entrepreneurial attitudes were thus introduced and established earlier than they would have been under a regime of free trade and peace.

Crouzet's key example concerns the development of the textile industries—mainly, though perhaps not exclusively, the cotton industry—behind a wall of tariff protection (and at times of import prohibition) and supported by a supply of liquid capital and entrepreneurial energies no longer tied up in overseas trade. Unfortunately, Crouzet presents little evidence here which would support this argument. First of all, there are some doubts in my mind as to the character of the new textile firms created: were they as "large-scale," "mechanized," and "integrated" as all that? Not by comparison with most British firms. Moreover, much of the new manufacturing edifice constructed—whatever its relative magnitude and specific characteristics—proved unremunerative after 1814 when faced by British competition, and it disappeared. In Saxony, for instance, there were numerous business failures, and cotton mills were sold at prices covering only a fraction of their original costs. Similar events took place in France. Then, too, one of the most progressive of Continental textile centers in the post-Napoleonic period was the old Duchy of Berg in the German Rhineland. This region was severely disadvantaged under the Continental system, and its subsequent success—beginning in the late 1820's—was closely linked to overseas markets. The Aachen region, in contrast, favored by and geared to the French market under the Napoleonic system, had great difficulty in the postwar period in getting reestablished in world markets. Its subsequent success, moreover, was in the "Atlantic sector" of the European economy. But citing examples is hardly enough. We need to know much more about the firms founded during the Napoleonic period, also a good deal more about their performance after 1814. Until we do, we can only compare impressions. At the moment, my impression is that the achievements in cotton and related branches of manufacturing were premature and at least two decades in advance of the conditions—one might speak here of external economies—necessary for sustained development.

Crouzet concluded his paper with some remarks on the inward-looking character of the Continental economy. It is true that the Continental countries had no transoceanic connections of the magnitude which blessed Great Britain after 1815, and it is true, particularly in the German case, that with the advent of the railroads and the Zollverein in the 1830's, rapid industrial development was perceptibly and increasingly a matter of internal markets. However, the lag between these developments and the Napoleonic period is considerable. Moreover, at least several centers of Continental textile development—one thinks once again of the German Rhineland, though it was not alone in this respect—owed much of their buoyancy and success in the 1830's to their transoceanic trading connections. Furthermore, if we admit the importance of export agriculture (which the Crouzet paper ignores), the picture of an inward-oriented Continental economy, at least for the first third of the nineteenth century, becomes still less plausible.

In summary, Crouzet has presented us with an interesting and very useful paper. I hope that further research concerning the questions his paper has raised will begin to provide us with some convincing answers.

A. Gerschenkron

Economic Backwardness in Historical Perspective

A HISTORICAL approach to current problems calls perhaps for a word of explanation. Unlike so many of their predecessors, modern historians no longer announce to the world what inevitably will, or at least what ideally should, happen. We have grown modest. The prophetic fervor was bound to vanish together with the childlike faith in a perfectly comprehensible past whose flow was determined by some exceedingly simple and general historical law. Between Seneca's assertion of the absolute certainty of our knowledge of the past and Goethe's description of history as a book eternally kept under seven seals, between the *omnia certa sunt* of the one and the *ignorabimus* of the other, modern historical relativism moves gingerly. Modern historians realize full well that comprehension of the past — and that perforce means the past itself — changes perpetually with the historian's emphasis, interest, and point of view. The search is no longer for a determination of the course of human events as ubiquitous and invariant as that of the course of the planets. The iron necessity of historical processes has been discarded. But along with what John Stuart Mill once called "the slavery of antecedent circumstances" have been demolished the great bridges between the past and the future upon which the nineteenth-century mind used to travel so safely and so confidently.

Does this mean that history cannot contribute anything to the understanding of current problems? Historical research consists essentially in application to empirical material of various sets of empirically derived hypothetical generalizations and in testing the closeness of the resulting fit, in the hope that in this way certain uniformities, certain typical situations, and certain typical relationships among individual factors in these situations can be ascertained. None of these lends itself to easy extrapolations. All that can be achieved is an extraction from the vast storehouse of the past of sets of intelligent questions that may be addressed to current materials. The importance of this contribution should not be exaggerated. But it should not be underrated either. For the quality of our understanding of current problems depends largely on the broadness of our frame of reference. Insularity is a limitation on comprehension. But insularity in thinking is not peculiar to any special geographic area. Furthermore, it is not only a spatial but also a temporal problem. All decisions in the field of economic policies are essentially decisions with regard to combinations of a number of relevant factors. And the historian's contribution consists in pointing at *potentially* relevant factors and at *potentially* significant combinations among them which could not be easily perceived within a more limited sphere of experience. These are the questions. The answers themselves, however, are a different matter. No past experience, however rich, and no historical research, however thorough, can save the living generation the creative task of finding their own answers and shaping their own future. The following remarks, therefore, purport to do no more than to point at some relationships which existed in the past and the consideration of which in current discussions might prove useful.

THE ELEMENTS OF BACKWARDNESS

A good deal of our thinking about industrialization of backward countries is dominated — consciously or unconsciously — by the grand Marxian generalization according to which it is the history of advanced or established industrial countries which traces out the road of development for the more backward countries. "The industrially more developed country presents to the less developed country a picture of the latter's future." [1] There is little doubt that in some

[1] Karl Marx, *Das Kapital* (1st ed.), preface.

broad sense this generalization has validity. It is meaningful to say that Germany, between the middle and the end of the last century, followed the road which England began to tread at an earlier time. But one should beware of accepting such a generalization too wholeheartedly. For the half-truth that it contains is likely to conceal the existence of the other half — that is to say, in several very important respects the development of a backward country may, by the very virtue of its backwardness, tend to differ fundamentally from that of an advanced country.

It is the main proposition of this essay that in a number of important historical instances industrialization processes, when launched at length in a backward country, showed considerable differences, as compared with more advanced countries, not only with regard to the speed of the development (the rate of industrial growth) but also with regard to the productive and organizational structures of industry which emerged from those processes. Furthermore, these differences in the speed and character of industrial development were to a considerable extent the result of application of institutional instruments for which there was little or no counterpart in an established industrial country. In addition, the intellectual climate within which industrialization proceeded, its "spirit" or "ideology," differed considerably among advanced and backward countries. Finally, the extent to which these attributes of backwardness occurred in individual instances appears to have varied directly with the degree of backwardness and the natural industrial potentialities of the countries concerned.

Let us first describe in general terms a few basic elements in the industrialization processes of backward countries as synthesized from the available historical information on economic development of European countries [2] in the nineteenth century and up until the beginning of the First World War. Thereupon, on the basis of concrete examples, more will be said on the effects of what may be called

[2] It would have been extremely desirable to transcend the European experience at least by including some references to the industrialization of Japan. Unfortunately, the writer's ignorance of Japanese economic history has effectively barred him from thus broadening the scope of his observations. The reader must be referred, however, to the excellent study by Henry Rosovsky, *Capital Formation in Japan, 1868–1940* (Glencoe, 1961), in which the validity of this writer's approach for Japanese industrial history is explicitly discussed.

"relative backwardness" upon the course of industrial development in individual countries.

The typical situation in a backward country prior to the initiation of considerable industrialization processes may be described as characterized by the tension between the actual state of economic activities in the country and the existing obstacles to industrial development, on the one hand, and the great promise inherent in such a development, on the other. The extent of opportunities that industrialization presents varied, of course, with the individual country's endowment of natural resources. Furthermore, no industrialization seemed possible, and hence no "tension" existed, as long as certain formidable institutional obstacles (such as the serfdom of the peasantry or the far-reaching absence of political unification) remained. Assuming an adequate endowment of usable resources, and assuming that the great blocks to industrialization had been removed, the opportunities inherent in industrialization may be said to vary directly with the backwardness of the country. Industrialization always seemed the more promising the greater the backlog of technological innovations which the backward country could take over from the more advanced country. Borrowed technology, so much and so rightly stressed by Veblen, was one of the primary factors assuring a high speed of development in a backward country entering the stage of industrialization. There always has been the inevitable tendency to deride the backward country because of its lack of originality. German mining engineers of the sixteenth century accused the English of being but slavish imitators of German methods, and the English fully reciprocated these charges in the fifties and sixties of the past century. In our own day, Soviet Russia has been said to have been altogether imitative in its industrial development, and the Russians have retorted by making extraordinary and extravagant claims. But all these superficialities tend to blur the basic fact that the contingency of large imports of foreign machinery and of foreign know-how, and the concomitant opportunities for rapid industrialization with the passage of time, increasingly widened the gulf between economic potentialities and economic actualities in backward countries.

The industrialization prospects of an underdeveloped country are frequently judged, and judged adversely, in terms of cheapness of labor as against capital goods and of the resulting difficulty in substituting scarce capital for abundant labor. Sometimes, on the contrary, the cheapness of labor in a backward country is said to aid greatly in the processes of industrialization. The actual situation, however, is more complex than would appear on the basis of simple models. In reality, conditions will vary from industry to industry and from country to country. But the overriding fact to consider is that industrial labor, in the sense of a stable, reliable, and disciplined group that has cut the umbilical cord connecting it with the land and has become suitable for utilization in factories, is not abundant but extremely scarce in a backward country. Creation of an industrial labor force that really deserves its name is a most difficult and protracted process. The history of Russian industry provides some striking illustrations in this respect. Many a German industrial laborer of the nineteenth century had been raised in the strict discipline of a Junker estate which presumably made him more amenable to accept the rigors of factory rules. And yet the difficulties were great, and one may recall the admiring and envious glances which, toward the very end of the century, German writers like Schulze-Gaevernitz kept casting across the Channel at the English industrial worker, "the man of the future . . . born and educated for the machine . . . [who] does not find his equal in the past." In our time, reports from industries in India repeat in a still more exaggerated form the past predicaments of European industrializations in the field of labor supply.

Under these conditions the statement may be hazarded that, to the extent that industrialization took place, it was largely by application of the most modern and efficient techniques that backward countries could hope to achieve success, particularly if their industrialization proceeded in the face of competition from the advanced country. The advantages inherent in the use of technologically superior equipment were not counteracted but reinforced by its labor-saving effect. This seems to explain the tendency on the part of backward countries to concentrate at a relatively early point of their industrialization on promotion of those branches of industrial activities in which recent technological progress had been particularly rapid; while the more advanced countries, either from inertia or from unwillingness to require or impose sacrifices implicit in a large investment program, were more hesitant to carry out continual modernizations of their plant. Clearly, there are limits to such a policy, one of them being the inability of a backward country to extend it to lines of output where very special technological skills are required. Backward countries (although not the United States) were slow to assimilate production of modern machine tools. But a branch like iron and steel production does provide a good example of the tendency to introduce most modern innovations, and it is instructive to see, for example, how German blast furnaces so very soon become superior to the English ones, while in the early years of this century blast furnaces in still more backward southern Russia were in the process of outstripping in equipment their German counterparts. Conversely, in the nineteenth century, England's superiority in cotton textile output was challenged neither by Germany nor by any other country.

To a considerable extent (as in the case of blast furnaces just cited), utilization of modern techniques required, in nineteenth-century conditions, increases in the average size of plant. Stress on bigness in this sense can be found in the history of most countries on the European continent. But industrialization of backward countries in Europe reveals a tendency toward bigness in another sense. The use of the term "industrial revolution" has been exposed to a good many justifiable strictures. But, if industrial revolution is conceived as denoting no more than cases of sudden considerable increases in the rate of industrial growth, there is little doubt that in several important instances industrial development began in such a sudden, eruptive, that is, "revolutionary," way.

The discontinuity was not accidental. As likely as not the period of stagnation (in the "physiocratic" sense of a period of low rate of growth) can be terminated and industrialization processes begun only if the industrialization movement can proceed, as it were, along a broad front, starting simultaneously along many lines of economic activities. This is partly the result of the existence of complementarity and indivisibilities in economic processes. Railroads cannot be built unless coal mines are opened up at the same time; building half a railroad will not do if an inland center is to be connected with a port city. Fruits of industrial progress in certain lines are received as external economies by other branches of industry whose progress in turn accords benefits to the former. In viewing the economic history of Europe in the nineteenth century, the impression is very strong that only when industrial development could commence on a large scale did the tension between the preindustrialization conditions and the benefits expected from industrialization become sufficiently strong to overcome the existing obstacles and to liberate the forces that made for industrial progress.

This aspect of the development may be conceived in terms of Toynbee's relation between challenge and response. His general observation that very frequently small challenges do not produce any responses and that the volume of response begins to grow very rapidly (at least up to a point) as the volume of the challenge increases seems to be quite applicable here. The challenge, that is to say, the "tension," must be considerable before a response in terms of industrial development will materialize.

The foregoing sketch purported to list a number of basic factors which historically were peculiar to economic situations in backward countries and made for higher speed of growth and different productive structure of industries. The effect of these basic factors was, however, greatly reinforced by the use in backward countries of certain institutional instruments and the acceptance of specific industrialization ideologies. Some of these specific factors and their mode of operation on various levels of backwardness are discussed in the following sections.

THE BANKS

The history of the Second Empire in France provides rather striking illustrations of these processes. The advent of Napoleon III terminated a long period of relative economic stagnation which had begun with the restoration of the Bourbons and which in some sense and to some extent was the result of the industrial policies pursued by Napoleon I. Through a policy of reduction of tariff duties and elimination of import prohibitions, culminating in the Cobden-Chevalier treaty of 1860, the French government destroyed the hothouse in which French industry had been kept for decades and exposed it to the stimulating atmosphere of international competition. By abolishing monopoly profits in the stagnating coal and iron production, French industry at length received profitable access to basic industrial raw materials.

To a not inconsiderable extent, the industrial development of France under Napoleon III must be attributed to that determined effort to untie the strait jacket in which weak governments and strong vested interests had inclosed the French economy. But along with these essentially, though not exclusively, negative policies of the government, French industry received a powerful positive impetus from a different quarter. The reference is to the development of industrial banking under Napoleon III.

The importance of that development has seldom been fully appreciated. Nor has it been properly understood as emanating from the specific conditions of a relatively backward economy. In particular, the story of the Crédit Mobilier of the brothers Pereire is often regarded as a dramatic but, on the whole, rather insignificant episode. All too often, as, for instance, in the powerful novels of Émile Zola, the actual significance of the developments is almost completely submerged in the description of speculative fever, corruption, and immorality which accompanied them. It seems to be much better in accord with the facts to speak of a truly momentous role of investment banking of the period for the economic history of France and of large portions of the Continent.

In saying that, one has in mind, of course, the immediate effects of creating financial organizations designed to build thousands of miles of railroads, drill mines, erect factories, pierce canals, construct ports, and modernize cities. The ventures of the Pereires and of a few others did all that in France and beyond the boundaries of France over vast areas stretching from Spain to Russia. This tremendous change in economic scenery took place only a few years after a great statesman and a great historian of the July monarchy assured the country that there was no need to reduce the duties on iron because the sheltered French iron production was quite able to cope with the iron needs of the railroads on the basis of his estimate of a prospective annual increase in construction by some fifteen to twenty miles.

But no less important than the actual economic accomplishments of a few men of great entrepreneurial vigor was their effect on their environment. The Crédit Mobilier was from the beginning engaged in a most violent conflict with the representatives of "old wealth" in French banking, most notably with the Rothschilds. It was this conflict that had sapped the force of the institution and was primarily responsible for its eventual collapse in 1867. But what is so seldom realized is that in the course of this conflict the "new wealth" succeeded in forcing the old wealth to adopt the policies of its opponents. The limitation of old wealth in banking policies to flotations of government loans and foreign-exchange transactions could not be maintained in the face of the new competition. When the Rothschilds prevented the Pereires from establishing the Austrian Credit-Anstalt, they succeeded only because they became willing to establish the bank themselves and to conduct it not as an old-fashioned banking enterprise but as a crédit mobilier, that is, as a bank devoted to railroadization and industrialization of the country.

This conversion of the old wealth to the creed of the new wealth points out the direction of the most far-reaching effects of the Crédit Mobilier. Occasional ventures of that sort had been in existence in Belgium, Germany, and France herself. But it was the great eruptive effect of the Pereires that profoundly influenced the history of Continental banking in Europe from the second half of the past century onward. The number of banks in various countries shaped upon the image of the Pereire bank was considerable. But more important than their slavish imitations was the creative adaptation of the basic idea of the Pereires and its incorporation in the new type of bank, the universal bank, which in Germany, along with most other countries on the Continent, became the dominant form of banking. The difference between banks of the crédit-mobilier type and commercial banks in the advanced industrial country of the time (England) was absolute. Between the English bank essentially designed to serve as a source of short-term capital and a bank designed to finance the long-run investment needs of the economy there was a complete gulf. The German banks, which may be taken as a paragon of the type of the universal bank, successfully combined the basic idea of the crédit mobilier with the short-term activities of commercial banks.

They were as a result infinitely sounder financial institutions than the Crédit Mobilier, with its enormously swollen industrial portfolio, which greatly exceeded its capital, and its dependence on favorable developments on the stock exchange for continuation of its activities. But the German banks, and with them the Austrian and Italian banks, established the closest possible relations with industrial enterprises. A German bank, as the saying went, accompanied an industrial enterprise from the cradle to the grave, from establishment to liquidation throughout all the vicissitudes of its existence. Through the device of formally short-term but in reality long-term current account credits and through development of the institution of the supervisory boards to the position of most powerful organs within corporate organizations, the banks acquired a formidable degree of ascendancy over industrial enterprises, which extended far beyond the sphere of financial control into that of entrepreneurial and managerial decisions.

It cannot be the purpose of this presentation to go into the details of this development. All that is necessary is to relate its origins and effects to the subject under discussion. The industrialization of England had proceeded without any substantial utilization of banking for long-term investment purposes. The more gradual character of the industrialization process and the more considerable accumulation of capital, first from earnings in trade and modernized agriculture and later from industry itself, obviated the pressure for developing any special institutional devices for provision of long-term capital to industry. By contrast, in a relatively backward country capital is scarce and diffused, the distrust of industrial activities is considerable, and, finally, there is greater pressure for bigness because of the scope of the industrialization movement, the larger average size of plant, and the concentration of industrialization processes on branches of relatively high ratios of capital to output. To these should be added the scarcity of entrepreneurial talent in the backward country.

It is the pressure of these circumstances which essentially gave rise to the divergent development in banking over large portions of the Continent as against England. The continental practices in the field of industrial investment banking must be conceived as specific instruments of industrialization in a backward country. It is here essentially that lies the historical and geographic locus of theories of economic development that assign a central role to processes of forced saving by the money-creating activities of banks. As will be shown presently, however, use of such instruments must be regarded as specific, not to backward countries in general, but rather to countries whose backwardness does not exceed certain limits. And even within the latter for a rather long time it was mere collection and distribution of available funds in which the banks were primarily engaged. This circumstance, of course, did not detract from the paramount importance of such activities on the part of the banks during the earlier industrialization periods with their desperate shortages of capital for industrial ventures.

The effects of these policies were far-reaching. All the basic tendencies inherent in industrial development in backward countries were greatly emphasized and magnified by deliberate attitudes on the part of the banks. From the outset of this evolution the banks were primarily attracted to certain lines of production to the neglect, if not virtual exclusion, of others. To consider Germany until the outbreak of World War I, it was essentially coal mining, iron- and steelmaking, electrical and general engineering, and heavy chemical output which became the primary sphere of activities of German banks. The textile industry, the leather industry, and the foodstuff-producing industries remained on the fringes of the banks' interest. To use modern terminology, it was heavy rather than light industry to which the attention was devoted.

Furthermore, the effects were not confined to the productive structure of industry. They extended to its organizational structure. The last three decades of the nineteenth century were marked by a rapid concentration movement in banking. This process indeed went on in very much the same way on the other side of the English Channel. But in Britain, because of the different nature of relations between banks and industry, the process was not paralleled by a similar development in industry.

It was different in Germany. The momentum shown by the cartelization movement of German industry cannot be fully explained, except as the natural result of the amalgamation of German banks. It was the mergers in the field of banking that kept placing banks in the positions of controlling competing enterprises. The banks refused to tolerate fratricidal struggles among their children. From the vantage point of centralized control, they were at all times quick to perceive profitable opportunities of cartelization and amalgamation of industrial enterprises. In the process, the average size of plant kept growing, and at the same time the interests of the banks and their assistance were even more than before devoted to those branches of industry where cartelization opportunities were rife.

Germany thus had derived full advantages from being a relatively late arrival in the field of industrial development, that is to say, from having been preceded by England. But, as a result, German industrial economy, because of specific methods used in the catching-up process, developed along lines not insignificantly different from those in England.

THE STATE

The German experience can be generalized. Similar developments took place in Austria, or rather in the western sections of the Austrian-Hungarian Empire, in Italy, in Switzerland, in France, in Belgium, and in other countries, even though there were differences among the individual countries. But it certainly cannot be generalized for the European continent as a whole, and this for two reasons: (1) because of the existence of certain backward countries where no comparable features of industrial development can be discovered and (2) because of the existence of countries where the basic elements of backwardness appear in such an accentuated form as to lead to the use of essentially different institutional instruments of industrialization.

Little need be said with reference to the first type of country. The industrial development of Denmark may serve as an appropriate illustration. Surely, that country was still very backward as the nineteenth century entered upon its second half. Yet no comparable sudden spurts of industrialization and no peculiar emphasis on heavy industries could be observed. The reasons must be sought, on the one hand, in the paucity of the country's natural resources and, on the other hand, in the great opportunities for agricultural improvement that were inherent in the proximity of the English market. The peculiar response did not materialize because of the absence of the challenge.

Russia may be considered as the clearest instance of the second type of country. The characteristic feature of economic conditions in Russia was not only that the great spurt of modern industrialization came in the middle of the 1880s, that is to say, more than three decades after the beginning of rapid industrialization in Germany; even more important was the fact that at the starting point the level of economic development in Russia had been incomparably lower than that of countries such as Germany and Austria.

The main reason for the abysmal economic backwardness of Russia was the preservation of serfdom until the emancipation of 1861. In a certain sense, this very fact may be attributed to the play of a curious mechanism of economic backwardness, and a few words of explanation may be in order. In the course of its process of territorial expansion, which over a few centuries transferred the small duchy of Moscow into the huge land mass of modern Russia, the country became increasingly involved in military conflicts with the West. This involvement revealed a curious internal conflict between the tasks of the Russian government that were "modern" in the contemporaneous sense of the word and the hopelessly backward economy of the country on which the military policies had to be based. As a result, the economic development in Russia at several important junctures assumed the form of a peculiar series of sequences: (1) Basic was the fact that the state, moved by its military interest, assumed the role of the primary agent propelling the economic progress in the country. (2) The fact that economic development thus became a function of military exigencies imparted a peculiarly jerky character to the course of that development; it proceeded fast whenever military necessities were pressing and subsided as the military pressures relaxed. (3) This mode of economic progress by fits and starts implied that, whenever a considerable upsurge of economic activities was required, a very formidable burden was placed on the shoulders of the generations whose lifespan happened to coincide with the period of intensified development. (4) In order to exact effectively the great sacrifices it required, the government had to subject the reluctant population to a number of severe measures of oppression lest the burdens imposed be evaded by escape to the frontier regions in the southeast and east. (5) Precisely because of the magnitude of the governmental exactions, a period of rapid development was very likely to give way to prolonged stagnation, because the great effort had been pushed beyond the limits of physical endurance of the population and long periods of economic stagnation were the inevitable consequences.

The sequences just mentioned present in a schematic way a pattern of Russian economic development in past centuries which fits best the period of the reforms under Peter the Great, but its applicability is by no means confined to that period.

What must strike the observer of this development is its curiously paradoxical course. While trying, as Russia did under Peter the Great, to adopt Western techniques, to raise output and the skills of the population to levels more closely approaching those of the West, Russia by virtue of this very effort was in some other respects thrown further away from the West. Broadly speaking, placing the trammels of serfdom upon the Russian peasantry must be understood as the obverse side of the processes of Westernization. Peter the Great did not institute serfdom in Russia, but perhaps more than anyone else he did succeed in making it effective. When in subsequent periods, partly because of point 2 and partly because of point 5 above, the state withdrew from active promotion of economic development and the nobility emancipated itself from its service obligations to the government, peasant serfdom was divested of its connection with economic development. What once was an indirect obligation to the state became a pure obligation toward the nobility and as such became by far the most important retarding factor in Russia's economic development.

Readers of Toynbee's may wish to regard this process, ending as it did with the emancipation of the peasantry, as an expression of the "withdrawal and return" sequence. Alternatively they may justifiably prefer to place it under the heading of "arrested civilizations." At any rate, the challenge-response mechanism is certainly useful in thinking about sequences of that nature. It should be noted, however, that the problem is not simply one of quantitative relationship between the volume of the challenge and that of the response. The crucial point is that the magnitude of the challenge changes the *quality* of the response and, by so doing, not only injects powerful retarding factors into the economic process but also more likely leads to a number of undesirable noneconomic consequences. To this aspect, which is most relevant to the current problem of industrialization of backward countries, we shall advert again in the concluding remarks of this essay.

To return to Russian industrialization in the eighties and the nineties of the past century, it may be said that in one sense it can be viewed as a recurrence of a previous pattern of economic development in the country. The role of the state distinguishes rather clearly the type of Russian industrialization from its German or Austrian counterpart.

Emancipation of the peasants, despite its manifold deficiencies, was an absolute prerequisite for industrialization. As such it was a negative action of the state designed to remove obstacles that had been earlier created by the state itself and in this sense was fully comparable to acts such as the agrarian reforms in Germany or the policies of Napoleon III which have been mentioned earlier. Similarly, the great judicial and administrative reforms of the sixties were in the nature of creating a suitable framework for industrial development rather than promoting it directly.

The main point of interest here is that, unlike the case of Western Europe, actions of this sort did not per se lead to an upsurge of individual activities in the country; and for almost a quarter of a century after the emancipation the rate of industrial growth remained relatively low. The great industrial upswing came when, from the middle of the eighties on, the railroad building of the state assumed unprecedented proportions and became the main lever of a rapid industrialization policy. Through multifarious devices such as preferential orders to domestic producers of railroad materials, high prices, subsidies, credits, and profit guaranties to new industrial enterprises, the government succeeded in maintaining a high and, in fact, increasing rate of growth until the end of the century. Concomitantly, the Russian taxation system was reorganized, and the financing of industrialization policies was thus provided for, while the stabilization of the ruble and the introduction of the gold standard assured foreign participation in the development of Russian industry.

The basic elements of a backward economy were, on the whole, the same in Russia of the nineties and in Germany of the fifties. But quantitatively the differences were formidable. The scarcity of capital in Russia was such that no banking system could conceivably succeed in attracting sufficient funds to finance a large-scale industrialization; the standards of honesty in business were so disastrously low, the general distrust of the public so great, that no bank could have hoped to attract even such small capital funds as were available, and no bank could have successfully engaged in long-term credit policies in an economy where fraudulent bankruptcy had been almost elevated to the rank of a general business practice. Supply of capital for the needs of industrialization required the compulsory machinery of the government, which, through its taxation policies, succeeded in directing incomes from consumption to investment. There is no doubt that the government as an *agens movens* of industrialization discharged its role in a far less than perfectly efficient manner. Incompetence and corruption of bureaucracy were great. The amount of waste that accompanied the process was formidable. But, when all is said and done, the great success of the policies pursued under Vyshnegradski and Witte is undeniable. Not only in their origins but also in their effects, the policies pursued by the Russian government in the nineties resembled closely those of the banks in Central Europe. The Russian state did not evince any interest in "light industry." Its whole attention was centered on output of basic industrial materials and on machinery production; like the banks in Germany, the Russian bureaucracy was primarily interested in large-scale enterprises and in amalgamations and coordinated policies among the industrial enterprises which it favored or had helped to create. Clearly, a good deal of the government's interest in industrialization was predicated upon its military policies. But these policies only reinforced and accentuated the basic tendencies of industrialization in conditions of economic backwardness.

Perhaps nothing serves to emphasize more these basic uniformities in the situation and the dependence of actual institutional instruments used on the degree of backwardness of the country than a comparison of policies pursued within the two halves of the Austrian-Hungarian monarchy, that is to say, within one and the same political body. The Austrian part of the monarchy was backward in relation to, say, Germany, but it was at all times much more advanced than its Hungarian counterpart. Accordingly, in Austria proper the banks could successfully devote themselves to the promotion of industrial activities. But across the Leitha Mountains, in Hungary, the activities of the banks proved altogether inadequate, and around the turn of the century the Hungarian government embarked upon vigorous policies of industrialization. Originally, the government showed a considerable interest in developing the textile industry of the region. And it is instructive to watch how, under the pressure of what the French like to call the "logic of things," the basic uniformities asserted themselves and how the generous government subsidies were more and more deflected from textile industries to promotion of heavy industries.

THE GRADATIONS OF BACKWARDNESS

To return to the basic German-Russian paradigm: what has been said in the foregoing does not exhaust the pattern of parallels. The question remains as to the effects of successful industrializations, that is to say, of the gradual diminution of backwardness.

At the turn of the century, if not somewhat earlier, changes became apparent in the relationship between German banks and German industry. As the former industrial infants had grown to strong manhood, the original undisputed ascendancy of the banks over industrial enterprises could no longer be maintained. This process of liberation of industry from the decades of tutelage expressed itself in a variety of ways. Increasingly, industrial enterprises transformed connection with a single bank into cooperation with several banks. As the former industrial protectorates became economically sovereign, they embarked upon the policy of changing alliances with regard to the banks. Many an industrial giant, such as the electrical engineering industry, which could not have developed without the aid and entrepreneurial daring of the banks, began to establish its own banks. The conditions of capital scarcity to which the German banks owed their historical position were no longer present. Germany had become a developed industrial country. But the specific features engendered by a process of industrialization in conditions of backwardness were to remain, and so was the close relation between banks and industry, even though the master-servant relation gave way to cooperation among equals and sometimes was even reversed.

In Russia the magnificent period of industrial development of the nineties was cut short by the 1900 depression and the following years of war and civil strife. But, when Russia emerged from the revolutionary years 1905–1906 and again achieved a high rate of industrial growth in the years 1907–1914, the character of the industrialization processes had changed greatly. Railroad construction by the government continued but on a much smaller scale both absolutely and even more so relatively to the increased industrial output. Certain increases in military expenditures that took place could not begin to compensate for the reduced significance of railroad-building. The conclusion is inescapable that, in that last period of industrialization under a prerevolutionary government, the significance of the state was very greatly reduced.

At the same time, the traditional pattern of Russian economic development happily failed to work itself out. The retrenchment of government activities led not to stagnation but to a continuation of industrial growth. Russian industry had reached a stage where it could throw away the crutches of government support and begin to walk independently — and, yet, very much less independently than industry in contemporaneous Germany, for at least to some extent the role of the retreating government was taken over by the banks.

A great transformation had taken place with regard to the banks during the fifty years that had elapsed since the emancipation. Commercial banks had been founded. Since it was the government that had fulfilled the function of industrial banks, the Russian banks, precisely because of the backwardness of the country, were organized as "deposit banks," thus resembling very much the type of banking in England. But, as industrial development proceeded apace and as capital accumulation increased, the standards of business behavior were growingly Westernized. The paralyzing atmosphere of distrust began to vanish, and the foundation was laid for the emergence of a different type of bank. Gradually, the Moscow deposit banks were overshadowed by the development of the St. Petersburg banks that were conducted upon principles that were characteristic not of English but of German banking. In short, after the economic backwardness of Russia had been reduced by state-sponsored industrialization processes, use of a different instrument of industrialization, suitable to the new "stage of backwardness," became applicable.

IDEOLOGIES OF DELAYED INDUSTRIALIZATIONS

Before drawing some general conclusions, a last differential aspect of industrialization in circumstances of economic backwardness should be mentioned. So far, important differences with regard to the character of industrial developments and its institutional vehicles were related to conditions and degrees of backwardness. A few words remain to be said on the ideological climate within which such industrialization proceeded.

Again we may revert to the instructive story of French industrialization under Napoleon III. A large proportion of the men who reached positions of economic and financial influence upon Napoleon's advent to power were not isolated individuals. They belonged to a rather well-defined group. They were not Bonapartists but Saint-Simonian socialists. The fact that a man like Isaac Pereire, who contributed so much, perhaps more than any other single person, to the spread of the modern capitalist system in France should have been — and should have remained to the end of his days — an ardent admirer of Saint-Simonian doctrines is on the face of it surprising. It becomes much less so if a few pertinent relationships are considered.

It could be argued that Saint-Simon was in reality far removed from being a socialist; that in his vision of an industrial society he hardly distinguished between laborers and employers; and that he considered the appropriate political form for his society of the future some kind of corporate state in which the "leaders of industry" would exercise major political functions. Yet arguments of that sort would hardly explain much. Saint-Simon had a profound interest in what he used to call the "most numerous and most suffering classes"; more importantly, Saint-Simonian doctrines, as expanded and redefined by the followers of the master (particularly by Bazard), incorporated into the system a good many socialist ideas, including abolition of inheritance and establishment of a system of planned economy designed to direct and to develop the economy of the country. And it was this interpretation of the doctrines which the Pereires accepted.

It is more relevant to point to the stress laid by Saint-Simon and his followers upon industrialization and the great task they had assigned to banks as an instrument of organization and development of the economy. This, no doubt, greatly appealed to the creators of the Crédit Mobilier, who liked to think of their institution as of a "bank to a higher power" and of themselves as "missionaries" rather than bankers. That Saint-Simon's stress upon the role to be played by the banks in economic development revealed a truly amazing — and altogether "unutopian" — insight into the problems of that development is as true as the fact that Saint-Simonian ideas most decisively influenced the course of economic events inside and outside France. But the question remains: why was the socialist garment draped around an essentially capitalist idea? And why was it the socialist form that was so readily accepted by the greatest capitalist entrepreneurs France ever possessed?

It would seem that the answer must again be given in terms of basic conditions of backwardness. Saint-Simon, the friend of J. B. Say, was never averse to ideas of laissez-faire policies. Chevalier, the co-author of the Franco-English treaty of commerce of 1860 that ushered in the great period of European free trade, had been an ardent Saint-Simonian. And yet under French conditions a laissez-faire ideology was altogether inadequate as a spiritual vehicle of an industrialization program.

To break through the barriers of stagnation in a backward country, to ignite the imaginations of men, and to place their energies in the service of economic development, a stronger medicine is needed than the promise of better allocation of resources or even of the lower price of bread. Under such conditions even the businessman, even the classical daring and innovating entrepreneur, needs a more powerful stimulus than the prospect of high profits. What is needed to remove the mountains of routine and prejudice is faith — faith, in the words of Saint-Simon, that the golden age lies not behind but ahead of mankind. It was not for nothing that Saint-Simon devoted his last years to the formulation of a new creed, the New Christianity, and suffered Auguste Comte to break with him over this "betrayal of true science." What sufficed in England did not suffice in France.

Shortly before his death, Saint-Simon urged Rouget de Lisle, the aged author of the "Marseillaise," to compose a new anthem, an "Industrial Marseillaise." Rouget de Lisle complied. In the new hymn the man who once had called upon "enfants de la patrie" to wage ruthless war upon the tyrants and their mercenary cohorts addresses himself to "enfants de l'industrie" — the "true nobles" — who would assure the "happiness of all" by spreading industrial arts and by submitting the world to the peaceful "laws of industry."

Ricardo is not known to have inspired anyone to change "God

Save the King" into "God Save Industry." No one would want to detract from the force of John Bright's passionate eloquence, but in an advanced country rational arguments in favor of industrialization policies need not be supplemented by a quasi-religious fervor. Buckle was not far wrong when in a famous passage of his *History* he presented the conversion of public opinion in England to free trade as achieved by the force of incontrovertible logic. In a backward country the great and sudden industrialization effort calls for a New Deal in emotions. Those carrying out the great transformation as well as those on whom it imposes burdens must feel, in the words of Matthew Arnold, that

> . . . Clearing a stage
> Scattering the past about
> Comes the new age.

Capitalist industrialization under the auspices of socialist ideologies may be, after all, less surprising a phenomenon than would appear at first sight.

Similarly, Friedrich List's industrialization theories may be largely conceived as an attempt, by a man whose personal ties to Saint-Simonians had been very strong, to translate the inspirational message of Saint-Simonism into a language that would be accepted in the German environment, where the lack of both a preceding political revolution and an early national unification rendered nationalist sentiment a much more suitable ideology of industrialization.

After what has been just said it will perhaps not seem astonishing that, in the Russian industrialization of the 1890s, orthodox Marxism can be said to have performed a very similar function. Nothing reconciled the Russian intelligentsia more to the advent of capitalism in the country and to the destruction of its old faith in the mir and the artel than a system of ideas which presented the capitalist industrialization of the country as the result of an iron law of historical development. It is this connection which largely explains the power wielded by Marxist thought in Russia when it extended to men like Struve and in some sense even Milyukov, whose Weltanschauung was altogether alien to the ideas of Marxian socialism. In conditions of Russian "absolute" backwardness, again, a much more powerful ideology was required to grease the intellectual and emotional wheels of industrialization than either in France or in Germany. The institutional gradations of backwardness seem to find their counterpart in men's thinking about backwardness and the way in which it can be abolished.

CONCLUSIONS

The story of European industrialization in the nineteenth century would seem to yield a few points of view which may be helpful for appreciation of present-day problems.

1. If the spurtlike character of the past century's industrialization on the European continent is conceived of as the result of the specific preindustrial situations in backward countries and if it is understood that pressures for high-speed industrializations are inherent in those situations, it should become easier to appreciate the oft-expressed desires in this direction by the governments of those countries. Slogans like "Factories quick!" which played such a large part in the discussions of the pertinent portions of the International Trade Organization charter, may then appear less unreasonable.

2. Similarly, the tendencies in backward countries to concentrate much of their efforts on introduction of the most modern and expensive technology, their stress on large-scale plant, and their interest in developing investment-goods industries need not necessarily be regarded as flowing mainly from a quest for prestige and from economic megalomania.

3. What makes it so difficult for an advanced country to appraise properly the industrialization policies of its less fortunate brethren is the fact that, in every instance of industrialization, imitation of the evolution in advanced countries appears in combination with different, indigenously determined elements. If it is not always easy for advanced countries to accept the former, it is even more difficult for them to acquiesce in the latter. This is particularly true of the institutional instruments used in carrying out industrial developments and even more so of ideologies which accompany it. What can be derived from a historical review is a strong sense for the significance of the native elements in the industrialization of backward countries.

A journey through the last century may, by destroying what

Bertrand Russell once called the "dogmatism of the untravelled," help in formulating a broader and more enlightened view of the pertinent problems and in replacing the absolute notions of what is "right" and what is "wrong" by a more flexible and relativistic approach.

It is, of course, not suggested here that current policies vis-à-vis backward areas should be formulated on the basis of the general experience of the past century without taking into account, in each individual instance, the degree of endowment with natural resources, the climatic disabilities, the strength of institutional obstacles to industrialization, the pattern of foreign trade, and other pertinent factors. But what is even more important is the fact that, useful as the "lessons" of the nineteenth century may be, they cannot properly be applied without understanding the climate of the present century, which in so many ways has added new and momentous aspects to the problems concerned.

Since the present problem of industrialization of backward areas largely concerns non-European countries, there is the question of the effects of their specific preindustrial cultural development upon their industrialization potentialities. Anthropological research of such cultural patterns has tended to come to rather pessimistic conclusions in this respect. But perhaps such conclusions are unduly lacking in dynamic perspective. At any rate, they do not deal with the individual factors involved in terms of their specific changeabilities. At the same time, past Russian experience does show how quickly in the last decades of the past century a pattern of life that had been so strongly opposed to industrial values, that tended to consider any nonagricultural economic activity as unnatural and sinful, began to give way to very different attitudes. In particular, the rapid emergence of native entrepreneurs with peasant-serf backgrounds should give pause to those who stress so greatly the disabling lack of entrepreneurial qualities in backward civilizations. Yet there are other problems.

In certain extensive backward areas the very fact that industrial development has been so long delayed has created, along with unprecedented opportunities for technological progress, great obstacles to industrialization. Industrial progress is arduous and expensive; medical progress is cheaper and easier of accomplishment. To the extent that the latter has preceded the former by a considerable span of time and has resulted in formidable overpopulation, industrial revolutions may be defeated by Malthusian counterrevolutions.

Closely related to the preceding but enormously more momentous in its effects is the fact that great delays in industrialization tend to allow time for social tensions to develop and to assume sinister proportions. As a mild example, the case of Mexico may be cited, where the established banks have been reluctant to cooperate in industrialization activities that are sponsored by a government whose radical hue they distrust. But the real case in point overshadowing everything else in scope and importance is, of course, that of Soviet Russia.

If what has been said in the preceding pages has validity, Soviet industrialization undoubtedly contains all the basic elements that were common to the industrializations of backward countries in the nineteenth century. The stress on heavy industry and oversized plant is, as such, by no means peculiar to Soviet Russia. But what is true is that in Soviet Russia those common features of industrialization processes have been magnified and distorted out of all proportion.

The problem is as much a political as it is an economic one. The Soviet government can be properly described as a product of the country's economic backwardness. Had serfdom been abolished by Catherine the Great or at the time of the Decembrist uprising in 1825, the peasant discontent, the driving force and the earnest of success of the Russian Revolution, would never have assumed disastrous proportions, while the economic development of the country would have proceeded in a much more gradual fashion. If anything is a "grounded historical assumption," this would seem to be one: the delayed industrial revolution was responsible for a political revolution in the course of which the power fell into the hands of a dictatorial government to which in the long run the vast majority of the population was opposed. It is one thing for such a government to gain power in a moment of great crisis; it is another to maintain this power for a long period. Whatever the strength of the army and the ubiquitousness of the secret police which such a government may have at its disposal, it would be naive to believe that those instruments of physical oppression can suffice. Such a government can maintain itself in power only if it succeeds in making people believe that it performs an important social function which could not be discharged in its absence.

Industrialization provided such a function for the Soviet government. All the basic factors in the situation of the country pressed in that direction. By reverting to a pattern of economic development that should have remained confined to a long-bygone age, by substituting collectivization for serfdom, and by pushing up the rate of investment to the maximum point within the limits of endurance of the population, the Soviet government did what no government relying on the consent of the governed could have done. That these policies, after having led through a period of violent struggles, have resulted in permanent day-to-day friction between the government and the population is undeniable. But, paradoxical as it may sound, these policies at the same time have secured some broad acquiescence on the part of the people. If all the forces of the population can be kept engaged in the processes of industrialization and if this industrialization can be justified by the promise of happiness and abundance for future generations and — much more importantly — by the menace of military aggression from beyond the borders, the dictatorial government will find its power broadly unchallenged. And the vindication of a threatening war is easily produced, as is shown by the history of the cold-war years. Economic backwardness, rapid industrialization, ruthless exercise of dictatorial power, and the danger of war have become inextricably intertwined in Soviet Russia.

This is not the place to elaborate this point further with regard to Soviet Russia. The problem at hand is not Soviet Russia but the problem of attitudes toward industrialization of backward countries. If the Soviet experience teaches anything, it is that it demonstrates *ad oculos* the formidable dangers inherent in our time in the existence of economic backwardness. There are no four-lane highways through the parks of industrial progress. The road may lead from backwardness to dictatorship and from dictatorship to war. In conditions of a "bipolar world" this sinister sequence is modified and aggrandized by deliberate imitation of Soviet policies by other backward countries and by their voluntary or involuntary incorporation in the Soviet orbit.

Thus, conclusions can be drawn from the historical experience of both centuries. The paramount lesson of the twentieth century is that the problems of backward nations are not exclusively their own. They are just as much problems of the advanced countries. It is not only Russia but the whole world that pays the price for the failure to emancipate the Russian peasants and to embark upon industrialization policies at an early time. Advanced countries cannot afford to ignore economic backwardness. But the lesson of the nineteenth century is that the policies toward the backward countries are unlikely to be successful if they ignore the basic peculiarities of economic backwardness. Only by frankly recognizing their existence and strength, and by attempting to develop fully rather than to stifle what Keynes once called the "possibilities of things," can the experience of the nineteenth century be used to avert the threat presented by its successor.

Commercial Expansion and the Industrial Revolution

Charles P. Kindleberger*
Massachusetts Institute of Technology

I

The problem to be examined is the relationship between the commercial expansion of the 17th and 18th centuries and the industrial revolution. In particular it may be asked whether a commercial revolution, i.e., a discontinuously rapid increase in commerce, was necessary to, sufficient for or merely contributed to the industrial revolution. If commercial expansion was not needed in advance of an industrial revolution, was it required as a concomitant? Are different types of merchants needed at different stages of industrial evolution or revolution? In particular, is there a possibility that at a fairly advanced stage of industry, the mercantile function may best be taken over by the industrial firm which undertakes direct buying and direct selling; and continued performance of the middleman function by the merchant community may inhibit technical change by acting as a filter between producer and consumer, blocking communication between them?

* I have been helped in writing this paper by comments and suggestions from Rondo Cameron, François Crouzet, Robert Forster, Donald N. McCloskey, Franklin F. Mendels, Joel Mokyr and R. G. Wilson. They are, however, absolved of all responsibility for errors of fact or judgement.

Charles P. Kindleberger

In the early stages of the industrial revolution, the merchant has been thought to have contributed to industrialization in various ways: by extending the market for outputs; by increasing the range and reducing the cost of available inputs; through capital accumulation; « to some extent » as a source of entrepreneurial ability; by directly and indirectly increasing purchasing power; by inducing rationality in business procedures and widening economic horizons [Minchinton, 1969, p. 51]. Additional points made in other lists, and somewhat less compelling, are the import of foreign techniques [Bairoch, 1973, p. 544] and serving as a cause of growth of large towns and industrial centers [Deane, 1965, p. 68].

The method employed here is to study the role of merchants in a number of countries which are not widely dissimilar. We first analyze the activities of merchants in general, and then examine the ties between commerce and industry in four countries: Holland, Britain, France and Germany. In particular, we examine the hypothesis that the commercial revolution produced the industrial revolution in Britain; a commercial revolution in Holland failed to lead to an industrialization process; and industrialization occurred in Germany based on direct selling and buying by industrial firms, without the need for or the benefit of a flourishing commercial network; and see which of these general views, if they are verified, applied in France. Limitations of time and capacity prevent extending the examination to the Italian city-states with their commercial expansion of the 12th to 15th centuries, or to Spain engaged in colonial trade via Seville and Cadiz in the 16th and 17th centuries. Mere mention is made of Swiss experience in textiles and watches.

Commerce may be foreign, and within foreign, continental and overseas; or domestic, and within domestic, national, regional, or local. The term "merchant", moreover, covers a wide spectrum of individuals engaged in various activities which overlap, but the central focus of many differs in significant ways. The literature mentions the great merchant or Merchant Adventurer, the merchant banker, merchant-shipper, merchant-manufacturer, stapling merchant, wholesaler, retailer. In French, distinctions, not always identical, are made among *négociant*, *armateur* (ship-outfitter), *marchand, détailleur, boutiquier*; in German among *Grosskauf-mann, Kaufmann, Händler*; in Holland among First Hand, Second Hand, and Third Hand. Differences among various writers in the same country and *a fortiori* between countries make precision about numbers difficult if not impossible. In the main, we are interested in overseas trade on the one hand, and the wholesaler at the national level on the other, and are specifically not concerned with retail shopkeepers.

The functioning of the merchant must of course be specified in historical time since his role has been continuously evolving. N.S.B. Gras traces this evolution from the travelling, unspecialized merchant of the Middle Ages who played a role especially in moving goods as well as financing, buying and selling them, and the sedentary specialized merchant who followed him and who operated from his counting house, and perhaps a warehouse, increasingly specialized by commodities, by function (export, import, shipping, finance, insurance) and by area [Gras, 1939, *passim*]. Specialization by commodities was linked to that by function since when a sedentary merchant was also a shipowner or outfitter, for example, he found it necessary to provide different commodities for both outbound and return voyage, whereas to specialize in commerce without shipping made it possible to deal with limited numbers of products down to one. The specializing merchant, however, was exposed to greater risk as commodity quality changed, or old markets were closed off, leaving his knowledge obsolete and him without a continuing function. After the 18th century, direct buying and selling reduced the roles of merchant and merchant nation.

The mercantile function can be divided analytically into two parts which we may represent by ideal types: the "gains-from trade" merchant who buys cheap and sell dear in arbitrage fashion, and the "value-added" merchant who concerns himself especially with the quality of goods, and among particular aspects of quality, delivery date or timeliness. The distinction of course is overdrawn, since transport and finance in transit add value to goods in the arbitrage case, and gains from trade are reaped where the merchant is deeply concerned with standardization, quality control and prompt delivery. Moreover the merchant who brings primary products from the Baltic, Mediterranean, East Indies and America to Amsterdam for stapling in "gains-from-trade" operations generally improves the quality of the goods, refining sugar, curing tobacco, roasting coffee, repacking grain to prevent spontaneous combustion, breaking bulk and the like. The distinction is nonetheless a useful one and runs between joining markets and altering markets by upgrading goods, standardizing them, economizing on inspection and handling. In both cases, the merchant provides information, in the first instance on where goods exist and where they may be needed, in the second of what kinds of goods should be produced to satisfy consumer demand at minimal cost.

Most emphasis in the commercial revolution has been on the "gains-from-trade" merchant who uses his information and courage to obtain large rents when he is successful. The value-added function has perhaps been neglected. To avoid the risk that the goods bought and financed will be rejected by the purchasing merchant or the ultimate consumer, the merchant first inspects, say, textiles closely, not only at the fair or in the cloth hall, but in his warehouse, under strong light, and then undertakes to full, shear, bleach, size, dye or otherwise finish the cloth for the market. New qualities are sought to displace old standards and penetrate new markets.

The distinction between the gains-from-trade and the value-added merchant, while loose, is broadly that between the First Hand in the United Provinces of Holland which dealt, bought and sold goods from overseas, and the Second Hand which sorted them, inspected, arranged for bleaching and dyeing when needed and then repacked. It is not far from Adam Smith's separation between the speculative-merchant who can sometimes acquire a considerable fortune by two or three successful speculations, and the slow accumulator in a single industry who seldom makes a great fortune in consequence of a long life of industry, frugality and attention [Smith, 1776, pp. 113-114]. Adam Smith claimed that the only difference between a competent retailer and a great merchant was the quantity of capital:

> « Besides possessing a little capital, he must be able to read, write and account, and must be a tolerable judge too of, perhaps, fifty or sixty different sorts of goods, their prices, qualities, and the markets where they are to be had cheapest. He must have all the knowledge, in short, that is necessary for a great merchant, which nothing hinders him from becoming but the want of sufficient capital ». (*Ibid.*, p. 112)

This underrates the qualities which Smith detects in the speculative merchant: boldness and foresight, including the boldness in the "value-added" merchant of urging new varieties of products on producers and consumers.

When the knowledge used by the gains-from-trade merchant has been widely diffused, his contribution is ended. Direct buying and selling replace his intermediary role. Similarly the specialized merchant who develops a particular product for a particular market faces a crisis when a new, better quality displaces his. He many sell the old product in new markets, or more likely a new set of merchants may take over selling the old product in new markets. On this score, the fact that merchants in many societies seek to make a fortune and withdraw from economic life into politics, gentle life or rural pursuits is helpful. Capital accumulated in commerce may be invested in land, government stock, canal, turnpike and later railroad securities, used to build spacious houses or chateaux, and consumed, and hence to be unavailable for industry. Some will be diverted to banking and recycled into other commerce. But the social aspirations of most successful merchants which induce them to withdraw from trade serve an economic and political purpose. Where they continue on, they block economic change and tend to perpetuate oligarchies which may be republican but are not democratic.

II

The commercial revolution for present purposes, ignoring the Hanseatic cities, the Italian city-states and Seville and Cadiz, began in the United Provinces of Holland. Favored by location in the days of sailing ships, at the crossroads of traffic between the

Atlantic (and Mediterranean) and the Baltic, and representing a « land of cities intimately united by an all-pervading network of waterways » [Marshall, 1920, p. 33], the Dutch developed colonies, shipbuilding, shipping, herring fisheries and the stapling trade. The linkage from trade to local production led to shipbuilding, and Dutch success in shipbuilding in the 17th century gave her an advantage in fishing for herring along the British east coast in the North Sea. Herring were also bought from Scandinavia, and packed with salt obtained from Portugal. Baltic grain was repacked in the stapling ports with Second Hand skill which prevented it from exploding in the heat of the Mediterranean. Linens from Europe exchanged against woolens from Britain were both bought and sold in Amsterdam, after having been finished at Haarlem with local supplies of acid whey for bleaching along with seaweed from the Mediterranean and ashes for lye obtained from many points. The First Hand bought, financed and brought the goods to Holland, later shipped them on to their ultimate destination. The Second Hand sorted, inspected, finished, repacked them. The Third Hand wholesaled them to the hinterland. All dealt in the highly developed commodity and financial markets of Holland [Wright, 1955, pp. 1-6].

Dutch shipping is said to have embodied few innovations, but made effective use of all existing techniques. There was specialization of design between naval and merchant vessels, so that the latter could be more lightly built — not having to carry cannon — and could be cheaper and more efficiently sailed. When protection was needed, as against pirates, they were convoyed. Dutch shipbuilding, it was said, would have provided Adam Smith with a more striking example of the division of labor than pinmaking [Haley, 1972, p. 19]. The Dutch *fluyt* was standardized, so that as many as one a day could be built in the yards of Saardam, which used labor-saving machinery, especially winddriven sawmills and great cranes to lift and move heavy timbers. Timber was imported from the Baltic duty-free, and large inventories were maintained — for as many as four or five thousand ships [Barbour, 1929-30, pp. 119-24]. With a crew of 10 against one of 30 for an equivalent English vessel, freight rates in Dutch ships were one third to one half lower in peacetime than those of any rival. Since it was always easy to find a ship bound for Amsterdam, stapling there was often cheapest even when it was the longest way around [*Ibid.*, p. 132].

Dutch stapling stimulated other industries beside shipbuilding and its related ropemaking, canvas, and naval supplies, and fishing, including herring, cod and whaling. There were cheese and gin, plus the finishing of colonial products and textiles. Haarlem, as mentioned, specialized in bleaching, Leiden in weaving, based on imported wool from Spain, and Amsterdam in silk. The Revocation of the Edict of Nantes in 1685 brought to the Netherlands from France Protestant immigrants estimated by Scoville at 50-75,000. Included among them was a number of entrepreneurs who undertook silk manufacture in Haarlem and Utrecht, as well as production of velvet and linen in the French manner. The stimulus was short-lived, however, as compared with that in commerce and finance, and France recaptured the market for these products after the Peace of Utrecht in 1713 [Scoville, 1960, pp. 341-47].

Dutch successes in the 16th and 17th centuries were not limited to commerce, exploration, shipbuilding, fishing, but extended to public works — especially land reclamation, and not only in Holland but in England, France, Sweden and elsewhere [C. Wilson, 1968, chapter 5] — and to agriculture. The Dutch entrepreneur, workman, farmer, and seaman seems to have been characterized by energy and innovation. This industry and vitality are the only basis that one can make sense of Adam Smith's remark that the « carrying trade is the natural effect and symptom of great wealth, but it does not seem to be the cause of it » [Smith, 1776, p. 354]. Or perhaps a distinction is being made between the carrying trade — shipping — as such, and commerce with large gains from trade. Smith recognized that Holland was the richest country of Europe — though there is some lag here as marks of decline in the Dutch and upsurge of the British were apparent by 1700 and clear by 1730. But in his identification of commerce with manufactures, and merchants with artificers, Smith does not recognize that unlike England, Holland was a country with commerce but not manufactures.

Charles P. Kindleberger

The decline of the Dutch economy after about 1730 is variously explained. Three wars against the British in the 17th century and two against the French in the early 18th were surely a major factor. On narrow economic grounds, these destroyed capital, cut trade routes, raised taxes, which were levied on consumption and thus raised wages. In broader sociological terms, they drained the economy of its energy and innovative capacity. Not only did trade decline, but also shipbuilding, textile production, fishing and agriculture. Dutch industry, sailors, innovators and capital went abroad. Many industries expired without a fight [Wright, 1955, p. 59]. In 1670 the Dutch considered themselves superior to the British in energy and ability as well as capital and material resources. In 1779 a Leiden clothier complained of the general lack of initiative among Dutch industrialists, and a deep-rooted aversion to experimenting with new methods . . . « a lack of initiative and enterprise in many industries and to some extent in agriculture in striking contrast to 100 years previously » [Boxer, 1970, pp. 245, 259]. Cipolla believes this decline in capacity to meet economic challenges is typical of all empires — a three — generations effect which applies to societies as well as to families. « The Dutch of the seventeenth century were great innovators. By the end of the 18th century they were incapable of keeping up with progress taking place outside their boundaries » [Cipolla, 1970, pp. 8-9].

Such sociological explanation is rejected by a recent study of the comparative industrial growth of Belgium and the Netherlands in the first half of the 19th century [Mokyr, 1974a, 1974b], which comes down squarely on the explanation of high wages in Holland as the cause of failure to industrialize [see also Smith, 1776, pp. 826-27, 837]. Other explanations relate to lack of mineral resources, especially coal, inadequate demand, and governmental policy, especially free trade, insisted on by the First Hand as a traditional protector of their gains from stapling [Mokyr, 1974b, chapter 6; Wright, 1955, *passim*]. High wages clearly would make it difficult for Dutch industry to compete in existing industries with existing methods with countries with lower wages, especially if Dutch capitalists provided capital to Dutch and foreign industry at equal rates. But Dutch wage rates were high only in the provinces of Holland and Zeeland, and low in Utrecht, Friesland, Overijessel, Gelderland, and Groningen. More fundamentally, the impact of high wages is damaging to industrialization only in a static model. In a dynamic model, high wages stimulate innovation as Habakkuk points out for the United States in the 19th century, and Crouzet for Britain in 18th [Habakkuk, 1962, *passim*; Crouzet, 1966, p. 288]. In the 16th century, the Dutch went into trade for lack of land [Gee, 1729, p. 128, quoted by Smith, 1776, p. 96]. In the early period a dynamic model applies; in the later, a static. The change from the dynamic to the static model may be more significant than parametric shifts in either.

Nonetheless the Dutch experience makes clear that commercial success is not sufficient for industrial success, and that flourishing and highly specialized commerce poses risks while it earns large gains from trade. England set out to challenge Dutch supremacy in trade, with the Navigation Acts and acts of war. Hamburg represented another threat. But the Dutch emporium declined under direct trading even without the aid of mercantile policy. Exeter serges which had first gone via London and Holland to Spain and Portugal cut out first London, and then Holland, after bringing the dyeing and finishing stages to Devon [Hoskins, 1935, pp. 69-74; C. Wilson, 1941, p. 39]. German linens ultimately skipped Haarlem on their way to British ports [C. Wilson, *ibid.*, pp. 55-56]. The proportion of Bordeaux exports which went to the United Provinces bound for Germany and the Baltic declined from 53% in 1717-21 to 21% in 1764-66 while that of exports direct for the "North" rose from 10% to 51% [Crouzet, 1968, p. 256]. From having been the "regulating granary" of Europe in grain, Holland by the 19th century was merely a dumping ground [Wright, 1955, p. 171].

Long-distance trade by sea need have no connection with domestic industry. It is not clear whether this is accidental: « the absence of any large export industry which could form the solid backbone of exports and domestic investment » [C. Wilson, 1941, p. 187] or in the nature of colonial trade. A reviewer of an eight-volume work on Bordeaux notes that « Just like Athens in the classical period, or Carthage in the Roman epoch, or Venice in the XVIth century or Amsterdam in the XVIIth, Bordeaux found itself operating, in the XVIIIth century, as a center for the deposit and distribution of a flood of goods which passed from one hemisphere (*sic*) to the other . . . left behind by all those ports like London, Le Havre or Marseilles which were connected to the great European industrial centers » [Olivieri, 1973, pp. 454-56]. « A merchant is not necessarily a citizen of any country » said Adam Smith. « It is in a great measure indifferent to him from what place he carries on his trade » [Smith, 1776, p. 395]. Smith's understanding of trade is not always complete: he believed that the merchant employed in exchanging corn from Koenigsberg for wine from Portugal brought both to Amsterdam because he felt uneasy separated from his capital, and wanted to see it enough to compensate for the double charge for loading and unloading [*Ibid.*, pp. 421-22]. This ignores the stapling functions of quality control through grading, packing, storing and the like. But « merchants knew perfectly well in what manner (commerce) enriched themselves. It was their business to know it. But to know in what manner it enriched the country was no part of their business » [*Ibid.*, p. 403].

To the extent that the entrepreneurship thesis is valid, it differs in Holland (and Hamburg and Marseilles) from the formulation applicable in Britain and most of France. Dutch merchants did not withdraw their capital to buy land and enter the nobility as the British and French sought to. There was too little land available for the purpose [C. Wilson, 1968, p. 35; Haley, 1972, p. 49; but for a contrary view see Geyl, 1961, p. 164]. The First Hand closed ranks, formed a tight oligarchy, resisted new entry and moved gradually out of trade into conspicuous consumption and banking. Moreover it was international, not domestic banking, and like Dutch trade unconnected with domestic industry. Charles Wilson overstates the case when he writes that finance became the handmaiden of the industrial revolution in England, but in Holland the mistress of a plundered and bankrupt household [C. Wilson, 1941, p. 188]. Holland's commercial role went into sharp decline by 1750; its financial role, despite crises such as those of 1763 and 1772, lasted until the Napoleonic Wars. In the long run, finance remains more concentrated, and is slower to cultivate direct relations outside the center than is commerce.

III

In the United Provinces, the commercial revolution was not followed by an industrial revolution. In Britain it was. But the British were not seeking to expand commerce as a basis for industrialization when at the end of the 16th century and throughout the 17th they deliberately set out on mercantilist lines to rival Dutch commerce. On the contrary, they were bent on taking over the Dutch monopoly as Florence had done to Venice [MacNeill, 1974]. The Duke of Albermarle said (later) « What we want is what the Dutchmen have » [Williams, 1970, p. 484]. Sir Josiah Child listed 15 respects in which the English should ape the Dutch [C. Wilson, 1965, p. 40] because with his compatriots he wanted to take over the rents inherent in Dutch gains from trade. To the extent that the commercial revolution in Britain determined the subsequent changes in industry — « perhaps determined » them [Mantoux, 1961, p. 92] it was accidental rather than by design.

Like its Dutch counterpart, the East India Company was established at the end of the 16th century. In 1614 a group of London merchants tried to escape the monopoly of textile finishing by forbidding the export of undyed cloth [C. Wilson, 1968, p. 31]. The effort broke down by 1622. A proclamation of 1615 urging the use of English ships found little favor with Yarmouth, Hull or London merchants [Barbour, 1929-30, p. 114]. Half a century later, however, these policies began to take effect. The Navigation Acts were adopted in 1651, strengthened in 1660, and supplemented in 1663 by the Staple Act which monopolized the carrying trade of the colonies [Williams, 1970, p. 484]. New Amsterdam was exchanged for Surinam after the Second Anglo-Dutch War and rechristened New York. Export restrictions were applied in wool in the time of Edward III and again in 1660; the earliest restriction on export of machinery was that of the stockingframe in 1698.

At the same time measures of protection of British industry were begun (and continued throughout the 18th century) in iron, linen, sailcloth, and to keep out calico in an effort to protect woollens.

A revolution in British trade consisted in the « sudden and rapid growth in re-export trade from negligible to one third of exports in 1699-1701 » [Davis, 1954, p. 94]. In the hundred years after 1660, « the English merchant class was able to grow rich, to accumulate capital, on middlemen's profits and on the growing shipping industry which was needed to carry cheap sugar and tobacco, pepper and saltpeter on the ocean routes. Although re-exports and these imports called for no investment in home industry, but only in trade, this could be called the commercial revolution » [*Ibid.*, p. 95].

As Dutch commercial vitality in colonial products diminished, that of Britain (and France) rose. The initial monopoly of London in the established trading companies and under the Methuen treaty of 1703 with Portugal made the outports restive. The West Coast ports were better located for Atlantic trade; Bristol, Liverpool, Exeter and Glasgow moved into triangular trade with West Africa and the West Indies and American colonies, and direct trade to the American colonies. Glasgow specialized in Virginia and Maryland tobacco, by 1750 outstripping London, Whitehaven, Bristol and Liverpool [Price, 1973, p. 590], and stapling it for Amsterdam and French ports by virtue not of value added in sorting, packaging and the like, but on the basis of the monopoly of British ports conferred by the Navigation Acts.

The great West-Indies merchants like John Pinney [Pares, 1950], or John Gladstone [Checkland, 1971] were "speculative merchants", forced into sugar plantations largely by the necessity to take over the properties of those to whom they sold slaves on credit or advanced funds for other purposes. The record shows little interest in improving such plantations, beyond hiring the most efficient overseer possible. John Gladstone never visited the Indies and John Pinney's "great pride" was "to be considered as a private country gentlemen" [Pares, 1950, p. 141]. Both had more money than they knew what to do with. The Pinney merchant house speculated in cotton and sugar, and once put £ 10,000 into the Great Western Cotton Works [*Ibid.*, p. 331]. For the most part they invested their profits and slave compensation money in English land, government securities, and railway shares [*Ibid.*, p. 319]. John Gladstone supported the industrial revolution to the extent of contributing to the education of artisans and workers through the Mechanics and Apprentices Library [Checkland, 1971, p. 176]. He toyed briefly with the idea of buying a country bank which had suspended payments in Gloucester in 1825 [*Ibid.*, pp. 182-83], but the deal fell through. For a man without pretensions to enter the aristocracy, there were not many investment outlets apart from houses and estates before the railroad boom and limited liability. « He knew little of industry, for the world of a Liverpool merchant prince was distinct from that of factory owner [*Ibid.*, p. 198].

The connections of gains-from-trade commerce with industry were limited. To a small degree Bristol went in for staple-based industries. It had 20 sugar refineries in operation in the middle of the 18th century, glass-making for home demand and exports, tobacco-processing, a chocolate industry established in 1731, and a range of metal manufacture [Michinton, 1969, p. 43]. But the city lost ground in the 18th century for a variety of reasons — delays in modernizing the harbor and the lack of a hinterland caught up in the industrial revolution [Minchinton, 1957b, p. xv]. This reverses the causation from industry to commerce. It savors of understatement to say « not all overseas trade had industrial consequences » [Minchinton, 1969, p. 44]. Glasgow was important in colonial trade and in manufacturing, but there was no connection between the tobacco fortunes and the cotton factories [Campbell, 1964, pp. 17-19]. Merchants became landowners [Devine, 1971, *passim*], bankers, or plantation owners overseas, not industrialists. Minchinton is forced to deny Eric Williams' claim that « profits from West African and West Indian trades provided one of the main streams of accumulation of capital in England which financed the industrial revolution [Minchinton, 1969, p. 47].

Recent research, to be sure, has concluded that the industrial revolution did not need large streams of capital from commerce apart from working capital furnished by merchants. Fixed capital needs were small, and expansion was financed by plowed back earnings [Crouzet, 1972, *passim*]. As we shall see presently, an exception existed for South Wales in iron and copper, but not in tinplate or coal. For the most part, however, the industrial revolution was not characterized by a discontinuous leap in rates of savings and capital formation, as economic historians have long thought.

At the other end of the analytical spectrum from the Merchant Adventurer seeking large gains from trade in overseas operations was the value-added merchant who operated internally to create a national market within Britain. Bit by bit a mercantile system evolved, starting with a single emporium or relay center, which like Amsterdam in the 17th century world, served as a hub for the system while information was a scarce input and transactions costs high, and gradually was replaced both by direct connections which bypassed the initial pivot, and by the development of new and more efficiently located markets in some commodities, such as Manchester for cotton, cotton yarn, cotton grey-goods and finished cottons.

London was the initial relay, for domestic as for foreign trade. The monopoly in foreign trade was man-made; that in domestic was an evolutionary construct. In 1677, London had 1686 merchants — although it is unclear what this means in the absence of unambiguous definitions. It was the entrepot of England, with the postal system at its inception in the 17th century radiating out along five or more great roads from London, and with one rate charged for a letter to or from London and double that for letters between other parts of the country, on the theory that all letters went to, from or through the capital [Westerfield, 1913, pp. 417, 419].

Trade was local, national and international. In the early stages both national and international trade went through London. International trade separated itself first to the outports [Court, 1938, chapter ix]. The merchants of Leeds and Wakefield finally gained full control of the Yorkshire export trade, and routed it through Hull, rather than London, forty years after 1700 [R.G. Wilson, 1971b, p. 147]. Inland traders were less important than exporters in the first half of the 18th century but seem to have used London still to a considerable extent [R.G. Wilson, 1971a, p. 82]. In import-competing goods with no export market, national marketing took place through London merchants, whether in silk as illustrated by Sir Thomas Lombe in London, or ribbons, manufactured in Conventry but sold through 12 merchant-manufacturers with warehouses in London and Coventry [Prest, 1960, p. 49].

The evolution of the process is seen in detail in cotton. In the first half of the 18th century, London was the center for the greater part of the sale of Manchester goods, home and foreign, and a merchant like Joseph Hague made a fortune in London as both importer of cotton for Lancashire and dealer in cotton goods [Wadsworth and Mann, 1931, p. 237]. The change to Manchester took place more rapidly in cloth than in raw cotton, where London remained the main port until 1795 — a decade after the industry had begun exporting cloth. London lost out in cloth slowly after 1780, and rapidly from 1790 when Liverpool, Manchester and Glasgow merchants took over. Ultimately Glasgow cloth was marketed through Manchester [Edwards, 1967, pp. 107, 151, 171, 176]. At the end, only surplus stock was sent to London [*Ibid.*, p. 181]. In the 19th century there began the movement of some manufacturers into the merchant function — buying cotton in Egypt and the United States, and dispatching cloth to commission agents at home and abroad [C. Wilson, 1955, p. 180]. The industry was one, however, in which the merchant system remained powerful and prevented wide development of direct selling. Woolens moved to direct trading [Allen, 1939, p. 260]. In cotton textiles, failure to eliminate the merchant led to an unnecessarily large number of separate qualities of product since the merchant lacked motive to induce the customer to change specifications, however arbitrary, and could transfer orders from one manufacturer to another until he found one who would fill them [Robson, 1957, pp. 92-95]. Like Amsterdam or London as a trading relay, the merchant is efficient at a certain stage of economic development, and disfunctional beyond it. In the functional stage, it is important to have trade develop with the scale of industry. Thereafter it helps to have the merchant ready to withdraw. The speculative merchant finds no difficulty in withdrawing provided that he is attracted into the high-prestige vocations of banking, or public service, or the high-prestige avocation of country life.

Between the pure types of gains-from-trade and value-added merchants broadly represented by the tobacco importer of Glasgow and the cotton-yarn specialist of Manchester lay a wide number who combined both functions. Perhaps the woolen trades provide the best illustration. In the 17th century Exeter was the scene of a flourishing export trade in woollens woven in Devon. The dependence of weaver on the merchant is illustrated in Tiverton's action when a merchant died in 1765 in sending to Exeter for another [Mantoux, 1961, p. 65]. Exeter first shipped its serges to London which relayed them to Amsterdam for finishing and distribution on the Continent. London took the finishing away from Amsterdam. Then Exeter shipped direct to Amsterdam. Later Exeter took over the finishing from London and Amsterdam and shipped direct to Hamburg, Oporto, Cadiz [Hoskins, 1935, chapters 2, 3]. While these changes were occurring, however, Norwich came up in worsteds and Devon serges began to decline. As early as 1714 Sir John Elwill, a leading Exeter merchant, writing to a Dutch client, noted that « mixt serges (were) not worn by Many Sort of people as formerly ». In comment, Charles Wilson suggests that it is not altogether fanciful to seek part of the explanation for failure in the relative weakness of the link between manufacturer and merchant. In Devonshire the two functions were often separate; in Norwich and Yorkshire they were often combined » [C. Wilson, 1955, p. 177]. After a brief glory, Profesor Wilson notes, Norwich which had benefited from London merchants gave way to Yorkshire where the link was most strong (Bowring put the point more generally in 1872: Exeter's decay he attributed to want of men of sufficient means and wide views to create new branches of trade when the old had gone [Hoskins, 1935, p. 53]).

To a considerable extent, however, the Yorkshire experience has been misinterpreted. A great deal of the difficulty has been caused by Herbert Heaton's attention to Benjamin Gott, a merchant who went into manufacturing and built Bean Ing, a vast mill in which all the processes from sorting the wool to packing the finished pieces were carried on under one roof. By 1800 the mill housed a thousand operatives. Heaton called Gott only an Industrial Half-Revolutionary because all his weavers worked handlooms and he supplemented their output by purchasing heavily from small domestic clothiers [Heaton, 1965, p. xix] (Compare Gras, who called Sir Thomas Lombe, a London merchant who financed a silk throwing mill on the Italian pattern on the Derwent river in 1718 the morning star of the industrial revolution, not the rising sun [Gras, 1939, p. 178] because it used waterpower instead of steam). There were others beside Gott who went from merchanting into manufacture, the Luptons [Heaton, 1965, p. xx], John Edwards and Law Atkinson [R.G. Wilson, 1971a, p. 60]. The outcry which this change of profession gave rise to has given a misleading impression. With very few exceptions, merchants did not enter manufacturing. When successful, they put their fortunes into land, annuities, government stock, canals and turnpikes, entered public life or retired to the country. Some like William Denison invested in land but did not retire to the country; he was interested in land for its safe return [R.G. Wilson, 1917b, p. 151]. But the gentlemen merchants of Leeds provided neither capital nor entrepreneurship for the industrial revolution.

They did, however, extend the market, and this by upgrading quality and pushing on delivery dates. Yorkshire's share of British exports of woolens rose from one-fifth in 1700 to one-third in 1772 and three-fifths in 1800. It was accomplished, states R.G. Wilson, by concentrating on the cheap end of the market in both woolens and worsteds and leaving to East Anglia (Norwich) the better worsteds, to the West Country (Cotswolds) the fine broadcloths, and to Devon (Exeter) the serges [*Ibid.*, pp. 6-7]. Cloth was bought in cloth halls, or purchased direct from the clothiers, but in any event was brought to the merchant's workshop after purchase and inspected [*Ibid.*, pp. 56, 75]. No one paid attention to the lead seal testifying to measurements. It was sheared, fulled, dyed and sized by the merchant in his workshop, or on order under his direction. Price was not an issue in exports as compared with quality. « William Denison bullied his clothiers about delays and inferior workmanship and occasionally threatened withdrawal of his customs, but never attempted to dictate prices » *[Ibid.*, p. 57]. Initially Yorkshire woolens were relayed by way of London, then Amsterdam, then sold direct to Europe. In due course the European market died down and expansion shifted to trans-Atlantic trade with the American colonies, and after interruption from 1770 to 1783 with the United States. As each major shift came in the trade, new merchant houses came to the fore and old either failed or withdrew. The shift from the German and Portuguese trade to America, for example, meant new risks, especially much longer extensions of credit and new names.

Extension of the market underlined by Adam Smith was far from unimportant, but it tended to move parallel to rather than in anticipation of increase in production. Davis claims that the process of industrialization in England from the second quarter of the 18th century relied importantly on colonial demands for « nails, axes, firearms, buckles, coaches, clocks, saddles, handkerchiefs, buttons, cordage and a thousand other things, "goods, several sorts"» [Davis, 1969, p. 106]. In nails, the nailmonger stood between nail-maker and export-merchant, serving especially the colonial trade, grew rich and turned banker [Court, 1938, pp. 199-206]. This did not involve financing industry directly, but rather more trade.

The major systematic provision of entrepreneurship and finance from commerce to industry is found in the iron and copper industry of South Wales. Here the industry was organized from the beginning in large units, and in those branches of industry which started small — coal and tinplate — there was little London and Bristol mercantile capital. In iron, however, the demands of war brought both merchants and their capital — Anthony Bacon to Cyfarthfa Furnace at Methrhy Tydfil in 1765, joined subsequently by the London merchants Richard Crawshay, Alexander Raby, Sir Benjamin Hammett, Henry Smithers, Thompsons, Harfords, Grenfells. John regards the process as an exception to British development. South Wales had raw materials or access to them — coal, iron, copper, tin — but lacked an industrial tradition comparable to that of the Midlands, Lancashire or Yorkshire.

When war cut off Britain from Russian and Swedish iron ore, an opportunity presented itself with few local resources to take advantage of it. London and to a much lesser extent Bristol merchants brought capital and entrepreneurship. The same opportunities did not present themselves elsewhere [John, 1950, chapters 2, 6].

In frequent other cases trade and industry were linked in the early 18th century, but nothing came of it. At an early non-specialized stage merchants were involved in all kinds of manufacture. The firm of Sutton and Cotesworth started out as tallow-chandlers and corn merchants in Newcastle, traded internationally in grindstones, glass bottles, lead and salt among exports, flax, hemp, madder, whalebone and wines among imports, manufactured candles and sword steel, mined coal and grindstones, and produced salt [Hughes, 1952), pp. 7, 51, 58, 59, 69]. Joseph Pease of Hull invested in trade, shipping, underwriting and whaling, founded a bank, but preferred to think of himself as an industrialist, engaged in oil-seed crushing, white lead, whiting, soap and paint-making, cotton spinning, and through his son-in-law distilling and copper mining. Jackson calls him « one of the great entrepreneurs of the 18th century » [1972, p. 101]. He can hardly have contributed to the industrial revolution.

There were of course revolutionary figures in trade: Strutt who financed Arkwright's patent for the roller spinner and subsequent patents [Fitton and Wadsworth, 1958, pp. 63-64], Robert Owen, David Dale, Samuel Oldknow, but they came from the country draper's shops, not from the merchant princes [Wadsworth and Mann, 1931, p. 239]. In *Fortunes Made in Business* [anon. 1883] only three of 28 industrial fortunes resulted from production of one who had started in trade. Sir John Brown was apprenticed to a cutlery merchant, having wanted to be a merchant as a boy despite family resistance, started to make his own cutlery, and gradually moved into other steel products such as files, and ended up making railway spring buffers. Titus Salt set himself up as a manufacturer to spin alpaca wool after trading in it because he could not induce Bradford spinners to spin that fiber. William Bass was a carrier of beer, then a trader, and finally a brewer. But these connections run between the humblest of value-added merchants, rather than the merchant bankers, and not at all between the gains-from-trade merchant princes and industry. David Dale who built the New Lanark cotton mills outside Glasgow was not one of the 140 pre-1776 Glasgow merchants making fortunes in tobacco, but started as an apprentice in a silk mercer's house, imported linen yarn, went then into banking and only then into cotton spinning [Fox Bourne, 1866, p. 189].

Trade thus grew with the industrial revolution, rather than starting it, as Bairoch also demonstrates [1973, *passim*]. In fact, it is just as well that merchants who do make great fortunes withdraw from trade. Gains-from-trade merchants have to make a fortune in a few years because their rents are likely to be competed away rapidly. Value-added merchants are usually capable of making only a few major improvements in the organization of trade or a change in product before they settle into a routine. To make a fortune and clear out leaves the coast clear for new families to enter and adjust to evolving conditions. Such new families may not exist, as noted earlier for Exeter.

If then the commercial revolution was the expansion of overseas trade in colonial goods — sugar, tobacco, tea and coffee — the merchant's role was not central to the industrial revolution. The value-added merchant aided industrialization in two ways — by urging improvements and expanding the market, and perhaps in a third way, in the usual case, by removing himself from the economic process after a limited period, and unblocking what would otherwise be a barrier to change. The process of building the market proceeds by stages — first through a relay point like London, and a relay institution like the merchant. As information becomes widespread and change slows down, to maintain the old circuits is inefficient.

IV

French foreign trade expanded somewhat faster than that of the United Kingdom during the 18th century, and trade of some ports — notably Bordeaux, the real exports of which grew at a compound rate of 4.1 percent per year, much faster. Fortunes were made not only in Bordeaux, but in Marseilles, Nantes, St. Malo, Rouen, Le Havre among ports, and by the great merchants — *négociants* — of inland cities, Paris, Lyons, Amiens, Beauvais, Le Mans and the like. Much of the trade of the ports, however, like that of Amsterdam, Bristol, Liverpool, Glasgow, was only loosely coupled with domestic production in France and thus had little to do with the industrialization process. Within France there was a commercial and equally an industrial evolution — without the discontinuity one finds in the Britain of the last quarter of the century. The *négociants* reaped large gains from trade and big fortunes. Some went into banking. Many bought *seigneuries* (landed estates) and *offices* (annuities) available from the King, including ennoblement. To a greater extent than in Britain, ports were populated by foreign merchants, or Protestants of French origin, usually nominally converted to Catholicism, whose loyalties ran partly or wholly to fellow-religionists and family members abroad. The internal market developed first through the relay of Paris, and to a lesser extent, Orleans and Lyons, which were finally overcome through direct shipments. Annual and semi-annual fairs gave way in the XVIIIth century to continuous distribution.

French economic historiography has given us a rich detailed picture of the position of the separate ports, and of the textile trades of Beauvais and Le Mans, based on archives and notarial records — Admiralty accounts, rolls for special taxation, and individual marriage contracts, testaments, and bankruptcy proceedings. The prosperity of the ports was due above all to overseas trade. Trade of the Middle Ages in wheat, salt and wine gave away after the middle of the 17th century, and especially in the last quarter, to preoccupation with sugar and slaves, and to a lesser degree coffee, cotton and indigo. Saint Malo and Nantes were deeply engaged in fishing for cod off Newfoundland, which was dried or salted for distribution to the Catholic peasantry of France, as well as, (along with salt beef from Ireland), to slaves in the Antilles. In the course of the 18th century, Bordeaux prospered especially because of her window on the Atlantic and her good relations, based on the ancient trade in wine, with Holland and the Baltic to which excess supplies of sugar, coffee and indigo were relayed.

Nantes (and its neighbor Lorient) had the privilege of the French East India Company (until 1769) and was a center of trade in spices. Its success, most marked in the first half of the century, was owing to its accumulation of capital from the Indian trade, and its three way trade in slaves and sugar. Well placed for the movement of high-value products like spices to the interior, its long-run disability lay in poor road and river connections with Paris [Meyer, 1969, pp. 66, 252]. The salt, cod, sugar, drugs and spices brought to Le Mans from Nantes had to be transhipped from boat to horses and cart at Malicorne [Dornic, 1955, p. 103]. Bordeaux was not much better situated. It could supply planters in Martinique and Saint-Dominique with flour, wines, brandy and salted food, cod transhipped from Brittany or beef from Ireland, but was far from manufacturing centers of France. Crouzet claims this was not important since the biggest export item was textiles and they were not heavy [Crouzet, 1968, p. 210]. Inward shipments were handicapped by high transport costs, administrative and customs obstacles, not to mention low incomes. They consisted of Spanish wool, salt and dried fish, and above all sugar and coffee. Soap, dried fruits, rice, drugs, perfumes and oils came by water from Marseilles [*Ibid.*, p. 275]. Elsewhere Crouzet suggests that if trade is regarded as a superficial phenomenon in France, so was it in the United Kingdom, and that Bordeaux, Nantes and Rouen furnished the colonies with textiles to clothe the slaves, materials for sugar mills, sails, cordage and cannon for ships [Crouzet, 1966, p. 265]. The list seems skimpy. Meyer produces a longer one for Nantes, nearer to the manufacturing centers than Bordeaux — luxury articles wanted by the planters — (furniture, silver, books, clothes) — food for slaves: dried cod from Newfoundland, salt beef and butter from Ireland and Brittany; food for the planters: flour, grain, wine, brandy; construction materials: bricks, lime, wood; cheap cloth for the slaves, etc. and calls them of "mediocre" or "feeble" value [Meyer, 1969, pp. 164, 225]. Commercial expansion was not a strategic factor in the growth of industry, nor did the international conjuncture spread to the peasantry — whose economic life was rather dominated by the state of the harvest well into the nineteenth century.

If overseas commerce did not help greatly with demand, there remains the question of capital formation. Fortunes were made at Nantes and Bordeaux in 10 to 15 years, Arthur Young said [quoted Meyer, 1969, p. 207]. Many of these were consumed in purchases of seigneuries and the building of chateaux. Others found their way into banking. Some went abroad. Foreign merchants never completely dominated the merchant communities of France, but they frequently led them. Bordeaux was a melting pot, and at the top of society were foreign merchants, living assembled at Chartrons and connected through the "Protestant international" to Huguenots and other Protestants not only at other French ports but at Amsterdam, London, Geneva [Poussou, 1968, p. 345; Crouzet, 1968, pp. 213-24]. Not all foreigners were Protestants; among them were Irish and English Catholics, themselves religious and political refugees as the Huguenots were from France. As the total number of *négociants* recorded by Carrière in Marseilles rose from 275 at the beginning of the 18th century to 750 near the end on the eve of the French Revolution, the proportion of foreigners went from 10 to 18 percent, and those of French from the Languedoc and Dauphiné, largely Protestants if nominal converts, from 9 to 28 percent [Carrière, 1973, p. 280]. Some Huguenot capital was exported from France following the Revocation of the Edict of Nantes and particularly in the waves of penalties against failure to abjure Protestantism begun 10 years before 1685 and lasting as late as 1730 [Scoville, 1960, chapter 2]. It was easy for merchants and shippers to export gold, silver, jewelry, furniture. Van Harzel, a Dutch Protestant, who possessed the second largest fortune in Nantes in 1715 — 400,000 livres — escaped in 1730 with his entire wealth [Meyer, 1969, p. 173]. Contemporaries and later scholars made a great deal of this loss, and of the dependence on Marseilles and even Nantes merchants on Swiss funds of the "Protestant diaspora" [Dermigny, 1960, p. 209], but this is denied [Carrière, 1973, pp. 930ff]. The Protestant infusion of the *négoce*, however important, owes a great deal to the movement of Protestants excluded from charges and *offices* and with inhospitable farmland, off the southern slopes of the Central Massif, to Bordeaux and Marseilles [Cavignac, 1967, p. 30, Carrière, 1973,

p. 279]. Foreigners took part when a port failed to maintain representatives abroad — as Marseilles, not deeply engaged in Northern trades, let Swiss, German, British, Dutch agents buy, rather than itself sought to sell [Carrière, 1973, p. 274] or in some views filled a vacuum left by French disdain for commerce [Mathorex, 1919, chapter II, section 1]. Or it could possibly be regarded as a supply response, with excess German potential merchants, limited by the closed ranks of the Hanseatic cities, and anxious to join the race for fortunes, finding a place in French ports and even in the West Indian islands. Jean-Jacques Bethmann (*sic*) was the richest shipowner and one of the greatest merchants in Bordeaux, of Frankfurt-am-Main origin, and there were many others [Cavignac, 1967, p. 30; Thésée, 1972, p. 10]. Spanish and Portuguese Jews expelled from their countries were not numerous, but some of them were among the most prominent. Abraham Gradis had many friends in high places, was owed money by the Marshall of Richelieu and the Duke of Lorges, and died, some years after a loss of 2,700,000 livres tournois, with a fortune of 10,000,000, second only to that at 15 millions in 1791 of the Languedocian, François Bonaffé, a Protestant who married a Creole in Guadeloupe [Poussou, 1968, pp. 349-50].

As in Britain, there are a few examples of close connection between commerce and industry. The Dolle brothers, originally from the Dauphiné, made money in commerce with the Antilles, via the fair of Beaucaire, and moved to buy plantations in the Islands, rather than land in France. In 1787 they explored the desirability of going into linen weaving in Grenoble to assure a supply of cloth for their slaves [Léon, 1963, p. 87]. This conveys a hint of market failure, that adequate supplies would not be available at appropriate times and prices when needed, although perhaps it is nothing more than an outlet for surplus funds. More apposite, but also more debated, is the Huguenot ship-owner, Antoine-Jean Solier, who took funds from his correligionists in the calico printing (*indiennes*) trade, whether in Switzerland to which they had been expelled by the Revocation of the Edict of Nantes, or after their return to France in the Oberkampf plant at Jouy [Dermigny, 1960, *passim*]. Solier et Cie was a *négociant* in general as well as a specialist in import of undyed calico from India, and it is not certain that Swiss participation in this last trade was more significant than any other.

One odd connection between trade and comerce led through land. The merchant retired to the country on an estate, but then entered industry as a *maître de forge* on the basis of iron discovered on the property and woods available for charcoal. Le Nicolais of Laval was a *négociant blanchisseur* who acquired the forge of Aron in this way [Dornic, 1955, p. 171]. Another was Narcisse L., associated with his father in the wine trade in Paris who, in 1833, took on interests in mines and blast furnaces [Daumard, 1970, p. 368]. The equivalent in Glasgow was the tobacco merchant whose land turned out, or was known in advance, to bear coal, and led its owner into the coal, salt or iron business [Price, 1973, p. 587; Hughes, 1952, chapter ii].

Rouen and Le Havre were more closely connected with industry, and Rouen more than Le Havre which participated in the colonial trade. Woolens and linens of Northern France were sold at fairs at Caen and Guilbray and shipped over Rouen when exported. More important was the growth of the cotton-textile industry, based on cotton imported from the Islands, and the putting-out system operated by Rouen merchants. "Siamoises", a mixture of linen and cotton patterned after the dress worn by the ambassador of Siam and the members of his suite in 1684, were one speciality. Total production rose rapidly in Rouen and its parishes from 50,000 pieces in 1717 to 489,000 in 1767, and were exported as well as sold within France [Dardel, 1966, p. 119]. As Britain mechanized, Rouen tried to follow suit, with the aid of British workers, merchants and manufacturers. Fortunes were perhaps made more quickly in the colonial trade either directly, or via Cadiz, than by industry, and in banking, which was closely associated with Paris. Rouen was more like Liverpool, Bordeaux like Glasgow, except that Liverpool was associated with an innovative textile center, Rouen with an imitative. It was badly hurt by the Eden treaty of 1786 which lowered duties on English textiles and earthenware.

Apart from Rouen, the *grand négoce* of the ports had a limited connection with the industrial evolution of France. The same was not true of the great merchants of inland cities, insofar as they have been studied. France is a hexagon, closed against the outside world, but connected internally by a communication network. Until the 18th century, this operated commercially with Paris as the entrepot, pivot, or relay. In the sixteenth century the brightly colored cloths of Paris and Rouen were being sold in the Languedoc by merchants travelling to fairs [Le Roy Ladurie, 1966, pp. 123-26]. At the end of the 17th and beginning of the 18th century, Paris began to be eliminated by direct selling. In 1650, Beauvais sold its linens two thirds to Paris and Lyons, with the rest local or exported over Rouen. In 1690 exports had gained at the expense of Paris, and direct relations had been established by Beauvais merchants with Montpellier, Toulouse and other inland cities. By 1730, the role of Paris had shrunk to very small proportions. Lyons sustained its trade. In the early period it had been a relay point for the south. Later it served as a stapling place for exports to Switzerland and Italy [Goubert, 1959, maps, pp. 29, 71, 103]. Over the eighty years and throughout the rest of the 18th century exports grew to a preponderant position because of the competition at home from cotton cloths. The high quality *demi-Hollandes* of the seventeenth century became an export staple because of the loss of home demand. By diverting the existing supply to new outlets, the Danse, Motte and other mercantile dynasties relieved the pressure on Beauvais manufacturers to keep up with technological change, and thus contributed indirectly to their fossilization and ultimate decline.

Paris remained the relay longer for the newer goods in greater demand. While Oberkampf at Jouy (after 1759) sold no *indiennes* in export markets, it used Paris as a center of consumption and for redistribution to distant points. Alongside this network, however, there developed marketing directly through the established textile centers: Abbeville, Amiens, Beauvais, Troyes, etc. and the establishment of depositories who handled shipments to fairs [Bergeron, 1973, p. 19]. There was of course specialization. One Parisian *indienneur*, Dubois, sold four-fifths of his output in the provinces; another, Jarry, mostly in Paris [*Ibid.*]. In silk, another high-grade product, as well as in printed cottons, Lyons sold through Paris throughout the century, apart from export sales. Two thirds of domestic sales went through the relay of Paris, rather than directly and the « overpowering superiority of Paris limited the direct intervention of Lyon *négociants* in the rest of the country » [Garden, 1973, p. 20]. Even in silk, however, Paris was a rival but did not dominate [*Ibid.*, p. 24].

There were some interesting features in the trade in *etamines*, a loosely woven light woolen cloth in which Le Mans and the surrounding cities specialized. The numbers of *négociants* in Le Mans engaged in this product was seven in 1712, a dozen toward the middle of the century [Dornic, 1955, pp. 45, 59]. It is stated these *négociants* followed the extension of the master manufacturers « or at least did not precede it », but became the masters of the trade by buying, finishing and shipping the cloth [*Ibid.*, p. 53]. Le Mans and its surrounding towns specialized in *etamines* dyed black, which was sold in large quantities to monks and nuns within France and to Italy as well. Domestic sales through fairs were handled largely by merchants from Paris or other towns, rather than by the *marchand/drapier mercier* of Le Mans who existed in large numbers or the Le Mans *négociants* [*Ibid.*, pp. 1, 43, 100, 105]. Like *demi-hollandes etamines* lost out to cotton textiles, and also suffered from the deterioration in the quality of wool. Paris merchants dropped out of the trade, leaving the great merchants of Le Mans only with overseas trade.

The account of textiles in Maine provides a footnote on Paris as a relay in the « curious commercial circuit » in linens. The Abbey of Le Beauvais in Le Mans bought linen for shirts, underpants and stocking from Alençon, 30 kilometers to the north and slightly west, each year from 1713 to 1730. These were delivered at the end of January or beginning of February, like other linens bought from Rouen for surplices, table linen and handerchiefs, by a certain Marigner, wholesale cloth merchant of Paris, 100 kilometers to the east of Le Mans [Dornic, 1955, p. 109]. Dornic suggests that the explanation lies in the state of the local roads in winter. Another possibility is that until direct selling took over especially in the declining products which lost out in the French luxury market, the relay of Paris remained critical. In foodstuffs, of course, it continued at Les Halles in Paris until the market was moved to Orly in the 1960s.

By the middle of the 18th century, Paris, like London and Amsterdam, was shifting from trade to finance. It continued to invest in trade, through the trade of the ports. One scholar states bluntly that Paris in the 18th century was a financial not a commercial city [Cavignac, 1967, p. 132], but this presumably would apply more fully to the second than to the first half of the century. But the funds were invested but little in industry. Le Couteulx, Paris banker and Rouen merchant, invested in textile manufactures in Rouen, but mainly trafficked in Spanish piastres through his Cadiz branch [*Ibid.*]. Paris was a relay for money which went back to the ports and to foreign trade, when it was not used to buy privileges from the court, rather than into industry. At the same time, the commercial network within France, if not the superficial but profitable trade of the ports, evolved side by side with the progress in production.

V

Germany has been characterized as an economy without highly developed commerce, and one which sprang directly from traditional society to industrialization without passing through an extended commercial development [Borchardt, 1971, pp. 89-90, 145-46; Kocka, 1974, p. 9]. In the early 19th century, a British observer characterized Münster as having « little commerce », claimed that the « commerce of Hanover (is) inconsiderable despite the connection of Bremen » and further that:

> « A body of merchants and manufacturers, who by their capital give employment to numerous workmen, and who like country gentlemen are too rich to be dependent on the smiles or frowns of the court, is not to be found in Prussia [Jacobs, 1820, pp. 94, 116, 225].

Bavaria was marked by a lack of a considerable group of great merchant families and great long-distance trading houses [Schremmer

1970, p. 594]. In Baden, merchants were mistrusted as swindlers and speculators and the task of industrial development (*Gewerbe-foerderung*) was to protect and strengthen the middle class against them [Fischer, 1962, p. 79].

This picture is perhaps one-sided, leaving out as it does the Hanseatic cities — Cologne, Bremen, Hamburg, Lübeck — the Rhineland, and Frankfort-am-Main, in which merchant communities were highly developed. Cologne had sunk back from a leading position in international commerce as Antwerp and Rotterdam proved better located for stapling. In its turn, Lübeck, the site of Thomas Mann's novel of a merchant dynasty, *Buddenbrooks* [1901], lost out to the better-located Hamburg and Bremen. These cities, however, developed their own merchant dynasties, with Hamburg especially strengthened by displaced Huguenots in the early 17th century and by refugees from the French Revolution at the end. Hamburg specialized in short-haul trade to the Baltic and above all London. It became known as the « English city ». Bremen developed long-distance traffic, especially trans-atlantic, in tobacco, cotton and rice. Hamburg dominated in sugar, coffee, hides, tropical wood and wines. Bremen complained that it lacked an export product, as Hamburg had grain and the Baltic ports grain and wood [Rauers, 1913, pp. 13-15]. Frankfurt was the gateway for British goods to southern German states.

Frankfurt and Hamburg differed profoundly in social and political ways. Frankfurt was a creation of the Prussian king, who established the Frankfurt fair there, at the major crossing on the east-west trade route of central Europe. When the Zollverein was formed under Prussian leadership, Frankfurt found membership irresistible. Hamburg was republican, oligarchic, devoted to trade, shipping and finance to the exclusion of all else. The leading merchants associated but little with bureaucrats from government, refused to accept titles, and frowned on intermarriage with Prussian officers [Eckardt, 1910, I, p. 200]. In contrast to Augsburg whose merchants were anxious to be ennobled, the city not only forbade ennoblement as a strict rule as early as 1276, a regulation renewed in 1658 and practised throughout the 17th century, but stated on the earliest of these occasions that no knight could live in the city [Schramm, 1969, p. 82]. Hamburg was a miniature Holland, populated by gains-from-trade merchants stapling commodities produced for the most part at a distance. It resisted incorporation in the Zollverein and the Reich, until the patriotic surge of 1870, and even then it remained a free port, rather than an integral economic member of the Empire, for another decade and half. Within Frankfurt, moreover, there was a difference of view between the patrician merchants who wanted to stay out of Zollverein, and the new group who succeeded in winning adherence [Böhme, 1968, *passim*]. The difference corresponds roughly to the gains-from-trade leading merchants, interested in overseas trade and especially British goods, and value-added men concerned with developing the local market.

Limited economic development in most of the German states, and especially poor transport connections among them, and political separation, led to stunted commercial growth outside the Hanseatic states, Frankfurt and the Rhineland. The Rhineland and Westphalia were different from the other German states even before guild restrictions were eliminated by Napoleon at the end of the century. Merchant networks large and small were much more fully developed. Barmen and Elberfeld concentrated in textiles, and especially bleaching and dyeing, with the former emphasizing production, the latter marketing. What is impressive about Barmen and Elberfeld is their capacity to shift as comparative advantage was lost — from linens to cottons and then to ribbons and embroidery. When high wages and English competition with machine-made cottons made it necessary, a plant was moved across the river to Gladbach-Rheydt. By the end of the 18th century, the mercantile elite with customers throughout Europe, and especially France, the Netherlands and London, was a close-knit oligarchy, cut off from the aristocracy as the English and French were not, and turned inward into a life of luxury — horses, servants, fiinishing schools for daughters, wine, jewelry and speaking French. Of 300 mercantile houses in Frankfurt, 240 were of German origin and in 1776 some 50 to 60 of them were from Barmen and Elberfeld [Kisch, 1972, pp. 347, 363, 383, 387]. Similar mercantile towns existed in Krefeld and Münster, not to mention Cologne.

Silesia was otherwise. The linen trade flourished from the 16th century only when Dutch and English merchants by-passed the guilds because they could not adjust to the requirements of foreign markets [Kisch, 1959, p. 543]. By the late 17th, early 18th century Silesian linen dominated the world, and produced a class of wealthy merchants, who loaned out their profits in mortgages of the aristocracy and, when they could, acquired estates [*Ibid.*, p. 552]. When linen lost its overseas market to cotton, the only response was lower wages and fraud and adulteration [Blumberg, 1960a, p. 99].

The merchant class in the Rhineland seems to have furnished entrepreneurs for industry, but the petty merchant class rather than the *Grosskaufmänner.* A study of Rhenish and Westphalian entrepreneurs notes that many of them in the 19th century came from merchant families of the previous century [Zunkel, 1962, chapter 11]. The author comments that whereas in Britain entrepreneurs came largely from artisans, in Germany merchants played at least an equal role [*Ibid.*, p. 25]. These small merchants, *Kaufmänner* rather than *Grosskaufmänner* of the sort found in Hamburg, went from putting-out into manufacturing, like J.G. Brugelmann who built the first mechanised cotton-spinning works at Ratingen near Düsseldorf in 1783, five stories high, with 1600 spindles, and partly but not entirely waterpowered [Adelmann, 1970, p. 86]. Until 1830-50, foreign trade and financing of foreign trade were largely in Dutch hands. Krupp, Stinnes, Haniel, Liebrecht, Ehrenbold also moved from trade to production in iron and coal in the first half of the 19th century [Wutzmer, 1960, pp. 148-50]. Mannesmann at Remscheid was a printer and merchant who started a file factory in competition with British imports [Zunkel, 1962, p. 26]. « It was not the successful handworker who made it big, but his son, who, converted from being a powerfully capitalized merchant, used the corporate form, and hired a technical man who remained in the second rank » [Aycoberry, 1968, p. 518]. On the other hand, in Switzerland in the textile industry, factory pioneers were technical people, not financial, trade or managerial [Braun, 1965, p. 78]. Elsewhere, however, Braun noted that the north of the Alps there formed a local stationary class of merchants united in guilds parallel to the handwork guilds of the city, who were interested in innovation, achievement-oriented, and busy not only in trade, traffic and finance but also in production. He cites the Geneva watch industry along with the Zurich textile industry, as linked to city merchants, who, once the restrictions of the ancient regime were lifted, put forth a tremendous burst of energy [*Ibid.*, pp. 263-64].

There is some slight evidence of merchants investing in industry. A table of professions of the founders (original stockholders) of 61 companies in Prussia in the 1850s, covering 480 individuals, showed 32 percent merchants, 14 percent manufacturers, 12 percent state officials, 11 percent bankers, 7 percent large landowners and 1.3 per cent military. The representation of merchants in mining in Silesia was lower than for Prussia as a whole, dominated by the Rhineland [Blumberg, 1960b, p. 196]. In one Cologne cotton spinning and weaving plant, 29 merchants, 8 bankers, 2 manufacturers, one mine-owner, one newspaper publisher and 2 state officials were investors; and in a Silesian mine, 3 mine-owners, 13 merchants, 13 mine officials, 22 independent artisans and tradesmen, 8 officials, 4 millowners, 2 teachers, 2 doctors, 2 innkeepers, 1 "investor", 1 printshop owner and eight other persons [*Ibid.*, p. 201]. The Stinnes coal and iron complex in 1850 had 41 living stockholders in Cologne consisting of 14 merchants, 7 bankers, 7 rentiers, 1 mine-owner, and one each of the occupations of director, building inspector, major general, chief accountant, judge, privy councillor, inspector, lawyer, doctor, lady, widow [Zunkel, 1962, p. 52]. The evidence is available only for the period after our primary interest, and is highly fragmentary. It suggests, however, that merchant wealth provided one of the major sources of company capital. It is assumed that these were the value-added type of merchant rather than gains-from-trade overseas traders.

The explanation given for underpresentation of merchants in Germany as contrasted with Britain and France has various elements: first, the late unification of the country and the failure at an early stage to build an entrepot or relay pivot apart from the fairs of Frankfurt and Leipzig; second, the disdain with which German social values regarded commercial pursuits, holding the merchant in lower status than the industrialist, in contrast with Britain where the opposite ranking obtained; and third, the early bureaucratization of the firm, with direct buying and selling combined with production in a single business entity.

The low standing of commerce in Germany (outside the Hanseatic cities and Frankfurt-am-Main) contrasted with the higher status of the nobility and the military of course, and of industry as well. Merchants were scorned as tradespeople, *Händler* as opposed to heroes, in Sombart's book *Helden und Händler*, contrasting the Germans and the British. The term contained anti-semitic prejudice as well as a sense of economic and social inferiority *vis-à-vis* the British [Kehr, 1965, p. 153]. The three-generations-from-shirtsleeves-to-shirtsleeves did not apply in Germany nor was there opportunity to gain acceptance in an open aristocracy [Zunkel, 1962, p. 13]. Purchase of land in Prussia was forbidden to all but the nobility until 1807 [Ucke, 1888, pp. 8, 11] although there was brisk trade in land along the coastal lands such as Mecklenburg and Holstein. After 1807 there is evidence of rich merchants buying estates [Kisch, 1959, p. 552] and farms which had been in the same family for two or three hundred years changing hands three, four or six times in ten to fifteen years [Abel, 1962, p. 304]. Sartorius von Waltershausen is quoted to the effect that in the 1820-1830 agricultural crisis 80 percent of aristocratic landlords lost their estates [Bramsted, 1964, p. 54], and presumably not all were broken up and sold off piecemeal to land-hungry peasants. Later estates were purchased. But the rule broadly held: while bourgeois were permitted to purchase estates after the Stein-Hardenberg reforms, and aristocrats to enter trade, for the most part th old barriers to social integration held.

Industrialization on a significant scale in Germany came in the middle of the 19th century, and relied on direct buying and merchants. There is some debate as to when German industrial "take-off" occurred [Hardach, 1972, pp. 67ff] but for our purposes the difference of 20 years is inconsequential. In mining, distribution of coal was regulated by the state, subject to price control, and did not need to rely on a mercantile structure [Jankowski, 1973, pp. 718ff]. Railroads were large-scale and so, quickly, were iron and steel producers. In the earliest stages of getting started, when the firm consisted of 35 workers, Albert Poensgen, of the steel tubing plant of Poensgen & Schoeller, took over all mercantile activities of the firm personally, especially supervising the routine purchasing of inputs by two employees after Dutch merchants failed to perform what he had expected from them [Hatzfeld, 1966, p. 204].

In the electrical and chemical new industries, bureaucratic organization took place from the beginning. Siemens and Halske was started by a military engineer, Werner Siemens, and quickly acquired a manager in another military man, Wilhelm Meyer. Initial sales were of telegraphic installations for the military forces of Prussia, and the Russian government. By 1872, Werner Siemens noted the necessity to write to customers in advance of undertaking production. « Only thus can our fabrication constructions be very cheap, good, and quickly delivered » [Kocka, 1969, p. 126]. The middleman was eliminated before he had taken hold in an effort to speed up technical progress. Kocka observes that the monopoly in education which the merchant class had had earlier was eroded by general universal education, and that rather than rely on a mercantile network, large companies enforced standardization, undertook communication of goods and prices, delivered goods to inventories of shops, sold by sending samples over distances, undertock direct contact between buyer and seller [*Ibid.*, p. 165]. Dye companies initially marketed abroad through export firms in the developing countries, and through wholesale firms in Europe and the United States. These agents were allowed, however to carry only one firm's lines. After 1885, they went to direct selling [Beer, 1959, p. 95]. The chemical industry requires intimate contact between producer and consumer to ensure proper use of frequently dangerous materials, but in any case products which must be employed in exact ways to achieve efficient results [Hohenburg, 1967, pp. 127ff].

Did the absence of a well-knit system of markets hurt? Political reality more than the absence of mercantile connections was responsible for the fact that prior to *Zollverein* and the railroad, Germany was not one country but a juxtaposition of a number of half-closed economies. With *Zollverein* and the railroad, a large German market quickly substituted for inumerable local markets [Benaerts, 1933, pp. 121-22]. It has frequently been claimed that Germany was poor in capital prior to 1950. Sombart observed that great fortunes were rare, and the man with a mere half-million talers regarded as rich [*Ibid.*, p. 123]. Hamburg at mid-century is said to have had 100 millionaires in mark banco (equivalent to 500,000 talers) as opposed to only seven or eight in Cologne [Aycoberry, 1968, p. 515]. But as Borchardt has made clear in a classic article, it is hard to give content to the widely held view that German's lag behind Britain in the early 19th century was the result of lack of capital. Capital markets, like goods markets, were fragmented and underdeveloped, but banking institutions (and foreign credits) responded quickly to the demand for capital for such large projects as railroads and iron foundries, and once started, they proceeded on reinvested profits [Borchardt, 1961, *passim*]. The so-called Germany way to industrialization may have been delayed by the absence of an antecedent capital market built on mercantile credit but not for long.

VI

No attempt has been made to compare numbers of merchants at various periods in Holland, Britain, France or Germany, or to evaluate such statements as that England in 1721 had two-thirds the numbers of merchants of all Europe put together, though Amsterdam in 1671 handled 10 times as much trade as London with one-twentieth of the retailers [Westerfield, 1915, p. 412]. The secondary sources are weak on the merchant communities of London and Paris because of their great size and complexity, and because of the destruction of the Paris archives in 1871 [Daumard, 1970, p. 8]. A few qualitative conclusions can perhaps be drawn, however.

The widely-held view of contribution of the merchant to industrialization seems clearly exaggerated insofar as the profits of "gains-from-trade" commerce are concerned. These fortunes are neither necesary nor sufficient to stimulate industry. They are typically consumed, invested in non-industrial outlets, including banks which finance primarily trade or foreign governments, and in land, government stock, annuities, *offices*. When industry catches hold, it largely uses trade credit, finances itself with ploughed back profits. In large-scale industry — South Wales iron and German joint-stock companies — one can find investor contributions from merchants and — especially in the latter case — banks. They are the exception.

Nor do merchants typically become manufacturing entrepreneurs, unless one limits the discussion to the putters-out of the proto-industrialization period. Again there are exceptions like Benjamin Gott among the "value-added" merchants. The great majority of merchants who made fortunes moved out of business into banking, politics or rural pursuits. Where there was no aristocracy to ape — perhaps the agriculture of the area did not produce a surplus which needed to be guarded, or which could be expropriated by the landed aristocracy and military — merchant oligarchies formed, tended to dig in, and ultimately lead to mercantile decay. The moving off of successful merchants served the economic function of removing what might otherwise have developed into a block to change.

The value-added merchant is a unit in an information network. In the early stages, information is needed to see what can be produced and what is needed or sought, and where. Surveillance of quality and delivery timeliness is one aspect of information control. Division of labour between producer and his intelligence-gathering collaborator permits maximum efficiency.

With success comes fossilization, at which stage the merchant network is a barrier to change, with the result that it is helpful to have him ready to retire. Where he clings to the old ways, decay ensues.

One general force reducing the role of the merchant was universal education. The rents earned by both types of merchants were initially based on his monopoly of calculation and languages. Apprenticeship was expensive and consisted in instruction at home and in the offices of associates abroad. Money and opportunity costs limited entry. But the spread of education brought down the rents of merchants through competition. With information cheap, direct connections rendered the middleman obsolete when he was not disfunctional.

Commercial Expansion and the Industrial Revolution

BIBLIOGRAPHY

GERHARD ADELMANN, *Structural Change in the Rhenish Linen and Cotton Trades at the Outset of Industrialization*, in F. Crouzet, W.H. Chaloner and W.M. Stern, eds., « Essays in European Economic History, 1789-1914 », New York, St. Martin's Press, 1970, pp. 82-87.

G.C. ALLEN, *British Industries*, London, Longmans Green, 2nd ed. 1939.

ANON., *Fortunes Made in Business*, Vols. I, II, III, London, no publisher stated, 1883.

PIERRE AYCOBERRY, *Probleme der Sozialschichtung in Köln im Zeitalter der Frühindustrialisierung*, in W. Fischer, Hgbr., « Wirtschafts- und sozialgeschichtliche Probleme der frühen Industrialisierung », Berlin, Colloquium Verlag, 1968, pp. 512-28.

PAUL BAIROCH, *Commerce international et genèse de la revolution industrielle anglaise*, « Annales, Economies, Sociétés, Civilisations », XXVIII, 3 (mars-avril 1973), pp. 541-71.

VIOLET BARBOUR, *Dutch and English Merchant Shipping in the Seventeenth Century*, « Economic History Review », Vol. II reprinted in Warren C. Scoville and J. Clayburn La Force, eds., « The Economic Development of Western Europe », Vol. II, The Sixteenth and Seventeenth Centuries, pp. 108-37.

JOHN JOSEPH BEER, *The Emergence of the German Dye-Stuffs Industry*, Urbana, University of Illinois Press, 1959.

PIERRE BENAERTS, *Les Origines de la grande industrie allemagne*, Paris, Editions F.H. Turot, 1933.

LOUIS BERGERON, *Paris dans l'organization des échanges intérieurs à la fin du XVIIe siècle*, a paper presented to the Colloque National de l'Association Française des Historiens Economistes at Paris, 6 October 1973, entitled « Aires et structures du commerce français au XVIIIe siècle ».

HORST BLUMBERG, *Ein Beitrag zur Geschichte der deutschen Leinenindustrie von 1834 bis 1870*, in Hans Mottek, Hgbr., « Studien zur Geschichte der industriellen Revolution », Berlin, Akademie-Verlag, 1960a, pp. 65-143.

HORST BLUMBERG, *Die Finanzierung der Neugründungen und Erweiterungen von Industriebetrieben in Form der Aktiengesellschaften während der fünfziger Jahrhunderts in Deutschland, am Beispiel der preussischen Verhältnisse erlaüert*, in Hans Mottek, Hgbr., « Studien zur Geschichte der industriellen Revolution in Deutschland », Berlin, Akademie Verlag, 1960b, pp. 165-208.

HELMUT BOHME, *Frankfurt und Hamburg*, Das Deutsches Reiches Silber- und Goldloch und die Allerenglishste Stadt des Kontinents, Frankfurt am Main, Europäische Verlagsanstalt, 1968.

KNUT BORCHARDT, *Zur Frage des Kapitalmangels in der ersten Hälfte des 19. Jahrhunderts in Deutschland*, in « Jahrbücher für Nationalökonomie und Statistik », Bd. 173 (1961), pp. 401-21.

C.R. BOXER, *The Dutch Economic Decline*, in Carlo M. Cipolla, ed., « The Economic Decline of Empires », London, Meuthen, 1970, pp. 235-63.

ERNEST K. BRAMSTED, *Aristocracy and the Middle-Classes in Germany: Social Types in German Literature, 1830-1900*, rev. ed., Chicago, University of Chicago Press, 1964.

RUDOLF BRAUN, *Sozialer und Kultureller Wandel in einem ländlichen Industriegebiet im 19. und 20. Jahrhundert*, Erlenbach-Zurich, Eugen Rentch Verlag, 1965.

Charles P. Kindleberger

R.H. Campbell, *An Economic History of Scotland in the Eighteenth Century*, « Scottish Journal of Political Economy », XI, 1 (February 1964), pp. 17-24.

Charles Carrière, *Négociants Marseillais au XVIIIe siècle*: Contribution à l'étude des économies maritime, Marseilles, Institut Historique de Provence, 1973.

S.G. Checkland, *The Gladstones, A Family Biography, 1764-1851*, Cambridge, Cambridge University Press, 1971.

Carlo M. Cipolla, *Editor's Introduction*, « The Economic Decline of Empires », London, Methuen, 1970.

W.H.B. Court, *The Rise of the Midland Industries, 1600-1838*, London, Oxford, 1939.

François Crouzet, Book II on *Economie et société (1715-1789)*, in François-George Pariset, ed., « Bordeaux au XVIIIe siècle », Bordeaux, Federation historique de Sud-Ouest, 1968, Vol. VI.

François Crouzet, *Croissances comparées de l'Angleterre et de la France au XVIIIe siècle, Essai d'analyse comparée de deux croissances économiques*, « Annales, Economies, Sociétés, Civilisations », XXI, 2 (mars-avril), 1966, pp. 254-91.

François Crouzet, ed., *Capital Formation in the Industrial Revolution*, London, Methuen & Co., 1972.

Pierre Dardel, *Commerce, industrie et navigation à Rouen et au Havre au XVIIIe siècle*, Rivalité croissant entre ces deux ports, Rouen, Société libre d'émulation de la Seine-Maritime, 1966.

Adeline Daumard, *Les Bourgeois de Paris au XIXe siècle*, Paris, Flammarion, 1970.

Ralph Davis, *English Foreign Trade, 1660-1700*, « Economic History Review », Second Series, VI (1954), reprinted in W.E. Minchinton, ed., « The Growth of English Overseas Trade in the 17th and 18th Century », London, Methuen, 1969, pp. 78-98.

Ralph Davis, *English Foreign Trade, 1700-1774*, « Economic History Review », Second Series, XV (1962), reprinted in W.E. Minchinton, ed., « The Growth of English Overseas Trade in the 17th and 18th Century », London, Methuen, 1969, pp. 99-120.

Phyllis Deane, *The First Industrial Revolution*, Cambridge, Cambridge University Press, 1965.

T.M. Devine, *Glasgow Colonial Merchants and Land, 1770-1815*, in T. Ward and R.G. Wilson, eds., « Land and Industry », The Landed Estate and the Industrial Revolution, New York, Barnes and Noble, 1971, pp. 205-65.

Louis Dermigny, *Cargaisons indiennes, Solier et Cie, 1781-1793*, Paris, S.E.V.P.E.N. Tome I, Text, Tome II Documents, 1960.

François Dornic, *L'industrie textile dans Le Maine et ses débouches internationaux (1650-1815)*, Le Mans, Editions Pierre-Belon, 1955.

Julius V. Eckardt, *Lebenserinnungen*, Leipsig Verlag S. Hirzel, Bands I, II, 1910.

Michael M. Edwards, *The Growth of the British Cotton Trade, 1780-1815*, Manchester, Manchester University Press, 1967.

Thomas Ellison, *The Cotton Trade of Britain*, first impression 1886, new ed. New York, Kelley, 1968.

Wolfram Fischer, *Der Staat und die Anfänge der Industrialisierung in Baden, 1800-1850*, Berlin, Duncker u. Humblot, 1962.

Commercial Expansion and the Industrial Revolution

H.R. Fox Bourne, *English Merchants*, Memories in Illustration of the Progress of British Commerce, London, Richard Bentley, 1866.

R.S. Fitton and A.P. Wadsworth, *The Strutts and the Arkwrights, 1758-1830*, A Study of the Early Factory System, Manchester, Manchester University Press, 1958.

Maurice Garden, *Le commerce lyonnais au XVIIIe siècle*, a paper presented to the Colloque National de l'Association Française des Historiens Economistes at Paris, 6 October 1973, entitled « Aires et structures du commerce français au XVIIIe siècle ».

Joshua Gee, *Trade and Navitagtion of Great Britain Considered*, London, 1729.

Pieter Geyl, *The Netherlands in the Seventeenth Century*, Part One, *1609-1648*, New York, Barnes and Noble, 1961.

Pierre Goubert, *Familles marchands sous l'Ancient Regime: les Danse et les Motte de Beauvais*, Paris, S.E.V.P.E.N., 1959.

N.S.B. Gras, *Business and Capitalism*, An Introduction to Business History, New York, F.S. Crofts, 1939.

H.J. Habakkuk, *American and British Technology in the Nineteenth Century*, Cambridge, Cambridge University Press, 1962.

K.H.D. Haley, *The Dutch in the Seventeenth Century*, London, Thames and Hudson, 1972.

Lutz Hatzfeld, *Kaufmannische Probleme der Rohrenfabrik Poensgen & Schoeller, 1844-1850*, in Karl Erich Born, Hgbr., « Moderne deutsche Wirtschaftsgeschichte », Köln-Berlin, Kiepenheuer & Witsch, 1966, pp. 203-13, from Lutz Hatzfeld, « Die Begründung der deutschen Röhrenindustrie durch die Fa. Poensgen & Schoeller, Manuel, 1844-1850 », Wiesbaden, Franz Steiner Verlag, 1962, pp. 60-75.

Karl W. Hardach, *Some Remarks on German Economic Historiography and its Understanding of the Industrial Revolution in Germany*, in « Journal of European Economic History », Vol. 1, No. 1 (Spring 1972), pp. 37-99.

Herbert Heaton, *The Yorshire Woollen and Worsted Industries*, Oxford, Clarendon Press, 2nd ed., 1965.

Paul M. Hohenberg, *Chemicals in Western Europe, 1850-1914*: An Economic Study of Technical Change, Chicago, Rand McNally & Co., 1967.

M.A. Holderness, *The English Land Market in the Eighteenth Century: The Case of Lincolnshire*, « Economic History Review », Second Series, XXVII, 4 (November 1974), pp. 557-76.

W.G. Hoskins, *Industry, Trade and People in Exeter, 1688-1800*, With Special Reference to the Serge Industry, Manchester University Press, 1935.

Edward Hughes, *North Country Life in the Eighteenth Century*, Vol. I, *The North-East, 1700-1750*, London, Oxford University Press, 1952; Vol. II, *Cumberland and Westmoreland, 1700 1830*, London, Oxford University Press, 1965.

Gordon Jackson, *Hull in the Eighteenth Century*, A Study in Economic and Social History, London, Oxford University Press for the University of Hull, 1972.

William Jacob, *A View of the Agriculture, Manufactures, Statistics and Society in the State of Germany and Parts of Holland and France*, Taken during a Journey through those Countries in 1819, London, John Murray, 1820.

Manfred D. Jankowski, *Law, Economic Policy and Private Enterprise: The case of the Early Ruhr Mining Region, 1766-1865*, « Journal of European Economic History », II, 3 (Winter 1973), pp. 688-727.

Charles P. Kindleberger

A.H. John, *The Industrial Development of South Wales, 1750-1850*, An Essay, Cardiff, University of Wales Press, 1950.

Eckhart Kehr, *Der Primat der Innenpolitik*, Gesammelte Aufsätze zur preussissch-deutschen Sozialgeschichte im 19. und 20. Jahrhundert, H.-U. Wehler, Hgbr., Berlin, Walter de Gruyter & Co., 1965.

Herbert Kisch, *The Textile Industries in Silesia and the Rhineland: a Comparative Study of Industrialization*, « Journal of Economic History », XIX, 4 (December 1959), pp. 591-64.

Herbert Kisch, *From Monopoly to Laissez-faire: The Early Growth of the Wupper Valley Textile Trades*, « Journal of European Economic History », I, 2 (Fall 1972), pp. 298-407.

Jürgen Kocka, *Unternehmungsverwaltung und Angestelltenschaft am Beispiel Siemens, 1847-1914*: Zum Verhältnis von Kapitalismus und Bürokratie in der deutschen Industrialisierung, Stuttgart, Ernst Klett Verlag, 1969.

Jürgen Kocka, *Expansion, Integration, Diversifikation: Wachstumstrategien industrieller Grossunternehmen in Deutschland vor 1914*, comment on Theme 2 of the Sixth International Congress on Economic History, Copenhagen, 1974, A.D. Chandler, Jr. and Herman Daems, « Investment Strategy in Europe, the United States and Japan », unpublished version, pp. 1-23 plus 1-10 of footnotes.

Emmanuel Le Roy Ladurie, *Les Paysans de Languedoc*, Paris, S.E.V.P.E.N., 1966.

Pierre Léon, *Marchands et speculateurs dauphinois dans le monde antillais du XVIIIe siècle*, Les Dolle et les Raby, Paris, « Les Belles Lettres », 1963.

Paul Mantoux, *The Industrial Revolution in the Eighteenth Century*: An Outline of the Beginnings of the Modern Factory System in Britain, New York, Harper & Row Torchbooks, rev. ed., 1961.

Alfred Marshall, *Industry and Trade*, A Study of Industrial Technique and Business Organization, and of Their Influences on the Conditions of Various Classes and Nations, London, Macmillan, 1920.

Peter Mathias, *The First Industrial Nation*: An Economic History of Britain, 1700-1914, London, Methuen, 1969.

J. Mathorez, *Les Étrangers en France sous l'Ancient Régime*, Histoire de la population française, Tome I, Paris, Edouard Chapinon, 1919.

William H. McNeill, *Venice*: The Hinge of Europe, 1081-1797, Chicago, University of Chicago Press, 1974.

Jean Meyer, *L'Armement Nantais dans le deuxième moitié du XVIII siècle*, Paris, S.E.V.P.E.N., 1969.

W.E. Minchinton, *The British Tinplate Industry*, A History, Oxford, Clarendon Press, 1957a.

W.E. Minchinton, ed., *The Trade of Bristol in the 18th Century*, Bristol, Record Society's Publication, XX, 1957b.

W.E. Minchinton, *The Merchants in England in the Eighteenth Century*, « Explorations in Entrepreneurial History », X (December 1957c), pp. 62-71, reprinted in Hugh G.J. Aitken, ed., « Explorations in Enterprise », Cambridge, Mass., Harvard University Press, 1965, pp. 278-95.

W.E. Minchinton, *Editor's Introduction*, in W.E. Minchinton, ed., « The Growth of English Overseas Trade in the 17th and 18th Centuries », London, Methuen & Co., 1969.

Commercial Expansion and the Industrial Revolution

JOEL MOKYR, *Industrial Growth and Stagnation in the Low Countries, 1800-1850*, doctoral dissertation, Yale University, 1974a, to be published by Yale University Press.

JOEL MOKYR, *The Industrial Revolution in the Low Countries in the First Half of the Nineteenth Century: A Comparative Case Study*, « Journal of Economic History », XXXIV, 2 (June 1974b), pp. 365-91.

ANGELO OLIVIERI, *An Urban Case History: Bordeaux*, « Journal of European Economic History », II, 2 (Fall 1973), pp. 447-56.

RICHARD PARES, *A West-India Fortune*, Longmans Green, 1950, reprinted New York, Archon Books, 1968.

J.-P. POUSSOU, *Demographic and Social Structures*, Chapter V in François-George Pariset, ed., « Bordeaux au XVIIIe siècle », Bordeaux, Fédération historique de Sud-Ouest, 1968.

JOHN PREST, *The Industrial Revolution in Coventry*, London, Oxford University Press, 1960.

JACOB M. PRICE, *France and the Chesapeake*, A History of the French Monopoly and of its Relationship to the British and American Tobacco Traders, Ann Arbor, University of Michigan Press, 1973, 2 vols.

FRIEDRICH RAUERS, *Bremer Handelsgeschichte im 19. Jahrhundert*, Bremen, Franz Leuwer, 1913.

R. ROBSON, *The Cotton Industry in Britain*, London, Macmillan, 1957.

PERCY ERNEST SCHRAMM, *Hamburg und die Adelsfrage (bis 1806)*, « Zeitschrift des Vereins für Hamburgische Geschichte », Bd. 55 (1969), pp. 81-93.

ECKART SCHREMMER, *Die Wirtschaft Bayerns vom hohen Mittelalter bis zum Beginn der Industrialisierung*: Bergbau, Gewerbe, Handel, München, Verlag C.H. Beck, 1970.

WARREN C. SCOVILLE, *The Persecution of Hugenots and French Economic Development, 1680-1720*, Los Angeles, University of California Press, 1960.

ADAM SMITH, *An Inquiry into the Nature and Causes of "The Wealth of Nations"*, Cannan ed., New York, Modern Library (1776), 1937.

FRANÇOISE THÉSÉE, *Négociants bordelais et colons de Saint-Domingle*, Liaisons d'habitations, La Maison Henry Romberg, Papst et Cie, Paris, Société française de histoire d'outre-mer, 1972.

RAY B. WESTERFIELD, *Middlemen in English Business*, Particularly between 1660 and 1760, New Haven, Yale University Press, 1915.

E.N. WILLIAMS, *The Ancien Regime in Europe*, Government and Society in the Major States, 1648-1879, New York, Harper & Row, 1970.

CHARLES WILSON, *Anglo-Dutch Commerce and Finance in the Eighteenth Century*, Cambridge, Cambridge University Press, 1941.

C. WILSON, *The Entrepreneur in the Industrial Revolution in Britain*, « Explorations in Entrepreneurial History », VII, 3 (February 1955), pp. 229-5, reprinted with omissions in Barry E. Supple, ed., « The Experience of Economic Growth »: Case Studies in Economic History, New York, Random House, 1963, pp. 171-88.

CHARLES WILSON, *England's Apprenticeship 1663-1763*, New York, St. Martin's Press, 1965.

CHARLES WILSON, *The Dutch Republic and the Civilisation of the Seventeenth Century*, London, Weidenfeld and Nicholson, 1968.

R.G. WILSON, *Gentlemen Merchants*, The Merchant Community of Leeds, 1700-1830, Manchester, Manchester University Press, 1971a.

R.G. WILSON, *The Denisons and Milneses: Eighteenth Century Merchant Landowners*, in J.T. Ward and R.G. Wilson, eds., « Land and Industry », The Landed Estate and the Industrial Revolution, New York, Barnes and Noble, 1971b, pp. 145-72.

H.R.C. WRIGHT, *Free Trade and Protection in the Netherlands, 1816-30*, Cambridge, Cambridge University Press, 1955.

HENIZ WUTZMER, *Die Herkunft der industrielle Bourgeoisie Preussens in den vierziger Jahren des 19. Jahrhunderts*, in Hans Mottek, Hgbr., « Studien zur Geschichte der industriellen Revolution in Deutschland », Berlin, Akademie-Verlag, 1960, pp. 145-63.

FRIEDERICH ZUNKEL, *Der Rheinische-Westfälische Unternehmer, 1834-1879*, Ein Beitrag zur Geschichte des deutsche Bürgertums im 19. Jahrhundert, Köln und Opladen, Westdeutscher Verlag, 1962.

Proto-industrialization: Theory and Reality

GENERAL REPORT

Franklin Mendels

University of Maryland Baltimore County
Catonsville, Maryland 21228
USA

If there is one subject that continues to sustain the interest of economic and social historians and to challenge them as teachers and as students more than ever, it is the theme of the Industrial Revolution. Explaining the Industrial Revolution, reconstituting the sequence of events that preceded and prepared the coming of the factory system and thus launched one society after another into a succession of astonishing structural transformations, that is the task that continues to defy us. The fact that the Industrial Revolution has been one of the great turning points in history is in itself enough to justify historical curiosity. But the complexity of the process, especially when contemplated in a comparative perspective, only adds to the challenge. There is great difficulty in reaching acceptable generalizations as the phenomenon contains some features that seem to occur independently in each country, while others involve imitation or adaptation across national borders; some that seem to repeat themselves over time, while others are specific to a particular date; some that seem to replicate themselves from place to place, and others that are place-specific. And yet we continue to seek generalizations because the subject has a persistent contemporary relevance. We can say that the blatant income disparities that exist among the various parts of the world today can be attributed to the fact that the spread of the Industrial Revolution has been truncated. Our responsiblity as economic historians at least to portray and depict the historical experience of industrialization as correctly and accurately as we can is enhanced by the common belief, as Marx wrote, that "the country that is the more developed industrially only shows, to the less developed, the image of its own future".[1] The historical experience of the now developed countries is commonly invoked to justify policy decisions.[2]

For a long time economic and social historians have accorded a considerable importance to technology and science. They constructed a scenario of economic expansion and transformation where the prime mover of change and development, whether continuous or not, was human ingenuity manifesting itself in the progress of knowledge and the growing capacity of the human race to harness the forces of nature. In this serene and self-confident description of economic and social evolution, an autonomous and leading role was played by great discoveries and inventions attributed to men of genius. As seen from Europe, the march of progress, the ascent of modern civilization, were marked out by the invention of the water wheel in the sixth century, the stirrup in the eighth, the horsecollar and the rudder in the ninth, the windmill in the twelfth, the printing press in the fifteenth, and so on.[3] In this historiographic tradition, the slightest technical improvement seemed a breakthrough with unanticipated and revolutionary consequences upon the economy, the social structure, and the conditions of work and of the working class. In this approach the agricultural and industrial revolutions of the eighteenth and nineteenth centuries received simple explanations. T. S. Ashton's famous schoolboy could write: "About 1760 a wave of gadgets swept over England," and everything else logically unfolded and followed.[4]

Before the discoveries of Henry Cort, James Watt, James Hargreaves, Richard Arkwright, or Edmund Cartwright, there were of course "earlier forms of industry," but they were mostly described in negative ways and relegated by historians to the earlier, pre-industrial way of doing things (Perlin*, Sonenscher*). They were all, "in one way or another, closely connected with agriculture," depending upon it for their seasonal labor and their raw materials, and technically stagnant and backward.[5] Historians did not place the growth of such industries, when it occurred, under the heading of "industrialization" because it did not take place in the organizational framework of the factory mill. To be sure, some historians spoke of "an industrial revolution of the thirteenth century" or of a "first industrial revolution" in 1540–1640.[6] But the former was predicated upon the observation that the use of the fulling mill, moved by water power, had become widespread; the latter upon the discovery, later called "nonsense" by D. C. Coleman, that the factory form of enterprise, or large-scale industry in establishments costing "hundreds, and in many cases thousands of pounds . . . owned by private capitalists, who employed in them dozens and sometimes scores and even hundreds of workmen, was not the novelty it was once believed to be."[7] The beginnings of industrialization might be pushed back, but only to the extent that capital and machinery were heavily utilized in the manufacturing process. Since the Industrial Revolution is almost congruent with the factory, earlier forms of industry and industrialization which preceded the factory and the machine age only seemed significant to the extent that they already prefigured the size, scale, and concentration manifest in the factory system, that is, if they took the form of mills, operated by water, wind, steam, or horses. Such antecedents of the Industrial Revolution were indeed to be found very far back in history.

Together with this emphasis on technology and invention, there has been a fixation among historians on capital as the strategic productive factor and on investment as the indispensable ingredient.[8] It is a verity that capital has indeed played an important role in what is also called the rise of capitalism. The Industrial Revolution was in part the application on a grand scale of techniques aimed at multiplying the productivity of labor by equipping workers with costly tools and machinery. The great theories of economic development, from Smith, Ricardo, and Marx to Schumpeter have brought this out. Soviet planners, such as Fel'dman in the 1920s, and Keynesian economists, such as Harrod and Domar in the 1940s, further reinforced the emphasis on capital and investment by building growth-theoretic models around these variables.[9] The growth of labor-intensive forms of traditional industry could not, and did not, receive any attention in this perspective, and the origins of modern capitalism were thus more likely to be found by historians within the pre-industrial era in new developments or institutions if they had presumably led to capital accumulation. For Earl Hamilton, for instance, the principal dynamic process at work in the period 1500–1700 was the price inflation that had allegedly caused a great accumulation of capital because rents and wages had lagged behind prices. This furnished the means to build up industrial equipment.[10]

The interest in the study of the Industrial Revolution was rekindled after World War II by the birth of newly independent but underdeveloped nations in the Third World. Economic historians tried to rewrite the history of the Industrial Revolution as the historical experience of economic growth.[11] Simultaneously, the Harrod-Domar model seemed to receive empirical support from the miraculous success of massive transfers of capital and technology from the United States in the postwar reconstruction of Europe and Japan. The Soviet plans were still emphasizing heavy, capital-intensive industry. There was a convergence on this point between Western and Communist views. It is within this intellectual and political atmosphere, the "Marshall Plan mentality," that Kuznets, Gerschenkron, and Rostow led American, English, and some European economic historians into the quantitative measurement and the analysis of savings, investment, and capital-output ratios, just as development policies advocated for third-world nations emphasized transfers of capital for the building of steel mills, aluminum plants, and hydro-electric dams, with putative trickle-down effects.[12]

However, the systematic application of quantitative methods to the historical study of industrialization, combined with a new interest in looking at "history from below," quickly started to undermine the dominant image of industrialization as here described. Statistical measures showed no sudden increase in investment as a proportion of the national product during the crucial decades of the Industrial Revolution.[13] Measures of fixed capital formation showed that its importance might have been heretofore overestimated.[14] The new research even revealed that agriculture was the capital-intensive branch of production, so that industrialization was capital-saving since it meant a reduction in the share of agricultural output in the [German] economy![14a] Aggregate measures showed that high rates of economic growth could be reached at earlier dates than formerly thought, before the factory system had become dominant.[15] The impact of the most spectacular innovations, the railroad and the steam engine, was not as great or as sudden as one might have thought.[16] In the meantime, historians of pre-industrial society were measuring the magnitude and growth of small-scale, intermittent, and decentralized industry before the Industrial Revolution, and demonstrating its dominance in terms of output and employment.[17] Social historians interested in the origins of the proletariat or in the impact of the Industrial Revolution on the working class, using new methods of identifying and numbering people at the local level, confirmed the importance of non-agricultural work in so-called pre-industrial societies, and were surprised to see how long these traditional forms of industrial work persisted after the onset of modern industrialization.[18] Not just in the lagging countries, but right in the leading regions of the Industrial Revolution, the traditional forms of industry, small urban workshops, urban domestic industry, rural farm by-employments have been dominant, typical, not only before the "wave of gadgets" began to sweep over the world, but through much of the nineteenth and indeed into the twentieth century. Whereas only a few thousand workers were employed in large scale manufactures in the various countries of Europe before the Industrial Revolution, hundreds of thousands of families were involved in the production of textile, metallurgical, and other goods produced for sale in distant markets in a most routine fashion, according to traditional manual methods, with a few hand tools and inexpensive implements (Spufford*). This meant that the level of income and its fluctuations, the organization of life, work, and leisure for as many families was affected, even determined, in part by the seasonal, cyclical, and secular fluctuations and opportunities of non-agricultural, that is, "industrial" work.

To be sure, the political arithmeticians and early statisticians of the eighteenth and nineteenth centuries were cognizant of the existence and numerical weight of traditional forms of industry. The *Encyclopédie*, in its article on the manufactures, provided a clear distinction between the *manufactures dispersées* and the *manufactures réunies* and presented an analysis of the circumstances under which one or the other would prosper;[20] but most economists believed that the only progressive form of industry was the concentrated one (Berg*). The ubiquity and banality of the traditional forms of industry made them less conspicuous, so that it was both more difficult and less necessary to give them a precise count. Furthermore, since much of traditional rural industry was carried out by persons and families whose social status was that of peasants, it was almost natural to lump those who were only engaged in rural manufacturing on a part-time basis into the same category as agriculturalists. By contrast, the spectacular nature, concentration, and sparseness of large mills and workshops made them much more amenable to precise enumeration.

Thus the confrontation of estimates made by Gregory King in 1688 and Joseph Massie in 1759, where the latter took account of dispersed manufactures while the former had ignored them, has revealed the magnitude of the traditional industrial sector in England and the extent of distortion that results from omitting it in drawing a picture of "pre-industrial society."[21] The error that is consequent upon the omission of traditional industry is magnified by its part-time, seasonal nature. Since domestic rural industry was generally carried out in the time interstices that were left vacant in the annual cycle of planting and harvesting, it took a disproportionately large number of people to manufacture whatever product is reflected in the trade and production statistics that historians can collect. The error is further compounded by the fact that the workload was commonly shared among family members, so that statistical inquiries on the occupation of household heads often missed the fact that spouses and children also belonged in the industrial labor force. Thus the mixed nature of work in pre-industrial industry and the cumulation of income from different sources poses a serious challenge to retrospective quantification and defies accurate counts according to economic categories—agriculture, industry, services, or primary, secondary, tertiary—that have been devised under modern circumstances and are only appropriate for the division of labor that has been prevalent only since the Industrial Revolution.

Hence it has become increasingly clear that it was a mistake to separate pre-industrial economic history and devolve it to the students of a static "world we have lost," while recognizing dynamic and progressive properties only in the economic organization that followed upon the wave of gadgets of the Industrial Revolution. It has become evident that an economy could achieve more output in other ways than by technological progress. A significant form of expansion or growth, not to be confused with modern economic growth, could take place when output increased by using more inputs, when people worked longer hours by engaging in market activities during seasons and hours which they would in earlier generations have set aside for non-market work or even for leisure, or when a larger proportion of the population participated in the labor force. This is not to deny the extraordinary importance of modern technology after the Industrial Revolution; but the image of European industrialization that is painted by concentrating research on technology and investment has been distorted. It is an image of industrialization as a transition from a static, backward, essentially agricultural society to one that is dynamic, advanced, and essentially industrial. The kind of industrial growth and economic expansion that took place with traditional techniques and no fixed capital investment in industry did not simply constitute preconditions for an eventual take-off. It means that in many places the process of industrialization had started before the Industrial Revolution. Ignoring this not only led to a mis-specification of the structure of economy and society before the Industrial Revolution, but also to erroneous views on the significance of the Industrial Revolution itself—how it changed the world, and how it affected those who went through it (Tilly*) as many debates concerning the impact of industrialization rest on assumptions, sometimes implicit, on the nature of society before it started.

In the late 1960s, therefore, the paradigm of the Industrial Revolution went through a crisis.[22] It was subverted. Some people began to feel that they could not longer ignore the proliferation of anomalies and discrepancies between the paradigm and the new facts emerging from new research. In the 1970s, a small but growing number of economic and social historians turned their efforts to the testing of one part or another of the set of hypotheses enunciated by Mendels in an article first published in 1972 in the *Journal of Economic History*.[23] The essay also included the definition of a term, "proto-industrialization", whose introduction had been aimed at focusing the discussion. The term itself, which, not unlike "pre-industrial", and especially "pre-industrialized", indicates a sort of retrospective projection, since its etymology signifies "first industrialization", has been used by others, notably Eric Jones and Stuart Woolf, but without an attempt to give it a great deal of thought, let alone a definition.[24] Herman Freudenberger and Fritz Redlich had used the term "proto-factory" to speak of early forms of factory organization before the introduction of modern machinery.[25]

A certain amount of confusion concerning the meaning of "proto-industrialization" and the exact nature of the hypotheses that were involved with it was injected into the subsequent discussion by the fact that the concept, whose paternity is traced back to Mendels, had been adopted by Charles and Richard Tilly as early as 1971, and then quickly expanded upon and elaborated by a number of economic historians before Mendels ever got to congeal his own research and thinking by publishing it in book form.[26] Thus in 1975 Hans Medick, Peter Kriedte, and Jürgen Schlumbohm published in Germany a very complex and interesting neo-Marxist expansion and discussion of proto-industrialization, which they attempted to situate within the "transition from feudalism to capitalism." It is now available in English.[27] Furthermore, the demographic aspects of the hypotheses were fully tested out by David Levine on some English parishes in a book published in 1977.[28] In the meantime, Mendels' own definition had gone through several stages of refinement. In 1969, the term had originally been defined in his dissertation as "an expansion of manufacturing activities principally in the countryside."[29] Mendels further added that manufacturing qualified as "protoindustry" if:

> the "overwhelming majority" of the labor force [was] composed of landed peasants and landless agricultural laborers. Whether landed or not, essential is that their summers were devoted to activities other than the manufacture of linen, generally agricultural.[30]

In 1971, concerned that the term had not received a definition that was operational enough, especially in view of the problem of identifying the onset of proto-industrialization in any concrete instance, Mendels expanded and modified its meaning. In a paper delivered that year, published the next, he made the definition more complex by emphasizing the agricultural dimension. Proto-industrialization was no longer defined as just "the rapid growth of traditionally organized but market-oriented, principally rural industry," but *also* as "accompanied by changes in the spatial organization of the economy,"[31] which involved the bifurcation (the term is lifted from

Eric Jones) of a commerically-oriented agriculture from a peasant sector which dovetailed agriculture and industry according to the rhythm of the seasons. Such an expansion of the definitoin was suggested by the need to differentiate proto-industrialization from the mere expansion of cottage industry at earlier times. The interactions between agriculture and industry observed in Flanders and previously discussed by Eric Jones[32] were now to be regarded as the differentiating criterion, and Mendels proceeded in this and in subsequent articles to explain why this link between agriculture and industry was to provide the disequilibrating force that propelled the regional economy along an involutive path, yet an expansion path that could lead to modern industrialization.[33] Hence what was first presented as parallel occurrences, namely, the growth of cottage industry on the one hand, and the bifurcation of agriculture into peasant and commercial zones on the other, was now promoted to the rank of a double characteristic, indeed a definition, of proto-industrialization. This was to make the hypotheses revolving around the concept empirically verifiable. By narrowing down his definition and demarcating the phenomenon, Mendels hopefully made his hypotheses falsifiable and refutable, and avoided the pitfall of tautology. He was responding to Hartwell's plea that:

> we should define problems in such a way that we know what an historian is investigating and can compare the work of different historians. We should also define criteria in such a way that conclusions can be tested and conflicing conclusions reconciled ... Then historical discussions will be meaningful and not confused by misunderstanding, by terminological confusion, and by comparing incomparables.[34]

When the Executive Committee of the IEHA decided to organize this Section and gave it the obviously challenging title of "Proto-industrialization: Theory and Reality," it is with the same objective of falsifiability that Pierre Deyon and Franklin Mendels prepared a preliminary text which was distributed to the participants in the Section. Perhaps unfortunately, there were again a few minor changes in the formulation of the hypotheses, compared with the earlier version, so that the confusion we wished to avoid was not totally eliminated from the debate. We wrote:

> By proto-industrialization we understand an essentially regional phenomenon, hence one that should be observed in regions, not at the national level, where some of the critical interactions that constitute the process of proto-industrialization can be masked.

This preliminary paragraph re-emphasized that the hypotheses relating to proto-industrialization should stand or fall on the basis of regional, not national or international analysis and data collection. Whereas regional economic history had been previously confined to the purview of local historians, fundamentally opposed to generalizations and comparisons, one of the premisses of the hypotheses surrounding the concept of proto-industrialization is that the nation-state represents an anachronism with regard to the choice of territorial unit for a comparative analysis of the Industrial Revolution in its early stages. This is not just a matter of convenience, research strategy, or data availability. The point is that the principal interactions between agriculture and industry, and the demographic adjustments to these interactions, are posited to work themselves out principally within the geographic space roughly coterminous with the traditional region as defined by geographers. In an age of limited mobility for labor and especially for cereals, the seasonal dovetailing of agriculture and industry that is so characteristic of proto-industrialization principally took place within a small radius around a regional capital. Further discussion of this matter was recently published elsewhere,[36] but the essential point was already contained in the 1972 article, although somewhat less explicitly.

Then Deyon and Mendels proceeded to *define* proto-industrialization as "the regional combination of three phenomena:"

> First, proto-industrialization involves a type of industry whose *market was located outside the region* and often even outside of the national borders. Hence production is submitted here to influences that are partly autonomous with respect to local economic conditions. One must distinguish proto-industrialization from the traditional handicrafts and the petty industry that had always supplied country dwellers with their clothing, shoes, or agricultural goods.
>
> Secondly, the most significant aspect of proto-industrialization concerns the participation of *peasant* populations in handicraft production for the market. These populations found in handicrafts the supplement of resources needed to assure their subsistence and the payment of various dues imposed upon them. Handicraft activity was seasonal, for it was intercalated within the rhythm of agricultural work—except however when proto-industrialization had reached its extreme or ultimate form and occupied entire families on a full-time basis. By proto-industrialization we thus understand an industry which provided employment in the countryside above all.

Here we note the inadvertent and unfortunate omission on our part of the dynamic element of the definition (Tilly*) that had been earlier provided by Mendels. What should have been said is that proto-industrialization is defined as a "*growing* participation of peasant populations in handicraft production for the market . . . By proto-industrialization we thus understand the *growth* of an industry which provided employment in the countryside above all."

> Yet it was associated in a variety of complex systems with diverse forms of urban handicrafts and centralized manufactures. It was often in the city that the last stages of product transformation were accomplished, those which required the highest level of skill on the part of artisans dispersed in towns or concentrated in workshops with a moderate investment in fixed capital. It was also in towns that the rural artisans usually came to find the outlet for their products (Kaufsystem) and that the merchant-manufacturers resided (Verlagsystem, putting-out system). Thus one should keep in mind that the entire handicraft sector was organized or coordinated from the town.

This represents somewhat of a modification of earlier definitions, which had, perhaps unwittingly, underestimated or at least not sufficiently spelled out the considerable role of the urban sector in the process of proto-industrialization. Certainly, some of the earlier writings on proto-industrialization had almost entirely restricted their analysis and description to the rural sector.[37] This gave one the impression that the regional towns were entirely exterior, even superfluous to the process of regional proto-industrialization, which would be just as extreme as considering on the contrary the cities as the exclusive locus of the dynamic forces in the regional economy, as earlier historians had done. To be sure, a recent study of the Lombard economy in the seventeenth century makes the point that, "throughout Europe, none of the old centers of early capitalism (whether Antwerp or Venice, Amsterdam or Genoa, Bordeaux or Florence) played a leading role in the advent of modern industrialization." In Lombardy the cities

> had few of the traits that we associate with modern industrialization and in fact some that were diametrically opposed to it. Their growth and early success had been made possible by the superiority of their craftsmen and merchants over a still "barbarous" Europe and had been powerfully assisted by the cities' privileged position vis-a-vis their hinterland . . . [T]he very precocity of their success had tended, in the long

> run, to breed complacency and an excessive reliance on traditional techniques and design. The cities were thus clearly ill-suited to serve as the cradle of large-scale industrialization; far from being the vanguard of the modern economy, they must be viewed as anachronistic relics of a rapidly fading past. To find the harbingers of the modern economy, it is to the countryside that we must turn . . .[38]

Furthermore, we also know that the demographic expansion of the cities of Europe was not as quick in the eighteenth century as that of the surrounding countryside, and of course that the growth of rural industry surpassed that of the urban crafts and manufactures. Still, the town was an essential element in the economic structure of the region during proto-industrialization, shall we say the capstone of the whole regional edifice? The town coordinated, even directed proto-industrialization through its merchants and markets. It was in the town that the merchants of the putting out system directed the manufacturing activity dispersed in the surrounding countryside. It was to the town market that the independent producers of the Kaufsystem delivered their wares. It was in the towns that the finishing stages of production, the most intensive in skill and capital, and hence often those that contributed the largest share of the value added into the product were carried out. It was from the town that the wholesale merchants shipped out the goods to other regions. It was to the town that the food surpluses from the commercial farms of the surrounding areas were delivered—and the urban population was dependent upon such surpluses.[41]

> Thirdly, by proto-industrialization we understand the association of commercial *agricultural surplus* producers with a peasantry which was unable to attain self-sufficiency on its small plots. Income from agricultural self-employment was not only augmented by handicraft activity but also by seasonal wage work on the big farms, at certain times of the year when a supplement of labor was required on the latter. Proto-industrialization is thus defined by the simultaneous occurrence of three ingredients within the framework of a region: rural industries, external destination, and symbiosis of rural industry with the regional development of a commercial agriculture.

Ditz* justly points out that there is in this paragraph a departure from earlier versions where the association of commercial agricultural surplus producers with small peasantry took the form of a spatial segregation into specialized geographic "regions": "some regions turned to industry, while others developed their commercial agriculture to supply the needed food surpluses."[42] (Later Mendels utilized the word "zones", e.g., "The juxtaposition within a region of contrasting zones that maintained different ecological balances was mutually beneficial."[43]) But this change was not inadvertent, it was meant to take into the category of proto-industrialization homogeneous regions where each village might have both a peasantry engaged in cottage industry and a class of a larger food surplus producing farmers.

Having thus *defined* what was meant by proto-industrialization, Deyon and Mendels proposed a set of hypotheses:

> Now, the above combination tended to propel regions in which it was found toward an industrial revolution, or at least it facilitated their adoption of mechanization.
>
> First, a consequence of proto-industrialization has been the break up of the self-regulating character of pre-industrial demography, which had hitherto adjusted the natural rate of growth of population to local means of subsistence. The adjustment had been carried out either by the peasants themselves or under the constraint of their lords or landowners or sometimes of ancient inheritance or succession customs. The breakup of the self-regulating (or "homeostatic") system facilitated a considerable expansion in the number of families and size of villages since supplementary handicraft incomes, no longer directly limited by the size of farms, were now added to the local sources of agricultural subsistence. It became possible to tolerate the fragmentation of farms from one generation to the next, the more so as food provisions were complemented within the region by the surplus producing farms.
>
> However, the regions where proto-industrialization had thus taken place soon began to encounter the diminishing returns of a dispersed industry whose expansion created rising difficulties for merchants in the collection of output and its quality control. As a consequence some of these merchants found it beneficial to gather at least part of their manpower in workshops and then to take advantage of the opportunities offered by labor-saving mechanical inventions.
>
> Secondly, the necesssary capital for the constitution of workshops and for the introduction of machines had been accumulated locally by the development of proto-industrialization. Merchants had been enriched by it, as well as commercial farmers and landowners who earned high rents from the small fragmented farms in which the artisans lived. In short, through proto-industrialization, the merchant as well as the landowning class had found new surpluses from which a reserve of capital could be constituted.

> Thirdly, proto-industrialization had contributed to the accumulation by merchants of technical knowledge and experience with inter-regional and international trade and markets. It became a training ground in which the early industrialists were recruited.
>
> Fourthly, the development of a regional commercial agricultural sector during the phase of proto-industrialization prepared that sector for the increased supplies it was to provide during the urbanization that accompanied the subsequent phase, and without excessive price increases.
>
> Such are the principal mechanisms which apparently facilitated the transition from proto-industrialization to the modern phase of industrialization. However, . . . regional proto-industrialization sometimes proved to be a dead end, even leading to de-industrialization in the region.

* * *

The authors and titles of the 46 papers we have received in response to our circular are listed in the Appendix. These papers offer the opportunity to engage in a confrontation of hypotheses enunciated so far with new independent empirical evidence. This report will now attempt to bring out the most significant facts and opinions that were found in the papers. Naturally, they were not written according to any standard format, and it shall therefore be necessary to sort their contents according to a classification that was prompted by a reading of the entire set of contributions.

The first type of comment dealt with the basic definitions themselves. Even though there was a confirmation of the seasonal character of cottage industry in many cases, including Korea and Taiwan (Ho*) and West Africa (M. Johnson*), in Japan Saito* has observed that there was not primarily a division of labor in the rural textile industries according to the seasons but according to the sexes. Rural textile by-employment was women's work. In Normandy as well, fluctuations in the textile trade and industry primarily affected women (Gullickson*), whereas men were primarily engaged in agriculture.

Clarkson and Collins* suggest that the definition of proto-industrialization does not make room for a community which, like Lisburn, was engaged in household industry, often in combination with other activities, and yet was a town. It must be observed, however, that their perspective is from a survey dated 1821, several years after the factory system was first established in that town, so that, strictly speaking, the phase of proto-industrialization was now over. However, like almost everywhere else, traditional forms of manufacturing organization survived and continued to coexist with new forms for a long time. Kriedte* discusses the complexities and dialectics of this question. Medick* and Enciso* bring out the fact that rural industry was not always free of guilds and corporations, which indicates that the advantages of rural over urban industry in terms of labor cost may have at times been less obvious than conventionaly believed.

The question of the significance of town industry also comes up with respect to periodization and chronological distinction between proto-industrialization and earlier phases of industrial growth. Gutmann* suggests that a very long view of the process of industrialization would represent proto-industrialization as the ruralization of industry that took place as a rational response of merchants and manufacturers to the losses inflicted upon them by the economic crisis of the seventeenth century. Somewhat in agreement, Hohenberg* considers that urban and rural industry alternated over long cycles as the advantage passed from one to the other with fluctuations in real wages. Mendels* essentially adopts this kind of analysis in his contributed paper, where a new explanation and identification of the onset of proto-industrialization in Flanders now focuses on relatives changes in wages, grain prices, linen prices, rents, and land prices. Von Stromer* and Cariacono* use a very long term perspective for the study of textile industries and find a remarkable persistence of certain regional centers over the centuries. Schultz* relates proto-industrialization to rural crafts. She argues that the growth of such crafts for local markets (butchers, bakers, candlestick makers) prepared the way for proto-industrialization. Diedericks points out that the prosperous but purely agricultural communities that could be found in Holland supported a diversified array of crafts catering to the needs of the local farmers.

Perlin* prefers to place proto-industrialization in a broader spatial framework. He suggests that the relevant unit of analysis for the study of the transition from feudalism to capitalism is the world economy, and that the concept we should use is "proto-capitalism", a term he hopefully thinks will facilitate our understanding of the development of pre-colonial India, with its incipient market economy.

The question of the beginnings of proto-industrialization comes up also in terms of the social and demographic circumstances that favor the growth of cottage industry. Saito* sees demographic pressure in Japan, generally much greater than in Europe, as an important explanation that is verified by a cross-sectional statistical analysis conducted on the basis of 1881 data. Ditz*, although observing that the population density was much lower in the United States than in Europe, nevertheless detects some evidence of rising population pressure upon the arable land toward the end of the eighteenth century, to the point that cottage industry was promoted as an alternative to emigration. In China, according to Chao*, rural spinning and weaving of cotton began to spread in the fifteenth century, which was also a time when population growth had resulted in land shortage. The more ancient Chinese silk industry had both an urban manufacturing branch with large workshops and factories, and a rural family workshop component. Competition drove out the former, but the latter had a greater survival power. In both silk and cotton, it is the growth of population that drove the marginal product of family labor below the marginal product of hired labor. Hired labor must at least receive a subsistence wage, which thus constitutes a floor for marginal product as well. There is no such floor for the marginal product of family labor, because family members are not hired and fired; labor supply is given, determined by family size and age distribution.

Mendels had originally argued that population pressure had been the ultimate cause for the growth of rural industry in Flanders. Vandenbroeke*, however, points out that the period 1700–1750 was for weavers and wage workers a time of very high real wages; they were at a level that had not been reached in Flanders since the late middle ages. This definitely contradicts the view according to which Flemish industry was "the daughter of poverty." Mendels*, however, continues to argue that weavers in Flanders were at or near the bottom of the rural social scale, while conceding that the whole scale was rising as incomes improved in this period. Mendels* proposes that the rising real wages caused a profit squeeze on the merchants, who then tried to recoup their losses by putting out the work to cheaper rural labor.

In West Africa, M. Johnson* sees rural industry as not just labor intensive: it was virtually all labor, no capital was involved. In Leicestershire, Mills* finds that the framework knitters certainly were above the socio-economic level of mere wage workers. At least during a first phase of proto-industrialization which he identifies in 1650–1750, the stockingers were independent entrepreneurs owning frames and tools, with incomes definitely above subsistence. But the knitting frame was in the upper range of complexity and cost among all machines in use among rural textile workers. Nevertheless in the period 1750–1850 the stockingers' status progressively declined as they became full-time weavers dependent upon baghosiers. In other words, the transition from Kaufsystem to Verlagsystem was accompanied by proletarianization and impoverishment. Deboutte* observes that the weapons and metal trades were carried out in the region of Liège by household manufacturers who considered themselves much above the common proletariat. Hudson* points out that the wealth and status of wool weavers depended upon whether they were independent landholders who owned their tools, as in the case of the woollen sector organized in a Kaufsystem, or depended on a putter out as was the case in the worsted manufacture of the same region of Yorkshire. Near Osnabruck, Schlumbohm* finds that the more land the peasants had, the more linen they sold (although he cannot tell how much of that linen was actually produced by the peasants' tenants); whereas in Bielefeld, the peasants with the most land had no looms, but peasants with some land had more looms than peasants with none. There were also wage-earning weavers owning no land whatsoever!

The general social organization and agrarian structure of a region have been pointed out as a relevant circumstance explaining the location of cottage industry and conducive to its growth and to the modernization (or backwardness) of agriculture. Mills* points out that the conditions propitious to the rise of industry were met in open villages, where manorial or landlord control was weak. He argues that close villages with strong control would tend, on the contrary, to repel industry. Badia* discusses the hurdles imposed upon rural society by the "foro" in Galicia. However, Freudenberger* points out that "proto-factories", at least, could be promoted by a traditional noble class, and Rudolph* argues that serfdom can very well be seen as a type of social and agrarian system that tended to promote industry since labor was a fixed input. Being like a piece of equipment, it was in the landlord's interest to utilize it at all times and in all seasons, not to leave it idle in the dead season. In such a highly seasonal agriculture as that of Russia, cottage industry was a handy way for landlords to keep their unfree labor force occupied and productive during the long winter season. For Chao*, it was not serfdom, but a very strong family organization that turned labor into a fixed cost in China, even after population pressure had driven its marginal productivity down. Ho's* theoretical model includes two interacting regions, one which institutionally controls its population, and one which does not.

Mager* observes that the growth of industry was sometimes a consequence of enclosures, when common lands became private. Similarly, the growth of peasant industry in Bulgaria (Palairet*) can be directly related and attributed to a land shortage consequent upon its appropriation by the

Ottoman rulers. When the indigenous inhabitants were barred from cultivating the rich lands, they retreated to the hills and started earning an income from industry. When the Ottoman rulers subsequently left and the land was returned the product of industry began to decline. Clearly, a low land/labor ratio was an inducement to seek industrial by-employment, but the reasons why this ratio was low, or fell, varied and did not matter: the effects were fundamentally the same whether the land shortage was institutionally imposed through prohibitions and expropriations, or whether it was the consequence of the population's own high rate of demographic reproduction.

The demographic and social consequences of proto-industrialization are probably the most unexpected and intriguing aspect of the whole scenario of proto-industrialization. According to the original model, whereas population growth had been the principal cause of impoverishment and peasants thought of industry as a means of alleviating their poverty, the poverty did not disappear because proto-industrialization in turn had the unfortunate feedback effect of promoting poverty by breaking down the old self-regulating demographic system. This happened primarily through marriages, which were facilitated by the expectation of a cash income from industry. Vandenbroeke* offers Belgian data that seem to contradict the model, by showing that the age of marriage did not go down during proto-industrialization. Mendels* shows that a more careful exploitation of nuptiality data leads to results that do support the model on this point. Impoverishment is denied by Vandenbroeke*, and also by Servais* for Eastern Belgium. Both see rural industry as a way for peasants to supplement their incomes and accede to landed property, even to reach an improved standard of living. However, Mendels* uses land prices, land rents, mortgages, interest rates, and food and linen prices to show that the relative evolution of all these variables is quite consistent with the hypothesis that the growing aggregate industrial and agricultural product of the Flemish economy during rapid proto-industrialization was to a large extent captured by landlords and merchant manufacturers. Indeed Servais* concedes that mortgage payments were so high that they could absorb the quasi-entirety, sometimes more, of the usufruct of the property. Furthermore Vandenbroeke's data show a decline in real wages as well as per capita product after 1750 or even after 1725.

In the Pays de Caux (Normandy), where, as mentioned above, manufacturing in the home was mostly carried out by women, much of the population growth can be explained by migration, according to Gullickson*. The average age of marriage of textile workers remained high. In fact the women who married at the youngest age were the daughters of wealthy peasants. When home spinning was displaced by machine-made yarn, nuptiality declined, but the demographic impact was mild, and Gullickson suggests that this may have been due to the fact that cottage industry was principally a female activity. "The impact was muted by the stability of male occupations." The research of Saito* on Japan yields results that in several interesting ways confirm the importance of the sexual division of labor as an intermediate variable that modifies the impact of industry on family formation. In Japan, proto-industrialization, although carefully assessed and identified, did not lead to a decline in the marriage age, perhaps even the opposite, and, as previously mentioned, cottage industry was mostly women's work. In addition, however, it must be added that the Japanese circumstances differ from the European in that female age of marriage was very low, whereas male age at first marriage was comparable to a European level. In the Japanese village T. C. Smith called Nakahara the mean age of first marriage for males in the eighteenth century was 27 years, for females 19.5.[42] A girl had no dowry, which is probably what made a big difference: the parents had no incentive to make her wait and be patient (except if her work in the parents' home made her very valuable), and the grooms had no reason to marry older women, unlike in Europe, where the latter could bring more into the marriage than younger ones. But Japanese women made up for this low age of marriage, which might have caused excessive population growth, by lowering their fertility rate, so that population almost stagnated in the eighteenth century. Finally, the Japanese families practiced infanticide and adoption, two methods of achieving optimal size. It thus appears that the whole logic of family organization, the links between family, marriage, birth, and death were not the same in Japan as they were in Europe.[43] Other statistical tests presented by Saito* yield results that give an impression of a low demographic responsiveness to agricultural or industrial changes, or to both.

In West Africa, M. Johnson* finds no demographic response, but she observes that since land was abundant in this region of the world, and there was no population pressure, a European marriage pattern was not to be expected, so that proto-industrialization could not disturb a non-existent self-regulating mechanism. Finally, Leboutte* finds, in the area around Liège, marriage ages which seem high and stable throughout the 18th and 19th centuries, a finding that is consistent with his overall conclusion of *embourgeoisement,* certainly more than with the opposite of proletarianization. Except for the village of Wandre, however, it must be said that Deboutte's evidence is a bit late for the purpose at hand. Most of it comes from the period when modern industrialization has begun.

The problem of the transition from proto-industrialization to modern industrialization has evoked much interest on the part of the contributors to this Section. In view of some of the comments that were made, it might be useful to go back to the formulation of the problem in the 1972 version. Proto-industrialization was then viewed as a

> first phase which preceded and prepared modern industrialization proper ... Phase two was facilitated by prior proto-industrialization ... prior proto-industrialization tended to induce the passage to modern industry

and *yet*

> nonetheless, some regions became "de-industrialized" ... It is thus clear that whereas proto-industrialization preceded factory industrialization where the latter occurred, and paved the way for it, there was nothing unavoidable or automatic in the passage from phase one to the next ... A similar development of rural industry in phase one could thus lead to broad differences in the extent and pace of industrialization in phase two ... Natural resources availability, the location of essential new fuels and raw materials and above all, the socio-political framework, determined the success with which the transition was effected under the tensions introduced by the industrial revolution in the advanced regions. The extent of capital accumulation as well as the political and social terrain in which it took place and the personnel in whose hands it fell determined very much the course of future (factory) industrialization.[44]

Since then, Peter Kriedte pointed out the possibility that proto-industrialization actually deterred future factory industrialization, that its very viability might have postponed the coming of the factory system.[45] Kang Chao* more precisely makes the point that the existence of a tradition of family-based cottage industry only deterred and slowed down the appearance of factories in China by so successfully competing against them. He argues that the competitiveness of the traditional household sector was in the last analysis the result of population pressure, that is, redundant labor absorbed by family farms. Lyons* argues that the growth of the factory sector in Lancashire was little affected by the handicraft sector. "Thus it may have been that in one set of circumstances protoindustry was a complement to factory growth, in another it was competition and in the case of the British cotton weavers it was neither." For Perlin*, "if we look carefully at the historical circumstances surrounding the growth of proto-industrialized regions in Europe and Asia, it is significant that stagnation eventually affected the vast bulk of them, and that the emergence of industrial capitalism was not at all predetermined by the particular locations of proto-industrial forms of production. . . Areas of proto-industry were generally destined for stagnation . . . [The] transitional role [of proto-industrialization] would be better seen as stimulating accumulation and reinvestment elsewhere." Vandenbroeke* sees the accumulation of capital in eighteenth-century Flanders leading to the industrialization of . . . Wallonia, but we feel that he underestimates here the significance of the Industrial Revolution in Ghent and Kortrijk. They were to become major cotton towns, just as Lille-Roubaix-Tourcoing were to become major textile cities on the other side of the border—drawing some of their labor force from Belgian Flanders.[46] Continuity of capital and labor between proto-industrialization and modern industrialization is emphasized by Dewerpe* and Bythell*.

In order to reach a conclusion on this matter, it should be appropriate to apply our triple criterion for the definition of proto-industrialization, and verify whether the latter "tended to induce the passage to modern industry" at least in the limited sense that there was a spatial correlation between the two. The Industrial Revolution is defined as the beginning of factory industrialization, that is, the moment when the first mechanized factory begins operating within the region. A pseudo-statistical test can be applied to the evidence brought forward by the contributors. The hypothesis will receive support from cases where (1) proto-industrialization was followed by an industrial revolution or where (2) an absence of proto-industrialization was followed by de-industrialization. It is, on the other hand, contradicted by cases where (1) proto-industrialization is not followed by an industrial revolution, or (2) an industrial revolution is not preceded by a phase of proto-industrialization.

The hypothesis appears to have been contradicted by the case of the Danish linen industry (Oxenbøll); the Japanese cotton industry (Saito*), where machine industrialisation occurred, but in a remote part of the country, and in the region studies by Nishikawa*, where there was such a considerable delay between the two phases that the link between them is questionable. We find in support of the hypothesis: Bologna (Poni*), where, as predicted, an urban silk industry turned out to be ephemeral; the same appears to be true for Norrköpping (Schön*). Support for the hypothesis can equally be found in Castille (Enciso*), Galicia (Badia*), other Spanish provinces (Aracil*), the cotton industry of northern Sweden (Schön), and Keyder's* description of the situation in the Ottoman Empire. In those instances, the cottage industry did not constitute proto-industrialization, either because its markets were local, or because an agricultural sector did not symbiotically modernize with it. Hence, as we anticipated, an absence of proto-industrialization led to no industrial revolution. Positive evidence toward the theory is found in the case of Holland (Diedericks*) where the agricultural sector modernized and prospered, but there was no cottage industry for external markets, no proto-industrialization.

Finally, support for the hypothesis is shown in those cases where one can ascertain that proto-industrialization occurred as defined, and that an industrial revolution subsequently took place. That is the case of Padrón (Badia*); the region of Verviers (Gutmann*); Liège (Leboutte*); Flanders, which, *pace* Vandenbroeke*, cannot be placed under the heading of de-industrialization in spite of the frightful crisis of 1845–50; the West Riding of Yorkshire (Hudson*), where both the woolen and the worsted sectors were eventually mechanized; the woolen knitting industry in Jutland (Oxenbøll*); the silk industry in Japan (Saito*); the north-Italian industrial triangle (Dewerpe*); the Pays de Caux, with its factories in Rouen (Gullickson*); the northeastern United States (Ditz*); the Belfast textiles, in spite of the complex shifts between linen and cotton (Sandel*); Ravensberg (Mager*), and Lancashire (Lyons*).

To be sure, some of the cases we have classified here as having experienced an industrial revolution and thus entered "phase two" may have subsequently undergone serious difficulties or decay. Indeed, the proto-industrialization hypothesis never pretended to explain or predict the complete trajectory of modern industrialization that was taken by each region after its second stage had started. As was earlier specified, there could be "broad differences in the extent and pace of industrialization in phase two,"[47] not only because of particular conditions in the supply of fuels, raw materials, or transportation services, or in the social and political environment—which could itself go through its own autonomous and determining evolution (C. Johnson*)—but also because once mechanization had started, an entirely new economic structure entered into play. At that point, a new model is in order for the study of urban-rural interactions and change. One where, for instance, fixed capital is finally called to play a very important, if not governing role.[48]

Evidence presented by the contributed papers has only shown three cases where the predictions of the theory concerning spatial and temporal continuity are contradicted by the facts. The overwhelming majority of the cases support the theory on this point.

De-industrialization is a central theme in the economic historiography of countries that have experienced a period of foreign colonization, particularly India, North Africa, and the Middle East, so that the perspective on European economic history that comes with the theory of proto-industrialization has started to reduce the chasm that previously separated historians of these different regions and opened a dialogue. Can the rural economy of India around 1800 or around 1700 be analyzed by using the same categories we employ for Western Europe? In a recent textbook, Maurice Garden ventured to suggest that "a Eurocentric vision could make us say that certain regions of the Ganges valley remind us of European Flanders."[49] That is certainly a change from an historiographical tradition which had been dominated by the presumption that

> the existence of large numbers of persons involved in family-economies, in which industrial and peasant pursuits were combined is a sign of non-development and tradition . . . where large numbers of textile producers can be observed they are treated as if they were evidence of isolated cases of change within the sub-continent, and ones that need not necessarily be viewed as having developmental implications . . . Attention remains confined to the question of the extent to which there was a division of labor . . . Application of the concept of proto-industry to this Indian historiographic context puts the terms of the debate onto a completely new plane because it implies asking questions on the extent to which production in India was subject to capitalistic relations before the advent of colonial rule . . . This means putting India onto the map of a developing world capitalism as early as the 17th and 18th centuries . . . This in turn implies another question . . . is it true that the origin of capitalism is only to be found in Europe . . . ? In the "Third World" proto-industrialization has inescapably revolutionary implications capable of subverting a whole historiographic and epistemological tradition concerning the character of pre-colonial life. (Perlin*)

In the Ganges valley, on a territory of some 40,000 square miles, lived a population with a demographic density comparable to that of East Flanders in 1800, and ten percent of the adults were then weaving, half of the women spinning.[50] The output of the Indian silk and cotton industry was exported around the world. In the 1830s, European textiles still accounted for 25 percent of imports into Senegambia, for instance, but Indian textiles amounted to almost 40 percent.[51] Until the English industrial revolution, there is no doubt that the Indian cotton industry was the largest in the world.[52] Toward the beginning of the sixteenth century, most of the great civilizations of Asia were technically and economically ahead of Europe. Two centuries later, the level of economic development of the majority of countries in today's Third World was still close, sometimes superior to that of Europe, as Paul Bairoch has pointed out.[53]

Lucette Vallensi compares the structure and organization of Tunis' *shashiya* (knitting) manufacture with the European model of proto-industrialization. This was an export industry, city based, but involving a complex network of rural putting out. By emphasizing the impressive scale and apparent dynamism of this industry, she also has to consider the causes of its decline, which she finds in endogenous circumstances: the backwardness and self-sufficiency of the rural population, the insecurity and lawlessness that was tolerated by a weak central government, and the conservatism of the merchants who dominated and controlled the manufacturing sector.[54] Her analysis is not inconsistent with that of Keyder*, who finds in the notion of an "Asiatic mode of production" an adequate description of a whole socio-political structure incompatible with capitalist economic growth.

China did not fall under complete colonial domination in the nineteenth century. Chinese handicrafts did not experience the kind of de-industrialization that India went through. The question that has been posed by Chinese industrial historians is: why was the factory system so slow in developing and why did the handicraft sector persist so long? Kang Chao* does not find the Asiatic mode of production applicable to China. What retarded factory industrialization was the intersection of a rising population density, reaching very high levels, with a very strong family spirit which prevented households from shedding their redundant manpower, no matter how unproductive it had become at the margin, and from turning them onto the labor market. Rather than facilitating the growth of modern industry by offering an unlimited supply of labor, as in the Fei-Ranis model, population pressure only led to self-exploitation *à la* Chayanov, a problem inherent to the "domestic mode of production" whenever the marginal product of labor in agriculture is driven below the competitive wage that can be earned elsewhere, if people are not willing or able to move away from the familial household.[55] In the *absence* of overpopulation, Chao expects that "the labor cost of the family industry would [have] not [been] cheaper than in any alternative production institution. Handicraft factories would [have] develop[ed] and eventually pav[ed] the way for the road to industrialization," since "handicraft factories are usually dynamic in the sense that they have the potential to breed new technology" (Chao*). By contrast, the Japanese *ie* [household], with the mobility it permitted, adoptions, numerous servants, gives much less importance to the kin group. Does this help to explain why the previous expansion of industrial by-employment in Japan seemingly did not act in the same manner as in China as a brake on the growth of concentrated industry, whereas Japan as well suffered otherwise from the symptoms of population pressure? By giving less weight to kinship than even England or America, by considering that, as people say, "one's neighbor is of more importance than one's relatives",[56] Japanese traditional society might not have been willing to accommodate its surplus progeny, so that as a consequence market forces were free to operate without the impediment of sentimental constraints. It would not be surprising in this view that in the Chōshū *han* "the marginal productivity of labor per day was [in the 1840s] approximately equal to the daily money wage for non agricultural work, at least in salt-making at Mitajiri *saiban*. This suggests that a competitive wage or income prevailed among forms of agricultural self-employment and non-agricultural by-employment in this region."[57] Ho's* abstract model follows Chao's point of view, however, since household production of industrial goods continues to be viable even after the introduction of "proto-factories" because farm household members producing handicraft goods "are in effect being subsidized by those producing agricultural goods." On the other hand, Ho also incorporates into his model the possibility that family solidarity might help the growth of centralized industry. That would happen if the rural household makes up the difference between the per capita family farm income and the (hypothetically) lower factory wage by giving the departing household member a subsidy.

Korea and Taiwan have been widely regarded as examples of countries which, thanks to a type of growth and industrialization that avoided at first the capital-intensive and labor-saving technology, showed that rapid economic growth is not incompatible with an improved relative share of the poor. Ho* presents evidence on these two countries in his empirical contribution. He shows the continuing importance of small-scale rural industries in the industrialization of both countries. The rural share of manufacturing even increased in Taiwan from 1956 to 1966. The small-scale industries at stake, however, included not only "sideline activity", i.e., part-time household by-employment, but also small rural factory production and handicrafts for the local markets. The long run trend of rural industries in Korea and Taiwan is now a declining one, but there seems to be no doubt that such industries have made a positive contribution to the development of the economy in both countries.

* * *

The debate surrounding proto-industrialization offers an illustration of an old problem which concerns the use of words. The Dutch historian Huizinga once called it "terminological inflation." It is a phenomenon that seems to be very common: a new word is coined, it refers to a clearly-circumscribed subject. Later on, however, the use of the word is broadened. With its specificity, the word loses its value. Thus the theory of proto-industrialization consists of careful definitions and specific hypotheses. These were circulated in advance. It has been made clear that proto-industrialization is not coterminous with the existence (or growth) of household rural industry. If it were, we would not have needed a new word for it, and the hypotheses concerning its disequilibrating impact as well as its continuity with modern industrialization could have been rejected handily.

Although this problem has made our debate somewhat more confusing and complicated, happily we are able to report that the debate has generated new light, thanks to the collective exercise in comparative regional history that was made possible by the 46 respondents to the challenge of our subtitle: "Theory and Reality." The contributed papers have permitted us to confront the hypotheses with evidence that was not available at the time of their formulation, and suggest that some revisions to the theory might be in order.

(1) The first revision that emerges from the discussion concerns the question of the starting point of proto-industrialization. This question, which had not been previously handled in a satisfactory manner, might be resolved by inserting proto-industrialization into a broader temporal horizon. Throughout Europe, the period 1650–1750 was marked by a complex set of trend reversals. Different regions successively entered into a period of demographic abatement, severe epidemics and subsistence crises, falling grain prices, and rising real wages. A more precise explication of the causes and consequences of these diverging and intersecting trends could be

formulated for each region in order to explain and identify the approximate date of the onset of proto-industrialization. It was done in the case of Flanders by Mendels* but it would be hazardous to presume that the same economic and demographic forces were acting in the same way in other regions. However, the large amount of quantitative evidence prepared by historians under the stimulus of the controversy surrounding the "seventeenth-century crisis" certainly provides an opportunity for analogous analyses elsewhere and for fruitful comparisons.[58] In the Flemish case, following Vandenbroeke*, we now propose that proto-industrialization started in a time of very high and rising real wages. Even though rural wages were rising as well, the squeeze on profits that this development inflicted upon urban industry promoted a search for cheaper labor in the industrial sectors where labor costs constituted the largest share of total cost. That cheaper labor could be found in the countryside surrounding the old urban industrial centers.

(2) At the close of our debate, there is still a good deal of evidence showing that the "average" household which responded to the opportunity of engaging in rural industry—if such an average could ever be discovered—would be low on the social and economic scale. Nevertheless, the papers that have been presented demonstrate that a certain amount of variance would be found around that hypothetical average. They also show the danger of extrapolating from the experience of textiles, particularly linen weavers, to all other rural industries and workers. It seems that the status of those engaged in cottage industry was lower than full, self-sufficient peasants, and above that of propertyless wage workers. Impoverishment was not structurally built into cottage industry. It was, rather, the consequence of conjunctural difficulties or it was, after the Industrial Revolution, the consequence of competition from machine-made substitutes. The actual position of individuals and households around the notional average mentioned above was probably determined by the nature of the industrial commodity they produced, more precisely its production function. The contributed papers suggest that the more capital and skill-intensive the commodity, the higher the income and status of those who were engaged in its production. The more labor intensive, the lower their income and status.

(3) No matter what the initial position of peasant-workers might be, the demographic consequences of the growth of industrial by-employment in the countryside tended to disturb the former homeostatic system that kept

the population in check with the amount of disposable land, if and when such a self-regulating system existed. There is no reason why we should ignore the earlier findings of Chambers, Braun, Deprez, Krause, Mendels, Levine, Almquist, Guignet, and others.[59] Early reports of a major forthcoming work on the historical demography of England indicate that marriages were the principal regulator of the English population in the early modern period, and that they were in turn correlated, with a 20–30 year lag, with income, as measured by wage rates.[60]

However, the impact of industry was modulated by the precise manner in which it was organized within the household. In a patriarchal society, when industry was women's work, its demographic consequences may not have been the same as when it was the work of the entire family, male and female alike. Furthermore, research emerging from and about Japan and Russia, and also what M. Johnson* has told us about Africa, suggests that proto-industrialization could bring with it different results in societies where the locus and the calculus of decisions relating to family formation, work, and leisure, and where the level of population pressure and land scarcity, were drastically different from those which prevailed in western Europe. This reaffirms, once more, the importance of Hajnal's hypotheses concerning the European model.[61]

As to the precise manner in which the homeostatic system was disturbed, evidence still seems to bring nuptiality to the fore, but fertility was instrumental in some cases (Hohorst*, Levine[62]), and several contributions detect differentials in the migration pattern and rates as well. There is no question that the relative importance of mobility in determining the growth of particular villages and areas during the process of proto-industrialization is a topic that deserves further research. What is at stake, as Charles Tilly pointed out,[63] is no less than the settlement of an old debate which already opposed Marx to the Classics, between those who see the European proletariat originating in the expropriation of the peasantry (i.e., the "primitive accumulation of capital") and those who see it coming out of the excess of progeny that was engendered by a reckless peasantry.

Contemplating the differences between the Japanese, Chinese, North-American, Russian, and Western European patterns of rural industrial-demographic interactions suggests that a critical importance should be accorded to the relationship between two variables: one is the emancipation of women and adolescents, and the other population pressure as expressed in the gap between what people could gain by working at home in the countryside and what they could earn elsewhere. In the United States, the emancipation of youngsters was probably very great and income on farms very high: the mechanized factory sector easily outcompeted the rural home industries at a very early date.[64] The opposite was true in China (Chao*). In Russia, until 1861 restrictions on mobility did not originate within the family, but from the estate owner. After 1861, restrictions on mobility came from the *mir* and the heads of household themselves. As the population was rapidly increasing and agricultural productivity was held back, the household-industrial sector should have shown, as in China, a great persistence. It did.[65] In Japan, the population pressure may have been extreme but the emancipation of youngsters from familial pressures reduced the intersectoral income gaps, which reduced the competitive potential of rural industries when modern factories were built. In England population pressure must have been greater than in the United States but less than in Japan or China. Emancipation from family constraints and thus labor mobility may have been, say, equal to that prevailing in Japan and the United States. We would thus expect rural domestic industry to have persisted longer than in the US, but less than in Japan.

We mentioned the emancipation of women. It is not that, like youngsters (male or female) who leave the familial household at an early age in some societies, they might terminate the marriage in response to economic opportunities. Rather, there is in different societies a difference in the wealth they are expected to bring to a new alliance at the time of its formation. This affects their marriage age as well as their position in the couple for the rest of their lives. In Japan, we saw that rural industry did not modify female marriage age and that this might be related to the female's subordination that is reflected in the absence of dowry and the low marriage age. In Normandy as well, women's work did not affect women's marriage age much. But within France, Normandy was unique in its inheritance practices. Within an inheritance system which was extraordinarily egalitarian—among males, that is—the females were totally excluded from the devolution of real property under the Old Regime.[66] We might speculate that this should have given Norman women a low bargaining position at the time of their marriage and ever after. There are in any case good reasons to presume that both the causes and consequences of the spread of domestic industry in Europe and elsewhere should be looked at in conjunction with the nature of domestic relations in various societies.[67]

(4) With the Industrial Revolution came a transition from dispersion to concentration, "an 'implosion' of industrial production into the cities"

(Tilly*), and a new division of labor, when rural became synonymous with agricultural and urban with industrial. The process was often so protracted that the term "implosion" may be misleading, but sooner or later, therefore, it is self-evident that particular villages and areas underwent de-industrialization, as industrial activity was taken from the peasants' homes to the factories. What is not so obvious and surely became socially costly and painful, is that sometimes the industrial activity was taken away from the region altogether, and the point of concentration located itself to a city within a region that looked as foreign to the peasants as a foreign country looks today. As Tilly* proposes, this kind of de-industrialization to an extra-regional concentration point is to be distinguished from ordinary de-industrialization for intra-regional concentration. Brittany underwent this kind of de-industrialization. Its rural linens and iron products had been exported afar, and it even possessed its own separate zone of surplus-producing and exporting grain agriculture. The population of that province had been growing faster than any other French province in the eighteenth century, its density had become one of the highest, as well as its marital fertility, which was also rising. The conditions had thus been right for an industrial revolution. Yet, Brittany de-industrialized. Its people had to emigrate.[68]

It appears that whereas the region seems the appropriate unit for the analysis of economic-demographic interactions during the phase of proto-industrialization, with the falling transportation costs that were caused by the building of canals and railroads, the improvement of roads, and the increase in the speed of ships, the region ceases being the appropriate unit for the study of the second, factory phase of industrialization and de-industrialization. The spatial framework of the analysis has to expand in accordance with the enlarged terrestrial domain within which the law of comparative advantage now operated to select the sites of concentrated industry. It would be misleading to continue using the same regional approach when analyzing industrialization and de-industrialization after the transportation revolution had occurred, as when the cost of grain still doubled when it was carted over 300 miles.[69] Regional societies had not been closed systems under proto-industrialization, since we have particularly emphasized that they were involved in exporting the products of their peasant industry, but the amount of inter-regional trade in "necessities" was kept minimal by the cost of moving bulky goods, by tolls and customs, and even by the concern of the authorities and the inhabitants in years of dearth.[70]

(5) A fortiori, therefore, if the de-industrialization of, say, Brittany cannot be analyzed by remaining within the confines of the single region, one should not suppose that the theory of proto-industrialization will do more for our understanding of the de-industrialization of India or Egypt than posing new questions by suggesting counterfactual scenarios. Proto-industrialization remains a paradigm in this context, but in the limited sense of "a mode of perception in which old facts take on a new perspective."[71] There are many people who consider that the "Celtic fringe" of Great Britain and the non-French-speaking periphery of France entered into quasi-colonial relationships with respect to their respective national core regions.[72] John Hicks once remarked that "The English handloom weavers, who were displaced by textile machinery, could (in the end and after much travail) find re-employment in England; but what of the Indian weavers who were displaced by the same improvement? Even in their case there would be a favourable effect, somewhere, but it might be anywhere; there would be no particular reason why it should be in India. The poorer the country, the narrower will be its range of opportunities; the more likely, therefore, it is that it will suffer long-lasting damage, now and then, from a backwash of improvements that have occurred elsewhere."[73] We cannot be so sure, however, that the Breton weavers found much solace in the fact that they could find re-employment in Roubaix or in Lyon.[74] It may well be no more than a difference of degree that thus separates an analysis of the de-industrialization of India from that of Brittany or the Languedoc, and if we are ready to accept that the expansion of "traditional handicrafts" as sideline activities for peasants but with an export destination, constituted a primitive form of industrialization then we must remember that the Industrial Revolution in the advanced regions may have meant de-industrialization of large regions of the world, "which may, in fact, account for the majority of the population of the under-developed countries and, very nearly, the majority of the population of the world as a whole."[75]

(6) How much help this provides in discussing what could now, in some cases be dubbed the "re-industrialization" of Third World nations is not immediately self-evident. We can be hopeful that the research and debate on "proto-industrialization" is improving our understanding of "what really happened" in the past, and has moved us further away from the simplistic linear depictions of the historical process of industrialization as one where static, "feudal", pre-industrial societies moved into the industrial age of capitalist self-sustain growth. There are many similarities between the

constraints faced by European and non-European regions during proto-industrialization some 250 years ago, and those faced by developing countries today, but the world economy in which industrialization or re-industrialization may be starting today or tomorrow is quite different from the environment in which proto-industrialization took place then. The capital needed to create one workplace at the time of the Industrial Revolution in western Europe used to cost the equivalent of four to eight months of wage labor. It now costs about 350 months of wage labor to establish a workplace in modern industry in a developing country, and the minimum threshold plant size has also increased.[76] Many economists, consequently, are now reconsidering the position they took in the 1950s, when it was held that the appropriate goal of development should be the maximization of per capita output, or average income, at some future point in time; and that since the capital-labor ratio (or capital intensity) determines output per capita, capital intensive projects should be favoured even when capital is scarce.[77] There is a debate on the appropriateness of "intermediate technology" and the positive role of the "informal sector."[78] Decentralized, small-scale, rural industry has the great advantage of *not* encouraging migration to already overcrowded cities. Certainly, the Taiwan experience, among others, "suggests that rapid industrialization and spatially decentralized industrialization can occur simultaneously, and that decentralization has important beneficial effects on the rural sector," where the majority of the population still resides.[79] It is true that the small-scale rural industries that may be appropriate today are in some cases quite different from the home industry of peasants in eighteenth-century Europe, but in Europe itself, the "re-industrialization" of the textile region of Prato takes place in the context of family industry coordinated in a putting out system.[80] Whereas almost everywhere else in Europe, textile industries are succumbing to the competition from Asia, Prato multiplied its exports by a factor of 14 between 1963 and 1979. The work is dirty and dangerous. Eighty percent of the weaving workers are deaf upon retirement, but there is no lack of jobs in this area, and take-home earnings are very high.

Notes

Starred (*) references point to contributions to this Section. A list of the contributed papers has been placed in the appendix. I wish to thank Pierre Deyon for his skilful and selfless role in organizing this section, and for his support and patience toward me. Financial support for the preparation of this report and for the organization of and travel to preliminary meetings has been generously granted by the Maison des Sciences de l'Homme and the Werner-Reimers-Stiftung. Thanks to Clemens Heller, Maurice Aymard, and Konrad von Krosigk for making that possible.

[1] Karl Marx, *Das Kapital.* Preface to the first German edition, 1869, in Robert C. Tucker, ed., *The Marx-Engels Reader,* 2nd ed. (New York, 1978), 296.

[2] In Philadelphia, on October 15, 1981, President Reagan invoked the American historical experience as a model of boot-straps economics before flying to the Cancún North-South conference.

[3] Kenneth Boulding, *The Meaning of the Twentieth Century* (New York, 1965), 58.

[4] T. S. Ashton, *The Industrial Revolution 1760-1830* (London, 1948), 58.

[5] id., 34.

[6] E. M. Carus-Wilson, "An Industrial Revolution in the Thirteenth Century," *Economic History Review [EHR],* XI (1941); J. U. Nef, "The Progress of Technology and the Growth of Large-scale Industry in Great Britain 1540–1640," in Carus-Wilson, ed., *Essays in Economic History,* Vol. I (London, 1954).

[7] D. C. Coleman, *Industry in Tudor and Stuart England* (London, 1975), 36; Nef, "Progress of Technology," 91, 88.

[8] Ashton, *Industrial Revolution,* 11 and passim.

[9] Evsey Domar, "Capital Expansion, Rate of Growth, and Employment," *Econometrica,* XIV (1946), 137–47; Roy F. Harrod, *Towards a Dynamic Economics* (London, 1948).

[10] Earl J. Hamilton, "American Treasure and the Rise of Capitalism," *Economica* (1929), 338–57.

[11] H. J. Habakkuk, "The Historical Experience of the Basic Conditions of Economic Progress," in Barry Supple, ed., *The Experience of Economic Growth: Case Studies in Economic History* (New York, 1963), 111–27.

[12] Herbert Kisch, "Postscriptum," in Peter Kriedte *et al., Industrialization before Industrialization. Rural Industry in the Genesis of Capitalism* (Cambridge & Paris, 1981), 194–200; Simon Kuznets, "Quantitative Aspects of the Economic Growth of Nations," *Economic Development and Cultural Change,* V (1956), VIII (1960), IX (1961); Alexander Gerschenkron, *Economic Backwardness in Historical Perspective* (Cambridge, Mass., 1962); Walt W. Rostow, *The Stages of Economic Growth. A Non-communist Manifesto* (Cambridge, 1960).

[13] Phyllis Deane and H. J. Habakkuk, "The Take-off in Britain," in Rostow, ed., *The Economics of Take-off into Sustained Growth* (New York, 1965), 65–82.

[14] François Crouzet, *Capital Formation in the Industrial Revolution* (London, 1972).

[14a] Richard Tilly, "Capital Formation in Germany in the Nineteenth Century," in Peter Mathias and Michael M. Postan, eds., The Cambridge Economic History of Europe, Vol. VII, Part i (Cambridge, 1978), 398. This confirms Paul Bairoch, *Le Tiers-monde dans l'impasse* (Paris, 1971), 56.

[15] François Crouzet, "England and France in the Eighteenth Century. A Comparative Analysis of Two Economic Growths," in R. M. Hartwell, ed., *The Causes of the Industrial Revolution* (London, 1967), 139–74; Paul David, "The Growth of Real Product in the United States before 1840," *Journal of Economic History [JEH]*, XXVIII (1967), 151–97.

[16] Robert W. Fogel, *Railroads and American Economic Growth* (Baltimore, 1964); G. N. von Tunzelmann, *Steam Power and British Industrialization to 1860* (London, 1978).

[17] Barry Supple, *Commercial Crisis and Change in England 1600–1642* (Cambridge, 1959); D. C. Coleman, *The Economy of England 1450–1750* (London, 1977); Pierre Deyon, *Amiens Capitale provinciale* (Paris, 1967), 214ff; Deyon, "La concurrence internationale des manufactures lainières aux XVIe et XVIIe siècles," *Annales*, XXVII (1972), 20–32; Tihomir J. Markovitch, "L'industrie lainière française au début du XVIIIe siècle," *Revue d'histoire économique et sociale*, XLVI (1968), 550–79; Deyon, "Variations de la production textile aux XVIe et XVIIe siècles," *Annales*, XVIII (1963), 939–56; Domenico Sella, "The Rise and Fall of the Venetian Woollen Industry," in Brian Pullan, ed., *Crisis and Change in the Venetian Economy* (London, 1968), 106–26; Carlo Poni, "Archéologie de la fabrique. La diffusion des moulins à soie 'alla bolognese' dans les Etats vénitiens du XVIe au XVIIIe siècle," *Annales*, XXVII (1972), 1475–96; Waclawa Dlugoborskiego, ed., *Historia Slaska*, Vol. II 1763–1850, Part I 1763–1806 (Wroclaw, 1966); Mariusz Kulczykowski, "En Pologne au XVIIIe siècle. Industrie paysanne et formation du marché national,"*Annales*, XXIV (1969), 61–69; id., *Andrychowski ... XVIII i XIX w.* (Wroclaw, 1972); L. A. Clarkson, *The Pre-Industrial Economy in England 1500-1750* (New York, 1972).

[18] Charles Tilly, *The Vendée* (Cambridge, Mass., 1964); Joan Scott, *The Glass Workers of Carmaux* (Cambridge, Mass., 1974); Yves Lequin, *Les ouvriers de la région lyonnaise 1848–1914* (Lyon, 1977).

[19] See the earlier work of John Clapham, *Economic History of Modern Britain* Vol. 2 (1932), 24–29; Raphael Samuel, "Workshop of the World, Steam Power and Hand Technology in Mid-Victorian Britain," *History Workshop Journal* 3 (1977), 6–72; Jean-Pierre Houssel, "Essor des villes manufacturières de l'habillement et industrialisation spontanée dans 'l'Italie du milieu'," *Revue de géographie de Lyon*, XLVII (1972), 361–83.

[20] *Encyclopédie ou dictionnaire raisonné des sciences, des arts et des métiers* 1751–72, ed. Alain Pons (Paris, 1963), 420–26.

[21] Peter Mathias, "The Social Structure in the Eighteenth Century," *[EHR]*, X (1957), 30–45; See the forthcoming work of Peter Lindert and Jeffrey G. Williamson.

[22] This and the following analysis is borrowed from Thomas S. Kuhn, *The Structure of Scientific Revolutions* (Chicago, 1972).

[23] Mendels, "Proto-industrialization: The First Phase of the Industrialization Process," *JEH*, XXXII (1972), 241–61.

[24] D. C. Coleman, *The Economy*, 1, 2 uses "pre-industrialized" in quotation marks. F. Braudel, *Civilisation matérielle, économie et capitalisme* Vol. 2 (Paris, 1979), 259ff, uses "*préindustrie*". Eric L. Jones and S. J. Woolf, "The Historical Role of Agrarian Change in Economic Development," in Jones and Woolf, eds., *Agrarian Change and Economic Development* (London, 1969), 10, 11.

[25] Herman Freudenberger and Fritz Redlich, "The Industrial Development of Europe: Reality, Symbols, Images," *Kyklos*, XVII (1964), 372-402. That term was later adopted by S. D. Chapman.

[26] Charles and Richard Tilly, "An Agenda for European Economic History in the 1970s," *JEH*, XXXI (1971); Mendels, "Industrialization and Population Pressure in Eighteenth Century Flanders," Ph.D. dissertation, University of Wisconsin, now published, with minor revisions (New York: Arno Press, 1981).

[27] Peter Kriedte et al., *Industrialisierung vor der Industrialisierung* (Göttingen, 1975); Kriedte et al., *Industrialization before Industrialization* (Cambridge and Paris, 1981).

[28] David Levine, *Family Formation in an Age of Nascent Capitalism* (New York, 1977). Medick also published a widely-cited article, "The Proto-industrial Family Economy: The Structural Function of Household and Family during the Transition from Peasant Society to Industrial Capitalism," *Social History* 1 (1976), 291–315.

[29] Mendels, *Industrialization*, 5, 6.

[30] ibid., 7, 8.

[31] Mendels, "Proto-industrialization," 241.

[32] Eric Jones, "The Agricultural Origins of Industry," *Past and Present*, No. 40 (1968), 58–71.

[33] Mendels, "Aux origines de la proto-industrialization," *Bulletin du Centre d'histoire économique et sociale de la région lyonnaise* (1978) no. 2, 1–18; id., "Agriculture and Peasant Industry in Eighteenth Century Flanders," in Eric Jones and W. N. Parker, eds., *European Peasants and Their Markets* (Princeton, 1975), reprinted in Kriedte et al., *Industrialization*; id., "Seasons and Regions in Agriculture and Industry during the Process of Industrialization," in Sidney Pollard, ed., *Region und Industrialisierung* (Göttingen, 1980), 177–89.

[34] R. M. Hartwell, "Economic Growth in England before the Industrial Revolution. Some Methodological Issues," *JEH*, XXIX (1969), 13–31.

[35] Pierre Deyon and F. Mendels, "Proto-industrialisation: théorie et réalité," published in French in *Revue du Nord*, LXIII (1981), 11–19 and in *Quaderni Storici*, no. 44 (1980), 747–52.

[36] Mendels, "Seasons and Regions," and "Les temps de l'industrie et les temps de l'agriculture. Logique d'une analyse régionale de la proto-industrialisation", *Revue du Nord*, LXIII (1981), 21–33.

[37] David Herlihy. "Urbanization and Social Change," *Seventh International Economic History Congress Edinburgh* (Edinburgh, 1978), 59.

[38] Domenico Sella, *Crisis and Continuity. The Economy of Spanish Lombardy in the Seventeenth Century* (Cambridge, Mass., 1979), 136.

[39] Mendels, *Industrialization*, 26ff, 152ff, 180ff; Mendels, "Seasons and Regions".

[40] Mendels, "Proto-industrialization," 245.

[41] id, "Seasons and Regions," 183; "Aux origines".

[42] Thomas C. Smith, *Nakahara, Family Farming and Population in a Japanese Village, 1717-1730* (Stanford, 1977).

[43] Smith, *Nakahara*, 97–8; Carl Mosk, "The Evolution of Premodern Demographic Regimes: A Research Note," *Explorations in Economic History*, XVIII (1981), 199–209; Mosk, "Nuptiality in Meiji Japan," *Journal of Social History*, XIII (1980) 474–89; Mosk, "The Evolution of Pre modern Demographic Regime in Japan," *Population Studies*, XXXV (1981), 28–52; Akira Hayami, "Class Differences in Marriage and Fertility among Tokugawa Villagers in Mino Province," *Keio Economic Studies*, XVII (1980), 1–16.

[44] Mendels, "Proto-industrialization," 241, 245, 246.

[45] Kriedte, *Industrialization*, 135–60.

[46] Luc Schepens, *Van vlaskutser tot Franschman. Bijdrage tot de geschiedenis van de Westvlaamse plattelandsbevolking in de negentiende eeuw* (Bruges, 1973).

[47] Mendels, "Proto-industrialization."

[48] Mendels, ibid., 253, 257.

[49] in Pierre Léon, ed., *Histoire économique et sociale du monde*, Vol. 3 (Paris, 1978), 236.
[50] ibid, 238.

[51] P. Curtin, Unpublished lecture, Johns Hopkins University, Baltimore, s.d.

[52] Fernand Braudel, *Civilisation matérielle, économie et capitalisme*, Vol. 3 (Paris, 1979), 439.

[53] Paul Bairoch, *Le tiers-monde*, 9, 116. Cf. also his earlier *Révolution industrielle et sous-développement* (Paris, 3rd ed., 1969) and, "The Main Trends in National Economic Disparities since the Industrial Revolution," in P. Bairoch and Maurice Lévy-Leboyer, eds., *Disparities in Economic Development since the Industrial Revolution* (New York, 1981), 7–9.

[54] in P. Léon, *Histoire économique*, 255–73. Cf. also Valensi, *Le Maghreb avant la prise d'Alger* (Paris, 1969), 54–57.

[55] John C. Fei and Gustav Ranis, *Development of the Labor Surplus Economy* (Homewood, Ill., 1964). A. V. Chayanov, *The Theory of Peasant Economy* (Homewood, Ill., 1966). F. Mendels, "La composition du ménage paysan en France au XIXè siècle: une analyse économique du mode de production domestique," *Annales*, XXXIII (1978), 780–802. Kriedte et al., *Industrialization*, 40ff.

[56] Chie Nakame, *Japanese Society* (Berkeley, 1970), 7. Peter Laslett, ed., *Household and Family in Past Time* (Cambridge, 1972), 429–544. T. C. Smith, "Pre-Modern Economic Growth: Japan and The West," *Past & Present*, No. 60 (1973), 127–60. Kang Chao's views are developed in Chao, *The Development of Cotton Textiles Production in China* (Cambridge, Mass., 1977) and Chao, "The Growth of a Modern Cotton Textile Industry and the Competition with Handicrafts," in Dwight Perkins, ed., *China's Modern Economy in Historical Perspective* (Stanford, 1975), 167–201.

[57] Shunsaku Nishikawa, "Productivity, Subsistence, and By-Employment in Mid-Nineteenth Century Chōshū," *Explorations in Economic History*, XV (1978), 70.

[58] See the references in Mendels*.

[59] J. D. Chambers, *The Vale of Trent 1660–1800*, (Cambridge, 1957). Rudolf Braun, *Industrialisierung und Volksleben* (Zurich, 1960). Braun, "Proto-industrialization and Demographic Changes in the Canton of Zürich," in Charles Tilly, ed., *Historical Studies of Changing Fertility* (Princeton, 1978), 289–334; Paul Deprez, "The Demographic Development of Eighteenth Century Flanders," in D. V. Glass and D. E. C. Eversley, eds., *Population in History* (London, 1965), 608–30. John T. Krause, "Some Aspects of Population Change 1690–1790," in E. L. Jones and G. E. Mingay, eds., *Land, Labour and Population in the Industrial Revolution* (London, 1967), 187–205. Mendels, "Industry and Marriage in Flanders before the Industrial Revolution," in Paul Deprez, ed., *Population and Economics*

(Winnipeg, 1970). David Levine, *Family Formation in an Age of Nascent Capitalism* (New York, 1977). David Almquist, "Pre-Famine Ireland and the Theory of European Proto-industrialization," *Journal of Economic History* XXXIX (1979), 699–718. Philippe Guignet, *Mines, manufactures et ouvriers du Valenciennois au XVIIIè siècle* (New York, 1977). Philippe Toutain and Didier Terrier, *Le travail des hommes à Comines au XVIIIè s.* (Comines, 1978). Nils Friberg, "The Growth of Population and its Economic-Geographical Background in a Mining District of Central Sweden 1650–1750," *Geografiska Annaler,* XXXVIII (1956), 395–440.

[60] Richard Smith, "Three Centuries of Fertility, Economy and Household Formation in England," *Population and Development Review,* VII (1981), 595–622. E. A. Wrigley and R. S. Schofield, *The Population History of England 1541–1871* (Cambridge, 1982).

[61] J. Hajnal, "European Marriage Patterns in Perspective," in Glass and Eversley, *Population in History,* 101–46. See also Mendels, "La composistion du ménage paysan en France au XIXè siècle: une analyse économique du mode de production domestique," *Annales,* XXXIII (1978), 780–802; Mendels, "Notes on the Age of Maternity, Population Growth, and Family Structure in the Past," *Journal of Family History,* III (1978), 236–50; Mendels and L. K. Berkner, "Inheritance Systems, Family Structure, and Demographic Patterns in Western Europe (1700–1900)," in Charles Tilly, ed., *Historical Studies,* 209–24.

[62] Levine, *Family Formation.*

[63] Charles Tilly, "Demographic Origins of the European Proletariat," unpublished manuscript, CSRC Working Paper No. 207 (Ann Arbor, 1979).

[64] V. S. Clark, *History of the manufactures in the United States,* Vol. I (New York, 1949). Diane Lindstrom, "The Industrial Revolution in America," in S. Pollard, *Region und Industrialisierung,* 69–88.

[65] Anita B. Baker, "Bigger was not Better: An Economic Appraisal of Kustarnichestvo in the 1920s," unpublished manuscript, 1982. Baker, "Deterioration or Development?: The Peasant Economy of Moscow Province Prior to 1914," *Russian History,* V (1978), 1–23. Citing Adam Kaufman, *Small-Scale Industry in the Soviet Union* (New York, 1962).

[66] Jean Yver, *Egalité entre héritiers et exclusion des enfants dotés* (Paris, 1966), 100.

[67] Berkner and Mendels, "Inheritance Systems," John Habakkuk, "Family Structure and Economic Change in Nineteenth-Century Europe," *JEH,* XV (1955), 1–12. Lutz K. Berkner, "Rural Family Organization in Europe," *Peasant Studies Newsletter* I (1972), 145–55. Martine Segalen, *Nuptialité et alliance* (Paris, 1972).

[68] Maurice Le Lannou, *Géographie de la Bretagne,* I (Rennes, 1950), 1–14. II (Rennes, 1952), 69–93. Mendels, *Industrialization,* 40ff. Serge Chassagne, "Aspects des phénomènes d'industrialisation et de désindustrialisation dans les campagnes françaises au XIXè siècle," *Revue du Nord,* LXIII (1981), 35–58.

[69] Bairoch, *Le tiers-monde.*

[70] Louise Tilly, "The Food Riot as a Form of Political Conflict in France," *Journal of Interdisciplinary History,* II (1971), 23–57.

[71] Margaret Masterman, "The Nature of Paradigm," in I. Lakatos and A. Musgrave, eds., *Criticism and the Growth of Knowledge* (Cambridge, 1972), 65.

[72] Robert Lafont, *Décoloniser en France* (Paris, 1971). Michael Hechter, *The Celtic Fringe* (Berkeley, 1978).

[73] John Hicks, *A Theory of Economic History* (Oxford, 1969), 165.

[74] Cf. Pierre Jakez Helias, *Le cheval d'orgueil. Mémoires d'un breton du pays bigouden* (Paris, 1975). Eugen Weber, *Peasants into Frenchmen* (Stanford, 1976).

[75] "De-industrialization in India in the Nineteenth Century: Some Theoretical Implications," Amiya Kumar Bagchi, *Journal of Development Studies*, XII (1976), 135–64.

[76] Bairoch, *Le tiers-monde*, 291.

[77] W. Galenson and H. Leibenstein, "Investment Criteria, Productivity and Economic Development," *Quarterly Journal of Economics* (1955), 343–70.

[78] Richard Eckaus, ed., *Appropriate Technologies for Developing Countries* (Washington, DC, 1977). E. F. Schumacher, *Small is Beautiful* (New York, 1973). For a Soviet discussion, see S. A. Kuz'min, *The Developing Countries. Employment and Capital Investment*, transl. A. Schultz (New York, 1969).

[79] Samuel P. S. Ho., "Decentralised Industrialization and Rural Development: Evidence from Taiwan," *Economic Development and Cultural Change*, XXVIII (1979), 95. Cf. *Rural Enterprise and Nonfarm Employment. A World Bank Paper* (Washington, DC, 1978).

[80] Houssel, "Essor des villes manufacturières". Véronique Maurus, "Hongkong à l'italienne," *Le Monde* (Paris) August 31, 1980. Michel Battiau, "Comment fonctionne le modèle d'industrialisation dans le Sud-Est asiatique," *Le Monde* (Paris), Nov. 4, 1980, 27.

Appendix

List of the Contributed Papers

Rafael Aracil and García Bonafé, "La industria rural en España a fines del siglo XVIII."

Xan Carmona Badía, "L'industrie rurale en Galicie. XVIIIè et XIXè siècles."

Maxine Berg, "Proto-industry, Political Economy, and the Division of Labour 1700–1800."

Duncan Bythell, "From Handloom to Powerloom: Corporation and Co-operation in the English Cotton Industry in the mid-Nineteenth Century."

Kang Chao, "Textile Production in Traditional China."

Salvatore Ciriacono, "Echecs et réussites de la proto-industrialisation dans la Vénétie: le cas du Haut-Vicentin (XVIIè-XIXè siècles)."

L. A. Clarkson and Brenda Collins, "Proto-Industrialization in an Irish Town: 1820–21."

Alain Dewerpe, "Genèse protoindustrielle d'une région développée: l'Italie septentrionale (1800–1880).

Herman Diedericks, "Proto-industrialization in the Netherlands. Some observations."

Toby L. Ditz, "Proto-industrialization and the Household Economy in the American North: Inheritance Patterns in Five Connecticut Towns, 1750–1820."

Agustín González Enciso, "Economic Structure of Cameros' Dispersed Industry. A Case Study in Eighteenth Century Castillian Textile Industry."

Herman Freudenberger, "Proto-Industrialization and Protofactories."

Gay L. Gullickson, "Proto-industrialization, Proletarianization and the Sexual Division of Labor."

Myron P. Gutmann, "How Many Phases? Continuity and Change in the Development of European Industry."

Paul H. Hohenberg, "Toward a Model of the European Economic System in Proto-industrial Perspective, 1300–1800."

Samuel P. S. Ho, "Small-Scale Rural Industries in Contemporary Economic Development: The Cases of South Korea and Taiwan."

Samuel P. S. Ho, "The Simple Analytics of Protoindustrialization and Deindustrialization."

Gerd Hohorst, "Protoindustrialisierung im Übergang zum industriellen Kapitalismus: Die demoökonomische Entwicklung im Kreis Hagen 1817 bis 1863."

Pat Hudson, "The Environment and the Dynamic of Proto-industry: Some Considerations."

Christopher H. Johnson, "Proto-industrialization and De-industrialization in Languedoc: Lodève and its Region, 1700–1870." [May 1981, 49pp., was used].

Marion Johnson, "Proto-industrialization in West Africa."

Çağlar Keyder, "Proto-industrialization and the Periphery. A Conceptual Enquiry."
Peter Kriedte, "Die Stadt im Prozesz der europäischen Proto-Industrialisierung."
René Leboutte, "Industrialisation et évolution démographique. Approche d'une analyse démographique differentielle: Le cas de la région liégeoise au XIXè s."
John S. Lyons, "Competitiveness or Complementarity in the Survival of Proto-industry?"
Wolfgang Mager, "Gesellschaftsformation im Übergang: agrarisch-gewerbliche Verflechtung und soziale Dynamik in Ravensberg während der Frühen Neuzeit (16. Jh. - 1. Hälfte 19. Jh.)."
Hans Medick, "'Freihandel für die Zunft.' Ein Kapitel aus der Geschichte der Preiskämpfe im württembergischen Leinengewerbe des 18. Jahrhunderts."
Franklin Mendels, "Faut-il modifier le modèle flamand?"
D. R. Mills, "Proto-industrialization and Social Structure: The Case of the Hosiery Industry in Leicestershire, England."
Shunsaku Nishikawa, "Protoindustrialization in the Domain of Chōshū in the Eighteenth and Nineteenth Centuries."
E. Oxenbøll, "Proto-industrialization in Denmark."
M. R. Palairet, "Woollen Textile Manufacturing in the Balkans 1850–1911. A Study of Protoindustrial Failure."
Frank Perlin, "Some Central Problems Concerning the Proto-industrialization Thesis and Pre-colonial South Asia." [Third Draft, 97pp., was utilized].
Carlo Poni, "A Proto-industrial City: Bologna: XVI-XVIII Century."
Richard L. Rudolph, "Agricultural Structure and Proto-industrialization in Russia: Economic Development with Unfree Labor."
Debra A. Sandel, "Spinning a New Tale: The Belfast Cotton Industry."
Osamu Saito, "Population and the Peasant Family in Proto-industrial Japan."
Jürgen Schlumbohm, "Agrarische Besitzklassen und gewerbliche Produktionsverhältnisse: Groszbauern, Kleinbesitzer und Landlose als Leinenproduzenten im Umland von Osnabrück und Bielefeld." [18pp., s.d., was utilized]
Lennart Schön, "Swedish Textile Production 1820–1870: Protoindustrialization and Mechanization."
Helga Schultz, "Die Ausweitung des Landhandwerks vor der industriellen Revolution. Begünstigende Faktoren und Bedeutung für die 'Protoindustrialisierung'."
Paul Servais, "Industries rurales et structures agraires: le cas de l'Entre-Vesdre-et-Meuse aux XVIIIème et XIXème siècles."
Michael Sonenscher, "Solomon's Temple: Notes upon Work and the Historiography of Industry in the Eighteenth Century."
Margaret Spufford, "English Petty Chapmen and their Wares in the later Seventeenth Century."
Charles Tilly, "Flows of Capital and Forms of Industry in Europe, 1500-1900."
Chr. Vandenbroeke, "Analyse critique de la phase proto-industrielle en Flandre: L'évolution sociale et le comportement démographique, 17è siècle — début du 19è s."
Wolfgang von Stromer, "Textil-Landschaften vom Frühmittelalter bis an die Schwelle der Industriellen Revolution."

107

An Economic Theory of the Growth of the Western World[1]

BY DOUGLASS C. NORTH and
ROBERT PAUL THOMAS

THE object of this article is to present a new explanation for the economic rise of the Western world. While the model presented below has equally radical implications for the study of contemporary economic growth, our focus is upon the economic history of the nations that comprise the North Atlantic nucleus between 1100 and 1800. In capsule form our explanation is that changes in relative product and factor prices, initially induced by Malthusian population pressure, and changes in the size of markets induced a set of fundamental institutional changes which channelled incentives towards productivity-raising types of economic activity. By the eighteenth century these institutional innovations and accompanying changes in property rights built productivity changes into the system enabling Western man to finally escape the Malthusian cycle. The so-called "Industrial Revolution" is simply a later surface manifestation of innovative activity reflecting this redirection of economic incentives.

In subsequent sections of this article we shall (1) summarize (mostly from price history and population studies) the general contours of economic changes in Western Europe from 1100 to 1800, (2) provide an economic explanation for these changes, (3) develop a theoretical approach to institutional change, (4) integrate the analysis of economic change with the model of institutional change, and (5) explore briefly the implications of this general model.

I

There exists only fragmentary information on the economic contours of society between 1100 and 1800. We know that sustained economic growth emerged some time before the latter date and has continued ever since, although we are unaware of just when it began. We also know that the world prior to, say, 1500—if not

[1] This essay is a drastic condensation of a forthcoming book by the authors and of necessity has eliminated extensive bibliographical references. We are indebted to Stanley Engerman and to our colleagues Mary Eysenbach, John Floyd, and M. D. Morris for helpful comments on an earlier draft.

typical of what we would call a stationary state—still shows no evidence of having achieved continuously rising *per capita* income.

What follows is a brief summary of our knowledge—fragmentary, incomplete, and occasionally contradictory (particularly with respect to the demographic evidence). We should caution, to begin with, that regional variations marked Western Europe, so that the general contours of population change and price movements described above were not always parallel. Nevertheless, general patterns do emerge; and these appear to offer hard enough evidence to justify an attempt to construct a more general theoretical model with which to interpret this information.

Scholars generally agree that in the twelfth century Western Europe experienced notable expansion both of population and of settlers moving into still good arable land. *Per capita* income probably rose as output increased faster than population. Exactly when Europe became settled and diminishing returns to further increases in population set in is a matter of some controversy; but it appears likely that by the end of the twelfth century population pressure was a significant factor leading to changing relative prices with consequent rising rents and falling real wages. Also a matter of some dispute is the question of when population growth reversed itself, some scholars maintaining that the downward trend began as early as the first decade of the fourteenth century, whereas others hold to the classical explanation of the Black Death between 1347 and 1351 as the turning-point. The relative price data suggest that the former explanation is more credible: that is, real wages appear to be rising and the prices of agricultural products falling, beginning with the early fourteenth century. What is indisputable is that from the middle of the century on, population fell for a substantial period of time as recurrent plagues swept over Western Europe. In England the plague returned in 1360–1, 1369, and again in 1374, and regardless of the relative impact of each it is clear that collectively the plagues throughout the rest of the century were continually checking population growth and probably causing an actual decline over a substantial period. During this disturbed time land rents were falling, real wages were rising, and agricultural prices were relatively depressed.

Neither demographic data nor relative price data enable us to say exactly when population began again to expand. Since large amounts of good land were available to be put into production, population could grow for some time before relative price changes reflecting diminishing returns in agriculture would signal the onset of population pressure. Some time during the fifteenth century, therefore, a turning-point was reached, although the abundance of available land, and probably some increases in productivity, may for a time have maintained rising *per capita* real income in the face of a growing population; the evidence on this is not conclusive. The classic work of Hamilton has made clear, however, that a general rise in the price level did occur in the sixteenth century. This was accompanied by a significant relative price change, with agricultural prices rising much more rapidly than others, and by higher rents and lower real wages.

It would also appear that this was an era of population expansion accompanied by significantly diminishing returns in agriculture. John U. Nef has suggested that technological change in the latter part of this century brought about some advance in productivity. It is doubtful whether this was of a magnitude to overcome diminishing returns in agriculture. The more likely inference is that increased productivity served merely to damp the Malthusian checks, allowing a larger total population to exist before the inevitable population pressures resulted, in the seventeenth century, in another crisis with population outstripping the rate of growth in output. If population did not absolutely decline in Western Europe during this century, it appears at best to have stagnated.

During this period there is clear indication that productivity change occurred in a number of sectors, the most apparent being substantial improvements of productivity in transportation. Not only was the Canal du Midi constructed in the seventeenth century; more impressively, unequivocal evidence points to an improvement of efficiency in ocean transportation leading to falling real costs in the movement of goods. The extensive grant of patent monopolies in England in both the sixteenth and the seventeenth centuries probably paid off in productivity increase due to a variety of innovations brought over from the Continent.

Another turning-point appears to have come somewhere in the latter part of the seventeenth century—perhaps 1680 is as good a date as any. Again, from 1680 to the mid-eighteenth century substantial population growth occurred on the Continent. Prices remained relatively stable, and there appears to have been some rising *per capita* real income during the period. By now, one sees indisputable evidence of widespread improvement in productivity, not only in the form of continuing gains in transportation (particularly in shipping) but also as a result of widespread changes in the character of agriculture. These have been highlighted in many regional studies in agriculture, pointing, in the cases of Provence and Languedoc, for example, to shifts into viticulture and silk cultivation. Regional price differentials for agricultural commodities were narrowing, indicating an expansion from local to regional, to national, and finally to international markets. The growth was destined to extend throughout the eighteenth century, and it is perfectly evident that the productivity increase and expansion which were to become watchwords of the Industrial Revolution were already in effect. In short, sustained growth had begun in Western Europe.

II

An explanation of this demographic, price, and productivity behaviour can be summarized as follows. Before the middle of the twelfth century growth in population led to new colonization and settlement, and output grew as rapidly and probably more rapidly than population, so that *per capita* income increased. The urge to colonize and settle new areas would develop when relative overpopulation and diminishing returns in an existing settlement led to a falling off in well-being. This process could be slowed down by farming the given settlement more intensively, but as population continued to grow it would ultimately induce people to move out into new "frontier" lands. The spread of population through Western Europe produced a powerful inducement to expand trade since different parts of Europe had different resource endowments and, with different densities of population, also different factor proportions. To take but two examples, viticulture was possible in many new areas and fine wine increasingly in demand throughout most of Europe. The result was an expansion of the wine trade. Manufacturing—in particular of wool cloth—developed in the more densely settled areas and the cloth trade grew. Thus, interregional trade was a natural consequence of the differential factor endowments concomitant with the spread of settlement in Europe as a consequence of population growth.

However, as Europe became settled and population continued to increase, generally diminishing returns set in, real wages fell, and ultimately as the standard of life declined the population became increasingly susceptible to famine and disease. The result was the disastrous events of the fourteenth century with the bottom of the cycle reached some time in the fifteenth century and the whole pattern started all over again at that point. However, the second cycle is considerably dampened compared to the first, not only because settlement moves beyond the frontiers of Europe to the New World thereby easing the population pressure but also, more particularly, because an accelerating rate of increase in productivity now permits rapidly expanding output to ameliorate the diminishing returns in agriculture; the trough of the second cycle is consequently far less devastating. Indeed, by the time a third cycle of population expansion begins, productivity increase has become built into the system, and output continues to grow more rapidly than population.

The inception of this sustained increase in productivity can be accounted for only by a theory of institutional change, and this is to be developed in the following section of this essay; but the pattern of population change, relative price movements, and the growth of interregional trade can be explained in terms of a strictly economic model which we shall call a Malthusian staple-thesis model. While this model oversimplifies the evolution of Western Europe that we wish to describe, the addition of complicating factors (such as some rate of technological change) does not appear to affect adversely its explanatory power.

The model consists of two sectors, a non-agricultural sector subject to constant returns to scale and an agricultural sector subject to diminishing returns as population in any given area of settlement continues to grow. However, as long as there exists available empty land of equal quality, the build-up of population and consequent diminishing returns in the settled areas can be relieved by colonization and settlement of new land. The factor endowments of areas as population grows thus become increasingly differentiated. In old settled areas, labour and capital (both physical and human) are relatively abundant, leading to the expansion of the non-agricultural sector. In the newly settled areas, differential resource patterns (from varied climates, soils, etc.) produce varied agricultural and mining output and the gains from trade are increasingly enhanced. The consequent specialization and interregional trade will themselves result in productivity increase, but cannot account for the sustained growth of productivity needed to break out of the Malthusian trap when all the available land becomes settled and general diminishing returns set in. In order to break out of the Malthusian cycle, institutional changes must occur which by incentive direct man's efforts towards technological change and sustained productivity growth.

THEORY OF GROWTH 5

III[1]

Economic institutions, and specifically property rights, are generally considered by economists as parameters; but for the study of long-term economic growth they are clearly variables, historically subject to fundamental change. The nature of existing economic institutions channels the behaviour of individuals within the system and, in the process, determines whether the aggregate result is to be economic growth, stagnation, or decay.

Before we proceed farther, a definition is needed. Common usage of the term "institution" defies precise meaning since it has variously referred to an organization (such as a bank), to the legal rules that govern the economic relations between people (private property), to a person or position (king or monarch), and sometimes to a particular document (the Magna Charta). For our purposes, we shall define an "institution" or an institutional arrangement (which is a more accurately descriptive term) as an arrangement between economic units that defines and specifies the ways by which these units can co-operate or compete.

As in the more familiar case of the introduction of a new product or process, economic institutions are innovated because it appears profitable for individuals or groups in society to undertake the costs of bringing about such changes. The purpose is to capture for the innovator some profit unattainable under the old arrangements. The essential requirement for initiating an institution (or a product) is that the discounted expected gains exceed the expected costs of the undertaking; only when this condition is met would we expect to find attempts being made to alter the existing structure of institutions and property rights within a society. We shall explore successively the nature of potential gains and potential costs in such innovation, and then discuss the economic forces that would alter the ratio of these costs and gains through time.

Potential profit from a number of general sources can be captured by institutional innovation, whether of a voluntary sort or undertaken by use of the coercive powers of government. To accomplish voluntary organizational change, a group of individuals may choose to place themselves under a common set of rules governing their behaviour. Since each member retains the right to withdraw, the organization will exist only so long as it serves the interests of the group. Partnerships, corporations, and co-operatives are examples of such organization. Differing fundamentally are organizations under government aegis, since in this case the force or coercion of government may be exerted to require specific performance and the individual may no longer withdraw from membership and the coercive consequences of the policies.

Potential benefits that historically have induced institutional innovations and thereby actually raised the productivity of society result from (1) economies of scale, and (2) reduction of transactions costs, which can best be viewed as the costs of operating an economic system. The latter in turn can be separated into

[1] The intellectual origins of the theory set out in this section can be traced to W. F. Baumol, *Welfare Economics and the Theory of the State* (Cambridge, Mass., 1952); Anthony Downs, *An Economic Theory of Democracy* (New York, 1957); J. Buchanan and G. Tullock, *The Calculus of Consent* (Ann Arbor, Michigan, 1962); Harold Demsetz, 'Towards a Theory of Property Rights', *American Economic Review* (Proceedings, May 1967); Kenneth Arrow, 'Political and Economic Evaluation of Social Effects and Externalities', Universities National Bureau Conference on the Economics of Public Output (unpublished, cited by permission of the author).

the costs of externalities, of information, and of risk. Both the realization of economies of scale and the reduction of transactions costs have been significant sources of profit to groups undertaking institutional rearrangement.

Economies of scale are essentially a technological phenomenon and reflect the fact that the most efficient (lowest cost per unit) output may entail a size that requires more complex organization than the single proprietor. The growth of chartered companies, joint-stock companies, the corporation, and the legal provision of limited liability provide classic examples of how institutions and property rights were created to permit individuals to capture the profits inherent in economies of scale. Such organizations were first sanctioned by government, then later became voluntary, within a set of property rights designed to allow businesses to organize on a large scale.

Among the transactions costs mentioned above, externalities have received the most attention. By an externality we mean that some of the benefits (or costs) in a transaction accrue to other than the directly participating parties. In modern times we think of the downstream benefits of a dam, or, historically, of pirates and privateers preying upon commerce, as familiar examples. An even more significant historical example is the inability of an entrepreneur to capture the gains from an innovation without the legal protection of intellectual property. In each case, the profits to the potential dam-builder, trading merchant, or innovating entrepreneur are lower than they could be by appropriate institutional rearrangement. A change in tax structure or property ownership could permit the dam-builder to capture downstream benefits. A naval force could wipe out pirates, thereby seeing that merchants rather than pirates reaped the gains from trade, and an enforceable patent law would see that an innovator captured the gains from his innovation. The internalization of an externality by reorganizing property rights, institutional innovation, or the policing of markets not only will raise the private profitability of the activity but, as in the case of economies of scale, will also improve the efficiency and output of the society by equating the private and social profitability of an activity.

A less explored aspect of transactions costs is that of organizing and improving the flow of relevant economic information, yet this has historically been a major area of institutional innovation.

If there were no costs to acquiring information then prices in all markets would differ only by transport costs. In fact, information is costly and the widespread existence of purely local markets reflected in good part simply the lack of information about profitable trading opportunities in more distant areas. The development of such information meant that merchants would send their products to the markets where net price differentials (i.e. prices adjusted for transport costs) were highest; thus the development of institutions to increase the flow of information increased the merchants' profits. The development of the whole network of trading institutions—merchants, brokers, factors, supercargoes, early exchanges in coffee houses, stock and commodity exchanges—were all directed basically to this end. Eventually, it became profitable to produce and sell this information in such forms as price currents or trade association journals. Like the two previous sources of institutional innovation, these were not only privately profitable to the merchant and entrepreneur but by arbitraging markets were a major source of productivity increase and economic growth in the Western world.

The prevalence of risk—the inability to be certain of the terms on which future transactions can be made—is yet another cost limiting the extent of transactions. The possibility of very high losses on an individual transaction will inhibit such enterprises even though when averaged out over a sufficient number of undertakings the average rate of return may be very high indeed. The creation of institutions to spread risks would clearly make it possible for entrepreneurs to realize these higher rates of return without the risk of disastrous consequences when an individual venture failed. The early share-partnership in shipping and trade, the joint-stock company, and insurance associations (such as Lloyd's of London) were created precisely to capture the profits from such high-risk ventures. Since the high potential profits mirrored the social rate of return, the development of such institutions to spread risk also led to productivity increase. We should note in passing that many institutional innovations were created both to increase information and to spread risks (many types of financial institutions, futures markets, etc.).

The above types of innovation are associated with institutional change which also improves the efficiency of the economy. Yet, quite clearly, this is not the only kind of individually profitable institutional innovation that may occur. Institutional arrangements may also be created, either by voluntary groups or by the government, which are designed to capture gains for individual groups at the expense of others (or, in short, to redistribute income).[1] We may conceive of any group which is organized to control supply effectively as being able to redistribute income in its favour. The guild system may provide one example; or we can note various attempts through coercive power of government to undertake such activities: typical are the several attempts to fix wages following the era of the Black Death or, to take a different example, the Poll Tax of 1381 in England.

Institutional arrangements are not costless to create. The cost of forming both voluntary and governmental associations varies directly with the numbers involved; the greater the number that must be brought into agreement, the greater are the costs incurred.[2] Since voluntary organization presumably also implies the privilege of voluntary withdrawal, such organization must in effect exist under a unanimity rule.[3]

However, in considering organization by government one must look carefully at the potential costs. Until a more sophisticated theory of the state is developed by political scientists, we can only specify at discrete points in time the structure of political power and decision-making rather than provide a dynamic theory of political change. In a democracy the costs of effecting a particular policy will typically be those of obtaining the support of 51 per cent of the populace under majority rule. But when we go back in time to explore the issues that concern us in this essay, decision rules with respect to political change are more difficult to as-

1 Institutional innovations designed to realize the gains inherent in economies of scale or to reduce transactions costs will probably also redistribute income within the economy, but this effect is a by-product of the primary objective. On the other hand, institutional innovations designed primarily to redistribute income will generally not raise society's output and will usually lower it.

2 This very brief discussion of the costs of organization is amplified in Buchanan and Tullock, op. cit. ch. 8.

3 This means that the participant must view the cost of withdrawal as exceeding the cost of the policy with which he disagrees.

certain. This is not to deny that changes could be effected in political policy at some particular cost. The most obvious method that comes to mind is the bribing of kings and princes who were notoriously hungry for funds to run their governments. Historically, we observe many innovational changes accomplished by the use of discreet bribery, sometimes in the guise of loans. But for these earlier forms of political change we do not yet possess even the embryo of a general theory such as we are beginning to achieve for changes occurring under a political democracy.[1] Therefore, so far we must conclude that decision rules must have varied widely over time, and we are unable at this point to specify them accurately. We must simply assume the ability of groups in society through time to modify decision rules or to modify the political process.

We should note an additional cost incurred in the political process but absent in the voluntary system: once agreement has been reached to pass a law, all citizens become subject to that law whether they approve of it or not. These are the costs of coercive agreement. For example, when the peasants of feudal Europe entered into the manorial system—providing labour services in exchange for "protection and justice"—they were from that point on subject to the coercive authority of the lord and had not the right to dissolve the bond even though the initial benefits might have long ago ceased to exist. The familiar story of the protracted death of feudalism aptly illustrates the potential cost of a coercive agreement. The extension of the authority of government always has potential costs in terms of its coercive power being utilized in ways that may injure even those members of society who may originally have promoted the extension of government authority.

What will decide whether groups choose voluntary rather than governmental means of implementing institutional innovation? The answer must be the relative benefits and costs of the alternative methods, and our brief outline above suggests some of the benefits and costs that might exist. We may note that the social costs appear to be greater in general for governmental organization; but we should also realize that in many instances the benefits will be still greater. The most obvious reason for this is the coercive power which permits government to undertake policies even though they may be strongly objected to by a part of the society. Coercive power is exerted through the capacity to collect taxes, to enact regulatory legislation, and to institute eminent domain proceedings. The relative benefits and costs will be particularly influenced by the degree of development of government "infrastructure" as compared to the degree of development of the market. The era of "mercantilism" represents a period when the government structure was relatively well developed and the markets were small, fragmented, and often monopolized. Merchants and entrepreneurs turned to government to open up markets, to break down local monopolies, and to provide naval and military protection for trade; in short, to undertake the institutional innovations essential to internalizing externalities and reducing transactions costs.

So far our exploration of the basis for institutional innovation has not explained why such innovation tends to occur. But we must remember that presumably ex-

[1] Even for political democracies the theory is tentative and a long way from satisfactory; Buchanan and Tullock, op. cit. and Downs, op. cit. have made a beginning towards exploring aspects of this part of the problem.

ternalities or transactions costs exist at any given moment because no group in society judges it profitable to internalize the externality or to arbitrage out the market imperfections associated with other forms of transactions costs. Equilibrium among economic institutions exists when the costs of change (to the individual or group) exceed potential benefits. Therefore, to induce institutional innovation, some disequilibrium must occur in the system that adds weight to the profit side in the judgement of the decision-making individuals or groups, i.e. the perceived benefits must have risen, the costs fallen, or both.

Several factors appear to have created disequilibrium in the social structure of the Western world. The ones that we will subsequently focus on in this essay are: (1) long-run changes in relative product and factor prices, (2) changes in the size of the market, and (3) changes in the decision rules of government.[1]

Long-run changes in relative factor and product prices have led to fundamental institutional changes. For instance, changes in the relative value of land and labour altered the profitability of "owning" one versus the other and destroyed the basic *raison d'être* of feudalism. Rising agricultural prices increased the value of land and made it profitable to develop exclusive ownership of land (i.e. develop private property in land). With growing population and a relatively fixed supply of land, labour services became decreasingly valuable, thus providing the basis for the development of a free labour force. When relative prices reversed, the pressures for institutional change worked in the opposite direction. Changes in relative prices also led to an expanded basis for interregional trade and caused other changes, to be elaborated in Section IV below (pp. 10–16).

The increasing size of the market has historically weighed heavily as a force for institutional innovation. Transactions costs tend to be subject to economies of scale. The cost of getting additional information about prices may be independent of the size of the market involved, and a larger potential market makes it worth while to acquire information which would have been unprofitable in a very small market. This would appear to be true with respect both to improving information and to reducing externalities in such cases as the policing of piracy or brigandage.

For example, the cost of establishing a resident factor in a foreign port is approximately the same whether trade with that area is of a value of £1,000 or £1 million annually. The reduction in costs through better information on prices and quantities may not be worth while in the first case, but is more likely to be so in the second. Similarly, when trade with the West Indies was small, the costs of eliminating piracy were simply too great relative to the benefits to be undertaken; but at some point, the increased volume of trade warranted the expenditure.

Changes in decision rules of the government are a factor which can be referred

[1] While changes in factor prices resulting from population growth and changes in decision rules of government are exogenous, changes in market size are partly endogenous (from population growth and relative price changes) and partly exogenous (from political and military actions). Other forces which produced disequilibrium were changes in the norms of behaviour—such as the willingness of the populace to accept private property—and innovations. The acceleration of innovation is endogenous to our model, but some innovation would and did occur even in the absence of institutions which enabled the innovator to capture a large share of the benefits of the innovation. They were generally confined to chance discoveries or innovation that involved few search costs or risks. The pace of such innovative activity was so slow that the consequent productivity increase was subsequently swamped by population growth. The introduction of the three-field system is a typical case in point.

to only briefly—at least in this essay. But, clearly, these must play a large part in any overall examination of the evolution of the Western world. The relative power of the crown, of feudal lords, and of the towns—and changes in their relative power over time—were in fact changes in decision rules which basically altered the costs of calling on government to implement economic policy. At present, we simply lack any body of theory to explain these changes.

It should now be evident that we are talking about two different levels of institutions. There are the fundamental institutions that specify the basic "ground rules" such as the underlying "constitutional" basis of property rights and basic decision rules with respect to political decision-making, and then there are the secondary institutional arrangements which may be created without altering the basic institutions. The fundamental institutions may be specified in a constitution or may exist by legal precedent or perhaps only by custom. Sometimes the ways in which these fundamental institutions can be changed are specified, as in the rules for amendment of a formal constitution, but more often they are not. The innovation of secondary institutional arrangements may sometimes take place without any political or legal change, but others require political implementation. Historically, many secondary institutional developments have been entirely consistent with existing basic institutions; but the implications of others have been in conflict with the status quo of property rights or of political decision rules. As such secondary innovations take place, political pressures accumulate to induce a reorganization of the fundamental institutions.

As already emphasized, institutional innovation does not necessarily lead to growth. Some institutional changes lead to the redistribution of income and actually reduce output. The history of ancient civilizations as well as of most underdeveloped economies today bears witness that it has often been more profitable for entrepreneurs to develop institutions to redistribute income than to innovate productivity-raising institutions. The essential key to our essay is that cumulative changes in secondary institutional arrangements ultimately led to a restructuring of fundamental institutions in Western Europe over this period, in such a way that individuals in the society were now encouraged to strive for greater productivity in their economic undertakings. New incentives had arisen in the form of increased private profitability of innovation and improvement in the quality of factors of production. To put it more specifically, while the social rate of return on innovation and capital formation (both human and physical) had always been high, changes in organization and property rights increasingly brought the private rate of return accruing to individuals and groups nearer the social rate of return. Growing efficiency was the inevitable result.

IV

The implications of the preceding sections are that basic economic changes have created a potential for profitable opportunities which could be realized by the innovation of economic institutions. In this section, we shall explore these implications in what must, of necessity, be no more than a brief sketch. Our primary objective is to illustrate the promise of this model. The illustrations will mainly be drawn from English experience.

While our model has implications for the way in which feudal institutions evolved, we here examine only the way in which they declined and were replaced. The basic economic characteristics of the medieval world before 1100 were these: (1) abundant land was available for colonization and settlement, of approximately equal quality to that already settled; (2) labour was relatively the scarce factor of production; (3) of the small volume of trade, most was confined to local exchange of goods and services: what long-distance trade existed was based essentially on differentials of resource endowments between regions, occasionally including differentials in specialized human capital investments; (4) the rate of innovative activity appears to have been very low.

The force which upset this economy was population change. The fundamental institutions of feudalism had developed because of the scarcity of labour relative to land, which made it imperative to capture from labour the rents not existing from a ubiquitous supply of land.[1] Population pressure undermined the economic basis for the institutional organization of feudalism by reversing the relationship of prices as a result of diminishing returns and by expanding the size of markets. Increases in population relative to a fixed supply of good land led to agricultural prices rising relative to non-agricultural prices; this in turn increased the value of land and decreased real wages as the output per labourer fell. Growing population, colonization, and consequent different regional factor endowments led to expanding trade. The result was that landlords now found it to their interest to commute labour dues to payments in kind and in cash, and to lease the demesne lands in return for rent.

On the other side of the scale the rising value of land also produced a basic disequilibrium in the medieval world. Land now offered vastly higher returns if only they could be captured by individuals: a continuous pressure arose to eliminate common-property use of land and to achieve private exclusive ownership. The enclosure movement, undoubtedly the most dramatic of the institutional changes induced by this relative price change, resulted in a reorganization of property to permit exclusive ownership. Voluntary enclosures date from precisely this period.

Population pressure also exerted significant effect on the non-agricultural sectors of the economy. Because it was unevenly felt throughout Western Europe, and not felt at all in Eastern Europe, the result was to differentiate increasingly the relative factor endowments of different regions and this, in turn, led to greater reason for trade. However, potential profits from trade were obstructed by high transactions costs due to lack of information about potential markets and to the existence of piracy and brigandage on a widespread scale. Not surprisingly, merchant groups now undertook concerted action to reduce these costs. The development of merchant-trading groups, the establishment of resident factors in different cities to increase information flows, and a drive to eliminate or reduce piracy on major trade routes during this period were all designed to enable merchants to capture the ever-growing gains inherent in long-distance trade. Deposit banking,

[1] Prof. Evsey Domar, in an unpublished essay which he has been kind enough to show us, constructs a model which explains the perpetuation of serfdom in Eastern Europe (in contrast to its demise in Western Europe) in terms of a gentry having the ability to appropriate "rents" from labour when land is ubiquitous. In contrast, in Western Europe, when land becomes scarce, the gentry can appropriate the returns from land. This explanation is consistent with our more general model.

insurance, and predecessors of the corporation were just three examples of institutional arrangements developed to reduce information costs and risks during the period.

The expansion of trade led to two other significant institutional consequences. (1) Formal contracts came to replace the earlier informal agreements, as the size of the market made trade relationships more impersonal; the rise of contractual obligation was an important step in the development of property rights. (2) Existing political administrative units were increasingly inadequate to meet the needs of expanding trade which required a wider scope of fiscal policies, policing protection, and information; consequently, larger political units grew up or consolidated to keep pace with the growing market size.

As discussed in the preceding sections, we do not know exactly when a check in population growth tended to counter these processes. However, the effects of the long decline in population were striking. Relative prices reversed. Land now relatively more abundant became less valuable; labour became more scarce and, as a consequence of competition between lords to acquire labour, more dear. To recapture the loss in rents from the now more valuable labour, the lords attempted to use their political power to reimpose feudal obligations. Peasants and labourers, on the other hand, saw freedom from these obligations as a chance to capture more fully the returns of their own labour, which had now risen strikingly in terms of real wages.

Our model, in these circumstances, would predict a period of rising social tension and of conflict between these groups, requiring an exploration of the structure and distribution of political power of the times to foresee its resolution. To the extent that the political power resided with the lord, he would be able to reimpose feudal obligations; to the extent that it was more widely diffused among diverse political interests, he might be less able to capture again the rents from labour.

With respect to the non-agricultural sector of the economy, the decline in population served to reduce the aggregate volume of trade in two ways: it affected both the total size of the market and the difference between regions in Europe. Thus, in a world of high transactions costs, it reduced the bases for trade. While it is true that *per capita* income increased and the distribution of income shifted in favour of the working classes (which would result in a shift in the composition of demands), it would be surprising if the income elasticity of demand for imports would be sufficient to overcome the forces making for a decline in trade. A further trade-reducing factor would be the higher transactions costs associated with declining volume. In the first section, we sketched how merchants found it profitable to reduce transactions costs by stationing factors in a distant city to acquire information about prices and possible trading opportunities; as the volume of trade shrank, this was no longer expedient.[1] Information flows dried up and trade volume was further reduced. It is thus not surprising that economic historians have found depression (for them meaning a decreased total volume of economic

[1] The Hanseatic League developed in this period not only to consolidate information sources, but particularly as a defensive "institution" to perpetuate the dominance of these northern European cities in the face of declining trade and severe competition from outsiders. Our model would predict that in periods of contracting markets the institutions which would develop would be redistributive in character (such as guilds).

activity) even in the midst of this world where higher *per capita* income would presumably have followed the relatively increased real wage that peasant and worker must have been experiencing.

With the advent of the sixteenth century the changing pattern of relative prices leaves little doubt that diminishing returns had obviously once again set in. Real wages declined dramatically throughout the century. While the decline in real wages and in the value of labour tolled the final and ultimate demise of feudal obligations (since labour became relatively less valuable), it equally rang in the effective rise and fundamental institutionalization of private-property ownership as a consequence of rising land values. The enclosures taking place at this time occurred in areas precisely predicted by the theory we have sketched—in pastures producing raw wool and in areas suitable for truck farms. The former was in response to an expanding demand for raw wool and the second to increases in local demands for foodstuffs by the growing urban areas. The sixteenth-century enclosure movement was most extensive in the highland regions of England because the returns to enclosures were higher there than in the arable regions, for two reasons. First, the areas suited to pasture had a lower population density than did the more arable ones; hence, in terms of our model, fewer people had to reach agreement for the enclosure to occur. Second, and probably more important, the increase in the price of wool would have caused individuals holding land in common to use it inefficiently by each attempting to pasture more sheep. The cost to the individual of pasturing another sheep on the common approached zero, but the cost to society of everyone doing so was clearly positive. The common would tend to become overgrazed and the total output of wool would actually decline—providing a classic example of an externality associated with common-property resources. Individuals with the power to enclose the common could avoid this occurrence by enclosing areas and denying access to all others. The extinguishing of the rights of the common allowed the new owners to increase their private output (and that of society) at the cost of persons excluded from the common, who were obviously worse off. However, since the pasture was now more efficiently used, the costs and benefits to society were brought closer in line with those accruing to the individual.

The arable land areas of southern England were slower to yield to enclosure. The cost of achieving agreement to the process was obviously higher where many persons were involved. Also, potential external diseconomies in these areas were less onerous than in the grazing areas. For one thing, the problems attendant on a common-property resource were less pressing, since here individuals farmed their own strips and captured the benefits directly; it was only on land held fallow or between crops and on the wastelands that common rights existed. Additionally, the reclamation of swamp and wastelands may well have provided a less expensive alternative for expanding output than would the undertaking of arable enclosure during the sixteenth and seventeenth centuries. In any event, the complete enclosure of arable lands had to await the use of the coercive power of government in the eighteenth century as a means of reducing its cost.

Turning again to the non-agricultural sector, we find two basic forces making for a rearrangement of institutions in these centuries. The first, which had already existed in the earlier period, was the divergence in relative prices due to differentiation of factor endowment among different regions in Europe, as well as between Eastern and Western Europe (since the eastern area lacked the pressure and diminishing returns characteristic of Western Europe). Interregional trade flourished as the shifting comparative advantage of different areas offered entrepreneurs new profits. Another influence leading to substantial increases in trade was the inflationary effect of an influx of treasure from the New World. While the emphasis of Hamilton and subsequent economic historians has been on the resultant lag of wages behind prices (a hypothesis which is supported neither by theory nor by evidence), the real consequence of the treasure was its impact upon expansion in the size of markets. The importation of precious metals into Europe via Spain stimulated the growth of all the Western economies by increasing the incentive to trade. The consequence of the expansion of trade was an era of substantial institutional innovation which has commonly been described as a "commercial revolution". Typical of the institutional innovations were the joint-stock company, a variety of sophisticated forms of insurance, and the development of international financial markets—all devised in response to the profitable potential of reducing information costs or spreading risks. Convoys and naval expeditions were organized to combat piracy. Cities like Cadiz, Lisbon, Bordeaux, Rouen, Antwerp, Amsterdam, Bristol, and London gained stature as trading centres, developing commercial and financial institutions to increase the flow of trade. It must be emphasized at this point that the consequence of this was a substantial increase in productivity; that the growth of increased information, as well as the decline of piracy and of other obstacles to trade, lowered the cost of the movement of goods between areas in dramatic fashion. Indeed, just as we look upon the innovator and technological change as being the great sources of productivity change in the period of the Industrial Revolution, we have been remiss in not looking upon the merchant and the provision of increased information as having performed the same function in the era of commercial growth.

A classic example of an institution evolved under the impetus of such newly profitable opportunities is the joint-stock company. This arrangement not only served to spread risk by making shares available to individual owners but it also increased information about trade by stationing resident partners in various locations and at times even protected its participants by its own police power.

The international financial markets which grew up in Antwerp and Amsterdam were also a consequence of the growing profitability of international trade. Expansion and greater profits for the merchants would attend any increased knowledge of the supply and demand for capital. These institutions performed such a function, thereby lowering the cost of capital and expanding trade.

The expansion in the size of the market encouraged innovation in two basic respects: the first was a natural consequence of specialization and the division of labour in the classic sense that Adam Smith described the process. In terms of our model this simply lowered the costs of discovering an innovation. The second was that the expansion in the size of the market raised the potential returns to innovation. However, as long as the innovation was freely available to be adopted by other entrepreneurs, most of the benefits generated could not be captured by the innovator.

The granting of monopoly privilege by the Elizabethan crown to ensure that inventions introduced in Western Europe might profitably be imported to England was a governmental device to encourage the adoption of improved techniques. The way in which such monopoly grants gradually evolved into a patent system embedded in the legal structure of English common law in the Statute of Monopolies of 1624 has been told elsewhere and illustrates the way in which a cumulative set of secondary institutional innovations led to a change in the fundamental institutional structure.[1]

The implications of the Statute of Monopolies far outreached the statute itself. In an earlier period, the power of government had been the most feasible agent to break down such barriers to trade as the threat of piracy and the proliferation of local monopolies; the very basis of mercantilism had been a tacit banding together of government and merchants to realize profits from commercial venturing. By the end of the sixteenth century, markets had widened and the potential gains from expanded foreign and domestic trade had increased to the point where government intervention not only seemed unnecessary but was now actually costly and hence unwarranted. Officially created monopolies now thwarted merchants and traders bent on capturing the growing trade. Even when the granting of monopolies was first inaugurated some grants had been tossed as largesse to court favourites (and were therefore purely redistributive in effect rather than a stimulation to trade). Now all monopolistic grants became increasingly suspect as vehicles of privilege and as devices to redistribute income by denying free access to the profits inherent in expanding markets. Pressures arose to oust the government and its monopolistic grantee groups from their entrenched positions.

The fundamental political conflicts of the seventeenth century reflect this resistance to oppressive influences in trade and commerce; the evolution of the first voluntary business organization, in the form of the joint-stock company, was a logical development. In this context, the crises of the seventeenth century and the revolutionary period between 1640 and 1688 are fully consistent with the model we have outlined.

It is clear that by this time substantial productivity increase was already taking place in the Western world. We see it strikingly evidenced in the case of shipping, where the gains stemmed not from technological change but from the decline of piracy (allowing a decline in crew size) and from the growing size of markets which reduced turn-around time and port time for ships. All these phenomena manifested declining transactions costs of the kind we have described.[2] A productivity change similar to that described for shipping was also happening elsewhere, particularly in trade and commerce, where transactions costs were moving inversely to the growing volume of trade and commerce.

Sustained economic growth in the Western world required the creation of institutions and property rights that served to bring the private rate of return to individual activities more nearly in line with the social rate of return. This means that the individual's perception of his own gains from undertaking an activity

[1] 'Innovation and Diffusion of Technology: A Theoretical Framework', by Douglass C. North, is forthcoming in the Proceedings of the Fourth International Congress of Economic Historians at Bloomington, Indiana, and is to be published by the University of Indiana Press.

[2] See Douglass C. North, 'Sources of Productivity Change in Ocean Shipping, 1600–1850', *Journal of Political Economy*, LXXVI (1968).

would in practice closely approximate to the benefits that society would receive from that activity. This necessitates a set of property rights and institutions that ensures that the factors of production directly receive their economic value.

Because it lacked such provisions, the late medieval world was beset by population growing more rapidly than output. Much of the land was held in common (and all had ill-defined property rights); when it became scarce, no economic incentive assured its efficient utilization. Labour, tied by law and by custom to a specific activity, had neither the opportunity nor the incentive to increase its productivity, either by seeking its most efficient employment or by improving its efficiency in its existing occupation. The institutions and legal attitudes of feudal Europe generally were designed to channel the energies of individuals away from productivity-increasing effort.

In contrast, the English economy at the end of the seventeenth century had achieved or was in the process of creating within its social structure a set of institutions and a common law that jointly encouraged improvements in productivity. Each factor of production was able to receive the bulk of its economic value. Land was in the process of passing into private hands, with the owner having the right to exclude others from its use and freely to alienate his property. Any improvements made were captured by the owners, who therefore were motivated to use their property to obtain the highest returns. Labour had been freed (with minor restrictions) to seek its best employment; what skills a workman developed were his to sell as best he could, as he was no longer subject to arbitrary exploitation by some third person. The development of private property and the legal enforcement of contracts reduced the externalities and risks that had been present in the feudal world, thereby raising the rate of return on undertaking all kinds of economic activities. The development of a patent system and other laws protecting intellectual property (however imperfectly enforced) encouraged the growth of innovation. In addition to the patent system (which was always imperfectly enforced) prices, bounties, and subsidies were all used to encourage innovation and produced the same consequences in terms of our model.

In effect, a cumulative series of once and for all secondary institutional arrangements had gradually led to the reorganization of the fundamental institutional structure. This basic change was clearly evidenced by the characteristics of the property rights that had evolved. The social milieu of England at the end of the seventeenth century served to direct the individual efforts of Englishmen into channels that increased the productivity of society, where earlier the social structure had encouraged economic activities that mainly redistributed existing income. This change was a necessary prerequisite for Western man to escape the Malthusian trap.

V

The explanatory sketch outlined above suggests that the cause of the rise of the Western world was the redirection of incentives as a consequence of the development of institutions which made it more profitable to attempt to increase productivity within any economic activity. Such institutional innovations, as we have pointed out, are in direct contrast to those that throughout most of the history of mankind have served mainly to redistribute income. The two major forces making for this change in the Western world were variations in relative prices and expansion in the size of the market.[1]

Because the theory rests on response not only to market forces but also to non-market forces inherent in the coercive power of government, we cannot make any definitive statement about the inevitability of such an outcome; presumably some combination of political power under other circumstances could have thwarted the pressures for productive institutional changes—as they clearly have done in other societies up to the present day.

This point requires emphasis. There is no way to predict the outcome of conflicting political groups or coalitions, and therefore there is no certainty that political policies which will induce "productive" economic changes will be passed. We do not have a theory of the state that allows us to predict the outcome of such political conflicts, and indeed the theory of non-zero sum games suggests in fact that the results are indeterminate.

Nevertheless, in spite of these limitations on the explanatory sketch developed above, the implications and promise are revolutionary enough.

University of Washington

[1] There are striking similarities—and differences—between this model and a Marxian model. In both, the development of a system of private-property rights was a critical aspect in unleashing the productive forces of economic growth (capitalism in the Marxian terminology). There are other similarities in the analysis of the characteristics of the feudal economy and the capturing of rents (surplus value) from the labour of the serf. Both this model and Marxian writing on the seventeenth century look upon it as a century of turmoil in which a basic restructuring of political power was necessary to implement the institutional changes essential to economic growth (a capitalist system). The most critical difference is in the sources of disequilibrium in the system which induced change. In the Marxian model it is technological change. In our model it is the change in relative prices and expansion in the size of the market.

1 Transport and Economic Development in Europe, 1789–1914

PATRICK O'BRIEN

1. THE GROWTH OF THE ECONOMY AND THE DEVELOPMENT OF TRANSPORT

European economic growth, which accelerated in the second half of the eighteenth century, was accompanied by an expansion in the supply of transportation. Demand for transport services increased when industrialists and farmers purchased their inputs from a resource base which widened in space and as they sold a growing proportion of their output on markets at an ever greater distance from their enterprises. As commodity output went up, the share marketed increased even more rapidly because improvements in transport made it possible to sell further afield and because specialisation (a major impetus to economic growth between 1789 and 1914) led to more trade between firms, farms and industries. In the traditional economy of early modern Europe production tended to occur within integrated forms of enterprise geographically concentrated in well defined regions. But over the nineteenth century the co-ordination of production came to be achieved through organised commodity and input markets serviced by extended and increasingly efficient transport and distribution networks. Adam Smith's twin processes of market penetration and specialisation implied increased demands for transport.

At the same time a series of innovations raised the efficiency of Europe's transport systems which lowered the cost of inputs used by farmers and manufacturers and the prices of their outputs. Such innovations, which were diffused gradually throughout the continent, included: surfaced roads, improved water and railed ways, the embodiment of new materials such as iron, and later steel, in the capital stock of national transport sectors and above all, the replacement of wind, water, animal and man power by steam power. These innovations, chronicled by historians of transport, provided producers with speed and continuity of delivery, and lowered the costs of transporting almost all goods across time and space. They were particularly important for commodities heavy or bulky in relation to value –

transport-intensive goods such as coal, mineral ores, building materials, grain and metal goods. The effects of innovations within the transport sector appeared as changes in price differentials between commodities sectors and regions of the European economy. Such 'forward linkages' operated upon the structure of relative prices and fostered regional specialisation and trade. The concomitant effects on the allocation of resources, labour migration, urban development and rates of capital formation were complex and, for some economies, profound.

Furthermore, the new and expanded transport networks made direct demands for the products of several industries which supplied the companies, responsible for the carriage of goods and people by rail and water, with their iron, steel, coal, construction materials and engineering products. To appreciate the force of these connections (or 'backward linkages') economic historians have measured the share of total output delivered by particular industries to the transport sector, and have attempted to analyse the degree to which the pressures of such demand promoted economies of scale and generated more rapid diffusion of technical progress in supplying industries.

Accelerated rates of capital formation which took place in transport sectors throughout Europe in the nineteenth century also gave rise to a range of externalities, or spinoffs, which are not captured either in the declining real price for transport services or changes in the structure of relative prices. For example, railways (but also canals) trained labour - engineers, foremen and managers - whose skills (initially acquired in transportation) contributed to the development of other industries. First canals and later railways made voracious demands for capital over relatively short periods of time which prompted the expansion and improvement of financial intermediaries for the mobilisation of domestic and foreign savings. Once established such institutions continued to meet the needs of other sectors of the economy. By tapping and mobilising new sources of finance the railways pushed up average rates of saving and investment over the long run. And in some countries where the construction of railways occurred in a contracyclical manner, the timing of their expenditures on capital goods reduced instability which presumably encouraged more investment and a greater willingness to assume risks throughout the economy.

Economic historians must do more than simply point out the range of interconnections between transport and economic growth. By looking (as this book does) at the history of transport across several nations they can compare output for the whole transport sector country by country for bench-mark years from the late eighteenth century through to 1914. That output should be broken down into intermediate output (ton-kilometres of freight per capita) and final output (passenger kilometres per head of the population) for major European economies. The latter is not only an index of how rapidly the idea of travel diffused among Europeans but it is a useful indicator of labour mobility (1). Personal travel (kilometres per passenger) grew more rapidly than services to producers, and a rising percentage of receipts from the provision of the entire range of transport services emanated from the sale of this 'new good' (2). For example, in France it rose from under 10 per cent of transport revenues in the 1830s to over 30 per cent by the end of the century (3). Almost the whole of the addition to passenger output came from railways and the railway systems of Europe consistently derived more than half of total receipts from the purchase of travel for business and pleasure. They supplied a novel, safe, comfortable and speedy service for which there was no real alternative before the development of the internal combustion engine. Demand for travel seems to have been both price and income elastic. Even the Dutch, who had enjoyed the most efficient mode of passenger transportation by water in Europe, switched to rail as soon as that facility became available (4).

Levels of intermediate output also provide historians with a useful index of market penetration in different economies which also cross checks on levels of commodity output across Europe. There is a fairly close correlation between the growth of commodity output and the increased volume of goods handled by the transport sector, and it may be possible to estimate a range of coefficients to link the two series. For countries which lack adequate statistics on movements in output, cautious and qualified inferences might be drawn about possible rates of growth from transportation data, if that happens to be available. French, British and Belgium data indicate that freight output expanded at more than double the rate for commodity output as a whole between 1830 and 1913 (5). For other countries (such as Germany and Spain) less favourably endowed with roads and waterways in the early nineteenth century, the growth of freight transported was probably even more elastic with respect to output (6).

Two kinds of influence should be distinguished when we try to explain why the share of transportation to gross domestic product more than doubled in most European countries over the nineteenth century. First the income effects are reasonably clear and the high income elasticity of demand for travel has already been mentioned. Furthermore, rising incomes in particular regions and countries had for centuries generated demand for more varied consumption and stimulated international and interregional trade. Foreign and internal commerce expanded much faster than output as the diversified produce of remote countries and regions entered the consciousness and consumption patterns of new social groups. Studies of diet, dress and taste by historians suggest that the nineteenth century was a century when European families found spice and variety by incorporating the products of alien cultures and remote regions into their traditional consumption patterns. In this process the demand curve for transport and distribution shifted to the right.

Two sorts of price effects also stimulated demand for transport. First, as improvements in economic efficiency depressed the prices of final outputs to consumers and the costs of inputs to producers, markets widened. Furthermore, regional specialisation, based initially upon natural and increasingly upon uneven rates of capital formation and technical innovation widened the gaps in productivity and altered price differentials between the regions of Europe. Lags in the diffusion of technology, the accumulation of capital and organisational improvements between countries and regions generated greater demands for transport. New techniques reduced costs of production for several commodities such as iron, coal, building materials and grain, which could then afford to carry the transportation mark-ups involved in marketing goods, heavy or bulky in relation to their retail or final sales prices.

Secondly, technical progress within the transport sector also widened the market and stimulated additional demand for freight and passenger services. Although indices of changes in total factor productivity have been constructed for British, French and German railways, almost no econometric research of a similar kind has been undertaken for transport along European waterways and roads (7). But it is certainly evident that real costs per passenger kilometre and per ton-kilometre of transport services fell sharply with the diffusion of railways and steam locomotives. British and French data testify to the considerable reductions in price per kilometre, which the railways offered to travellers. For example, in nineteenth-century France it was two to three times more expensive for passengers to travel by road than by rail (8). And in the 1860s a British Royal Commission compared first-class coaching at two shillings per mile with first-class rail fares of between two and three pence a mile (9). When and where trucks drawn by steam-powered locomotives competed with carriages and carts pulled along roads by horses, oxen and mules, railways offered producers a far cheaper service per ton-kilometre of freight carried. Thus between 1855 and 1913 it was three to four times more expensive for French firms to send merchandise by road (10). As early as 1850 on average it cost nine centimes a ton-kilometre to despatch freight by road in Belgium compared to five centimes by rail (11). And in the 1860s, coal was carried by British railways at about one twentieth of the cost by road (12).

Whenever the new technology competed with waterborne transport the decline in costs per ton-kilometre of freight carried was, however, much less obvious, particularly where steam-powered boats competed with locomotives by carrying heavy and bulky commodities along inland and coastal waterways (13). In France, Belgium and Italy (but not in England) costs per ton-kilometre were lower by boat (14). Railways competed with waterways by offering producers a speedier and more reliable service but at a comparable or even higher tariff per ton kilometre. If speed was necessary (for example in the carriage of perishable produce) this service could be described as indispensable. Furthermore, the diffusion of railways compelled carriers who used waterways to lower their prices in order to compete; and by eroding the rents or monopolistic profits of carriers by water (and road) the railways made a further contribution towards extending the market and promoting economic growth.

The precise impact of the rise in total factor productivity in transport can only be properly assessed when it becomes possible to compare movements in the prices of transport services with trends in wholesale prices. More price indices are required to measure how far the relative price of transport declined between 1789-1914. The evidence for Britain, France and Germany suggests that this decrease was really significant (15). A significant decline implies that the increase in total factor productivity in the European transportation sectors was sufficient not merely to meet rising demands for its services from other sectors of the economy, at constant real prices, but made an 'autonomous' contribution to economic growth. Here lies the difference between transport operating to complement productivity growth achieved by industry and agriculture, and transport acting as a leading sector in the process of widening markets and altering the structure of relative prices for commodity output and inputs. The degree to which such an autonomous contribution occurred would be reflected in a fall in the average share of the transport margin in the selling prices of goods and services marketed throughout the continent. Such effects would vary from commodity to commodity, region to region, and sector to sector. They would depend on the share of transport embodied in the market price (and would thus be more significant for transport intensive commodities). The effects also depended upon the price elasticity of demand for particular goods. For example, although the transport margin embodied in the retail prices of flour and sugar is similar, a fairly small cut in sugar prices could perhaps widen the market far more than a comparable cut in the price of flour.

Although prices fell, studies on the pricing policies pursued by enterprises concerned with transport by road, rail and water, reveals how the monopolistic nature of the industry and government intervention (in many cases more concerned with public revenue than consumer welfare) operated to hold the prices charged by transport companies at levels which were suboptimal for the overall growth of the economy. For example, Fenoaltea's analysis for Italy reveals how the high tariffs charged by Italian railways constrained specialisation and the extension of the market while the more efficient French and German companies set their prices closer to marginal cost and were not burdened by onerous obligations to transfer a high percentage of their operating revenues to the state (16). But in other countries, including Britain, Belgium and the United States, actual and incipient competition from waterborne transport probably kept railway tariffs closer to competitive norms (17).

To conclude: transportation grew more rapidly than national output because demand for travel was income and price elastic and because the sector was called upon to carry a rising proportion of commodity output across greater and greater distances. For example, in France in the 1830s the average lead or distance travelled by freight from the point of production was 50-60 kilometres by road. By 1905-13 this average had risen to 190 kilometres by rail (18). Technical progress with rising factor productivity occurred in all forms of transport: in surfacing roads, improving inland waterways, in laying down railed ways, but above all in the application of steam power to carriers by rail and water.

Over the nineteenth century the growth of European output was not constrained as it had been in past centuries by increasing costs of transportation. On the contrary, extensions to and improvements in the transportation networks in most European countries made a positive and autonomous contribution to national rates of growth. Without the revolution in transport which raised its productivity a greater volume of scarce resources would have been diverted into that sector in order to cope with the rising demand for freight services (and for travel) between 1789 and 1914. If European economies had been confronted with rising real costs of transport, trade and specialisation would have been constrained and the overall growth of output would have been slower. If, in turn, historians could answer the question how much slower, then they would be in a position to measure the contribution to the growth rate made by improvements in the efficiency of transport networks. Interconnections between transport and other sectors of the economy can be specified, but historians are still a long way from 'measuring' the contribution of transport to economic growth between 1789 and 1914. They have not and perhaps cannot quantify the decline in the rate of growth of gross national product which would have occurred if total factor productivity in transportation had remained constant over time.

2. THE SIGNIFICANCE OF RAILWAYS: SOCIAL SAVINGS

New economic historians have, however, attempted to measure something more restricted, namely the impact of the diffusion of railways on economic growth. In focusing upon railways they selected the core element in the transport revolution of the nineteenth century. Although technical change occurred throughout the transportation sector and steam power was also utilised by waterborne carriers, the construction of networks of lines across Europe rising from nearly 3000 kilometres of track in 1840 to 362 000 kilometres by 1913 placed railway companies in a position to provide nearly all the additional passenger kilometres sold to consumers of travel between 1840 and 1910 (19).

They also met the demand for an extremely high proportion of the extra ton-kilometres supplied for the transportation of freight around Europe over the nineteenth century. For example, French statistics show that while output of roads, canals and coastal shipping remained roughly constant between 1830 and 1914 the ton-kilometres of transport services supplied by railway companies expanded to a point where they supplied nearly 70 per cent of total freight transportation in 1914 (20). No doubt statistics for other economies would also reveal the dominance of railways in the expansion of internal transportation over the nineteenth century. Thus in Belgium, railways moved from a position where they supplied 7 per cent of freight transportation in 1845-6 to where they provided 77 per cent in 1908-13 (21). Although railways clearly met most of the additional demand for transport which emerged after 1840 it cannot be assumed that railways alone possessed the techniques and capacity to obviate a potential 'transport bottleneck' to the growth of output over the nineteenth century. But can their contribution be measured or quantified?

One well-known approach to answering this question is to estimate social savings from railways some decades after national networks had been in full and efficient operation. This approach postulates that the importance of railways to any economy could be measured as the *costs* of coping without them over one year.

Table 1.1: *European railway development (territory and population in relation to kilometres of lines)*

	1860	*1880*	*1900*	*1910*
UK	44	66	69	69
Belgium	30	60	88	102
France	18	44	77	87
Germany	21	54	70	75
Sweden	3	32	61	76
Switzerland	28	63	79	88
Russia	1	10	21	24
Spain	6	23	40	58
Italy	6	23	38	38

Source: P. Bairoch, 'Niveaux developpement economique de 1810 a 1910', *Annales, ESC,* December 1965.

The method envisages a counterfactual situation in which a given national network is closed down for twelve months but producers continue to despatch exactly the same volume of freight to the same destinations as before by using roads and waterways. But, in order to maintain a constant level of transportation when denied access to railways, farmers and industrialists would clearly incur extra costs because:

(a) freight despatched by road and water had to pay higher tariffs per ton-kilometre;
(b) their goods would usually move along more circuitous and longer routes;
(c) the time taken between despatch and delivery would lengthen;
(d) transportation by wagons and boats was more risky, more subject to delay and loss than transportation by rail.

Assuming that the extra costs of sending commodities by water and road (instead of by rail - the preferred and cheaper method) could be estimated, new economic historians have ostensibly measured social savings which they define as equal to the *extra* or marginal benefit derived by a society from the normal operation of its railway system (22).

Estimates of social savings on ceteris paribus assumptions, first formulated for the United States, have now been manufactured for several European countries (including Tsarist Russia) and the statistics expressed as a ratio of Gross National Product are set out in Table 1.2.

Of course there are formidable problems connected with the quality of the data deployed to estimate social savings and a major theoretical controversy has arisen around the extent to which historical records of prices charged by road hauliers, inland waterways and railways provide adequate proxies for the 'real' or opportunity costs of the resources employed by transportation networks of Europe and America (23). Formidable index number problems inherent in the concept are also still unresolved. For example, what are the implications of assuming that enterprises would continue to despatch the same volume of freight with and without railways? If the cost differences between railways and other modes of transport are large and the derived demand for transport is price elastic the volume of freight transported is not independent from the introduction of railways (24).

Finally social savings estimates are sensitive to the assumptions made about the real opportunity costs of the resources 'released' to the economy from the transport sector through the operation of railways. If the rise in labour productivity simply adds to the volume of unemployment or to emigration (pace Italy) then the gains to the national economy are reduced (25). For Spain the availability of seasonally underemployed agricultural workers reduced the social savings from railways from 11.8 per cent to a feasible 7.5 per cent of GNP (26). Debate over the new economic history of railways continues and has been reviewed and summarised elsewhere (27). Let us presume that the available estimates of social savings provide rough outer-bound measures of the percentage falls in gross national products which might have followed from the closure of railways during the years specified in Table 1.2. It then becomes instructive to analyse the variations in social savings attributed to railways across countries and over time.

With all their theoretical and empirical imperfections, estimates of social savings do enable historians to dispose of claims that railways were 'vital' or 'indispensable' for economic progress, because they show that (apart from Spain and Mexico) closure of a fully operational railways network would have led to a fall of less than one tenth of national

Table 1.2: *Estimates of social savings on freight transported by railways*

	Date	*SS expressed as a share of GNP (%)*	
England and Wales	1865	4.1	(a)
England and Wales	1890	11.0	(a)
USA	1859	3.7	(b)
USA	1890	8.9	(c)
Russia	1907	4.6	(d)
France	1872	5.8	(e)
Germany	1890s	5.0	(f)
Spain	1878	11.8	(g)
Spain	1912	18.5	(g)
Belgium	1865	2.5	(h)
Belgium	1912	4.5	(h)
Mexico	1910	25-39	(i)

Note: These estimates all based on the ceteris paribus assumption are comparable.

Sources:

(a) G. R. Hawke, *Railways and Economic Growth in England and Wales, 1840-70* (Oxford, 1970) p. 196. The 1890 estimate has been calculated from the data presented on p. 89 and an estimate for the National Income of England and Wales for 1890 based upon the assumptions specified on p. 196.
(b) Calculated from A. Fishlow, *American Railroads and the Transformation of the Ante-Bellum Economy* (Harvard, 1965) pp. 37 and 52.
(c) This is Fogel's unadjusted estimate based on ceteris paribus assumptions and calculated from data in R. Fogel, *Railroads and American Economic Growth* (Baltimore, 1964) chapters 2 and 3.
(d) J. Metzer, 'Railroads in Tsarist Russia: Direct Gains and Implications', *Explorations in Economic History*, 13 (1976) p. 90.
(e) See p. 44.
(f) See p. 139.
(g) See p. 154.
(h) See p. 221.

Table 1.2 - *Continued*

(i) J. H. Coatsworth, 'Indispensable Railroads in a Backward Economy: The Case of Mexico', *Journal of Economic History* XXXIX (1979) p. 952, This estimate is expressed as a share of GDP.

outputs. Nevertheless a fall of 10 per cent of GNP (or even 5 per cent (for that matter) is still a very large magnitude. Railways were clearly important (and in the Spanish case very important) for the growth and maintenance of commodity production at observed levels. Just how important can only be assessed by comparing the social savings achieved by railways with the social savings for other major nineteenth-century innovations, such as steel, power looms, canals, chemicals, etc.

The increase over time in the benefits derived from railways (the rise in the ratio of social savings to national income) reflects the continuous improvement in the efficiency of railways which occurred through the steady diffusion of better techniques in railway engineering, the replacement of iron by steel rails and in business organisation (28). Thus freight rates on German railways declined over the nineteenth century to one quarter of their level for 1845 (29). In Belgium the rates per ton-kilometre declined from 10.8 centimes in 1845 to 3.6 centimes in 1913 and in France they fell by 50 per cent (over the same decades) (30). Improvements in productivity helped railways to hold their own against competition from waterborne transport which also became more efficient over the nineteenth century and which (in those countries well endowed with rivers or coastal shipping lanes) enjoyed advantages in the form of lower track costs per ton-kilometre carried. Over longer distances railways competed by offering producers speed and regularity of delivery. For large countries like Russia, the Austro-Hungarian Empire and the United States railways did not provide a much cheaper service than waterways per ton mile carried over long and comparable routes, and the social savings on that account could be fairly small, even negligible. Thus where trunk lines were constructed to compete with waterways over long distances, and the track accommodated to existing locations and patterns of internal trade, the initial gains to the economy would be positive but not dramatic. Belgium, France and Germany continued to build canals during the second half of the nineteenth century (31). Furthermore, it is difficult to speculate ('counterfactually') how waterborne transport might have developed without railways. It is not unreasonable to assume (pace Fogel) that without railways the efficiency of waterborne and road transport would have improved more rapidly than it did with competition from railways.

As time went on gains increased because railway companies expanded their lines outside and beyond the networks of trade already serviced by waterways. And they helped to create more trade and production in new locations. Social savings really went up when railways captured an ever increasing share of freight carried along roads over medium distances. The differential in average costs per ton-kilometre between railways and roads was already considerable in the 1840s and that differential widened over time because productivity in road transport failed to go up very much over the next seven decades (32). Although road networks expanded at a high rate technical progress in that subsector of transportation was minimal before the internal combustion engine provided a substitute for animal and man power used for cartage. Moreover, the opportunity costs of the principal inputs used to carry goods by roads (labour, animals and animal feed) rose with the development of agriculture and industry (33). For example, 32 per cent to 43 per cent of the hectares devoted to the cultivation of bread grains would have been required to cope with the extra demand for animal feedstuffs generated by the hypothetical shutdown of Spanish railways in 1878 (34). Thus the 'savings' in resources which emanated from the shift of freight from roads to railways was far greater than from the reallocation of merchandise from boats to trains and it increased over time. Considerations of this kind explain the relatively low ratios of social savings to national income for 1859 in the United States, 1865 in England and 1907 in Russia (see Table 1.2). Three or four decades later when the output of the transport sector had increased the ratio nearly trebled in Britain and America and the same tendency was certainly more pronounced in Russia because estimates for 1907 relate to the early stages of development. Between 1907-39 ton-kilometres supplied by rail multiplied four times to account for 82 per cent of Soviet transportation output. In 1907 84 per cent of freight services supplied by Russian railways could have been provided by waterways and that ceased to be true when transportation output expanded very rapidly over subsequent decades (35).

Similar observations follow from comparisons of social savings across countries. Thus the higher the percentage of freight taken off the roads and onto railway networks the greater the gains to an economy from its investment in railways. Countries badly endowed with navigable rivers, afflicted with a terrain hostile to the construction of canals, and with poor natural facilities and opportunities for coastal trade gained most from their investment in railways. Railways alleviated the adverse economic effects on internal trade of an unfavourable natural endowment. Thus over the long run Mexico, Spain, Russia, certain regions of France, the Upper Po valley in Italy and Germany obtained higher returns from the diffusion of railways than Britain, the rest of Italy, Holland, Belgium or the United States. The significance of this point can be measured by the percentage of freight which would be diverted from rail to water in the counterfactual case of a shutdown of the railway network. For example, during hypothetical closure of Spanish railways in 1878 coastal shipping and inland waterways supplied only 10 per cent of the ton-kilometres required to maintain a constant level of transportation; in France (1872) waterborne carriers provided 30 per cent (and in Britain (1865) they took about half of the traffic). Italian trunk lines were unprofitable because of the availability of coastal shipping (36).

To a more limited extent the relative significance of railways for different economies can also be discerned in terms of their effects on the levels of stocks held by farms and industrial firms before the construction of national networks. Railways supplied producers with transport which was faster, safer and far more immune to seasonal and geographical obstruction than roads or waterways. In the middle of the nineteenth century a barge took no less than 86 days to carry a load of coal from Mons to Paris and to return to its mooring (37). Where and when transport by road and water was both slow and irregular, firms could partially insure production against costly interruptions to inputs and sales by holding higher inventories of raw materials, finished goods and work in progress. Railways obviated the need to invest in stocks, and thereby enabled producers to convert circulating into fixed investment. This saving could be particularly important for economies like Spain and Russia where traditional transport of goods by road and water was highly seasonal (38). Nearly a quarter of the social savings achieved by Russian railways in 1907 came from a reduction in the cost of inventories (39). In Britain with its temperate climate, gentle terrain and specialised road and water-borne services, the gains from additional speed and regularity supplied by railways was less than 4 per cent of the social savings for 1865 (40).

To sum up: it appears that the traditional inefficiency and rather limited technical progress achieved in transportation by road made railways important in countries badly endowed with waterways. Over long and medium hauls road transport utilised more capital labour and other inputs per ton-kilometres than the two alternative modes of production, although road transport continued to dominate short distance transportation, particularly within cities it lost freight (first to waterways and then increasingly to railways on long and passenger hauls) (41).

3. RAILWAYS AND LONG-RUN GROWTH

Nowhere in Europe did railways make the difference between development and stagnation and economic historians are still a long way from measuring their contribution to the growth rates of particular economies. Although the social savings methodology represents a big step forward, static equilibrium analysis cannot deal effectively with the dynamic effects of railways over time. For such influences the relevant counter-factual is not the shutdown for one year (used with illuminating effect by Fishlow, Hawke, Metzer, Caron and Gomez) but a hypothetical refusal by a national government to establish a railway network at any time between 1830 and 1914. In this context and with the aid of general equilibrium models, the entire range of influences exercised by railways on savings and investment, location patterns, markets, the product mix, scale, technical progress and innovation, labour training, government policy, urbanisation, international flows of capital and labour, etc. could in theory be quantified. But even if models could be designed to capture the complex interconnections of railways to economic growth through time, empirically the task of quantification seems impossible - a point exemplified by the recent debate between Fogel and Williamson (42). Historians can, however, write more or less rigorously about the impact of railways on economic growth. They can also quantify some of the more important backward and forward linkages.

For example, several economic historians have analysed and measured the 'feedbacks' derived from the establishment and extension of railways. The great bulk of investment expenditures feedback to the construction industry. For example, between 1861-95 Italy's engineering and metalmaking industries absorbed only a tiny percentage of railway investment (43). For industry as a whole, even at peak levels of demand, railways absorbed far too low a percentage of total output to qualify as a 'leading sector' in the growing process. Industry normally delivered well over 90 per cent of its output to other producers and consumers.

Deliveries to railways did, however, exercise an important influence on the development of particular industries -

including iron and steel, coal, bricks, timber and engineering (see Table 1.3). Of course the relative significance of railway demand hinges entirely on the level of disaggregation used to define an industry. Thus for Italy: 'the growth in railway related production seems to have represented significant growth for the machinery and metal-making industries but relatively unimpressive growth for the entire metalworking group' (44).

Even for industries strongly stimulated by the establishment and maintenance of a railway system, the significance of 'feedback effects' was never extraordinary. Backward linkages were, moreover, concentrated over particular cycles of production, when a large proportion of the increment to output was delivered to railways. Thus demand for iron and steel, engineering products and other manufactured commodities could be important for the growth of a small range of industries over fairly short periods of time in the nineteenth century when the boost from the establishment of large-scale extensions to a national network pushed production up onto a higher plateau. Thereafter normal demand for replacement and extensions to the network and its stock of equipment accounted for a tiny proportion of the annual additions to industrial output.

During these boosts to total demand, firms and industries supplying railways might realise economies of scale and introduce improved technology more rapidly. Railway expenditures could even accelerate technical progress, particularly in metals and engineering. Historians have traced such influences on the South Wales iron industry; on the development of locomotive production and bar iron in Germany; on the steel industry in France and the United States, and upon specific sectors of the engineering industry in Belgium, Russia and Germany (45). Feedbacks to steel production stand out as significant for the expansion of capacity and the diffusion of technology in France, Germany, United States and Italy (in the late nineteenth century) (46).

Of course the impact of backward linkages varied from country to country and their importance for industrial development depended primarily upon the extent to which demand for capital goods required to establish the main network and its rolling stock was directed towards imported inputs and equipment. At the outset almost every country except Britain and Belgium laid down imported rails and hauled freight with foreign locomotives. But in some countries the process of import substitution quickly re-directed demand towards domestic industry. In others the lags, for a variety of reasons, were long, and the

Table 1.3: *Percentages of Output delivered to railways*

	Pig Iron (%)	*Steel* (%)	*Coal* (%)	*Bricks* (%)	*Wood* (%)	*Transport Equipment* (%)
USA	5-21 (1840-60)	50-87 (1867-80)	2-20 (1840-90)	-	3-10 (1840-60)	25 (1859)
England & Wales	5-13 (1835-69)	-	2-14 (1865)	30 (1840s)	-	-
Spain	6.0 (1890-1914)	8.5 (1890-1914)	18-29 (1865-1914)	-	-	-
Germany	22-37 (1840-59)	-	3 (1860s)	-	-	-
France	-	12-18[a] (1845-85)	-	13-18[b] (1845-85)	1 (1875-84)	-
Italy	-	12-13[c] (1861-1913)	-	16-23[d] (1861-1913)	-	5-11[e] (1861-1913)
Belgium	6 (1860-1913)	-	-	4-10 (1865-1913)	-	-

Note: Iron and Steel output delivered has been defined as a net flow of output.
(a) iron and steel; (b) building materials; (c) all ferrous metals;
(d) construction materials; (e) engineering products.

Source: See Table 1.2, O'Brien, *New Economic History of Railways* (1977) ch.5, pp. 38-9, 60-72, 123-32, 157-63, 209-14.

establishment of the initial networks exercised almost no impact upon local metallurgy, coal and engineering but contributed to the continued development of British, French and Belgium export industries. For example Spain, Italy and Tsarist Russia are all examples of countries where, until late in the nineteenth century, foreign suppliers derived the most of the benefits from orders for their rails and rolling stock (47). While in France and Germany rapid and decisive moves towards import substitution - initially in the rolling mill sector of the iron industry and in the construction of locomotives, but then followed by the shift into the manufacture of coke-smelted pig iron - imparted a significant stimulus to domestic metallurgy and to the transition from charcoal to coke-smelted pig iron (48).

Why did Tsarist Russia, Italy and Spain not take full advantage of 'potential linkages' to industrialisation, advantages which had been demonstrated during the construction of railways in other countries? One possible explanation is that their governments failed to impose tariffs on cheaper foreign rails and rolling stock, either because their statesmen were ideologically blinkered in favour of free trade or because the foreign capitalists and companies, who provided a large share of the finance required for the establishment of rail networks in Spain, Italy and Russia, insisted on directing the order for rails and equipment overseas (49). While such explanations are not without point, they seem to rest too heavily on the lack of foresight exhibited by statesmen and the bias of foreign investors towards their own national industries. At the time the decision presented itself as a problem of how quickly local capacity in metallurgy and engineering could be expanded to cope with a 'bunched' and cyclical demand for rails, locomotives, rolling stock and other manufactured inputs required by the railways. In Germany, France and Belgium the metallurgical and engineering industries had attained levels of development and efficiency from where import substitution could proceed smoothly. In other words, the fairly rapid reallocation of demand towards local suppliers which occurred in these economies did not lead to bottlenecks, intolerable delays and prohibitive costs for the establishment of railway networks (50). Furthermore (in contrast to Spain, Italy and Russia) these economies seem far less dependent on foreign capital to provide the finance required for investment in social overhead capital. Spain, Italy and Russia were short on domestic venture capital. Their levels of production trade and income rendered that investment a more expensive and riskier proposition. Their relatively 'difficult' terrains increased the ratio of capital to output in transportation.

Their existing plant and manpower in metallurgical and engineering industries compared unfavourably with France, Germany and Belgium. Only a handful of engineers worked in Russia in the 1860s (51). At the onset of railway construction in 1856 Spain possessed sufficient capacity to produce only 15 000 tons of iron compared with a demand for rails over the next five years of around 30 000 tons per annum. And Italy emerged from the Risorgimento with very limited capacity to produce pig and wrought iron (52).

In short, the preconditions for the kind of successful import substitution which occurred in Germany, France and Belgium were not present everywhere even in Western Europe and in some states the imposition of tariffs on imported railway equipment would surely have led to costly delays in construction and the emergence of a railway network saddled with considerable capital charges and compelled to charge high prices for freight and passenger services. Thus by implication the criticism of Cavour, Kankrin, Bunge, Luxan and other European statesmen for their failure to utilise protective tariffs to take advantage of opportunities for backward linkages is an argument for a railway system heavily subsidised by governments. The unexplored premise of that view is that the long-term social gains generated specifically by railway demand for the products of the metallurgical and engineering industries would more than compensate for the delays, higher costs and increased taxes necessitated by constructing and maintaining national railways networks with domestic rather than foreign rails, rolling stock and other equipment. Part of the costs of protection afforded to iron and steel would (as the Italian example shows) be borne by industries utilising metals, particularly engineering and shipbuilding (53).

For Spain, Italy, Russia and other parts of Eastern Europe the counterfactual scenario of tariffs, import substitution and railway construction combining to promote rates of industrial growth at far higher levels than those achieved with free trade seems altogether improbable, and the examples of France, Belgium and Germany are not relevant for countries with limited capacities in engineering and metallurgy at mid century.

Railways influenced the volume of capital formation as well as variations in rates of investment. For example, net investment in railways dominated investment in the Tewerbe, or modern sector of the German economy from 1851-79 (54). But to assess their influence on the long-term growth of particular economies historians need to specify a model which exposes interconnections between railway investment and other relevant variables in a closed and open economic system. Such models are, however, easier to adapt from economic theory than to use to quantify the influence of railway expenditure (55). Several historians have, however, analysed the extent to which investment expenditures on railways operated to impart greater stability to economies afflicted by fluctuations in the level of economic activity. Apparently in Britain and America such expenditures helped to induce booms and to alleviate depressions (56). But for Italy Fenoaltea's analysis 'suggests a mild boost to industry's cyclical instability' while French investment in railway companies also intensified the amplitude of cycles in the nineteenth century (57). But in Germany railway investment fluctuated in the approved contracyclical manner (58). The sequence of expenditures on railway networks is likely to vary from country to country and from period to period. No definite conclusions can, however, be drawn from these correlations until the connections between cycles of investment in railways and of economic activity are analysed and elucidated at the level of industries and even firms.

More important questions should be raised about connections between opportunities to buy shares in railway companies and the total volume of investible funds made available to an economy. If we assume inflexible rates of domestic saving as well as the availability of other projects competing for funds with railways, their influence on the level of capital formation will be seen as negligible. But if expenditures on railways attracted foreign funds or new savings to finance the networks, then it becomes possible to argue that railways raised the overall rate of investment by some positive percentage. Railroads apparently attracted both foreign capital and immigrants into the United States (59). Although sceptics maintain that capital migrates in search of profits not railways, there can be no doubt that European investors, banks and other financial institutions, weaned on the paper securities of British and French railways, were geared towards similar opportunities in Belgium, Italy, Spain and Russia (60). While the voracious demands for capital made by railway companies over short cycles of economic activity influenced institutional developments in the capital markets of Britain and France which not only mobilised funds but educated new social groups in impersonal investment and increased propensities to save (61). Debate on the precise importance of effects of this kind is likely to remain inconclusive.

Finally, railways widened markets which promoted economies of scale, encouraged the relocation of economic activity, stimulated competition and increased trade and specialisation. Williamson's important analysis of these 'forward linkages' from railways to the economy of the Midwestern United States demonstrates that they could be profound. Nothing like his general equilibrium model has been applied to the economic history of Western Europe (62). Prima facie it seems that for most parts of Europe broad patterns of interregional and international trade established before the advent of railways continued and the iron road carried forward the process of integrating local economies into national and international markets. Trade in the nineteenth century emanated basically from natural and traditional advantages. Comparative advantages of the kind stressed by Ricardo (and adumbrated by Heckshers and Ohlin) were becoming more important and railways reinforced tendencies towards industrial specialisation based upon favourable endowments of skilled labour, accessible supplies of capital and those 'temporary' gains which accrued to pioneers in new products and technology. Nevertheless the composition of commodities traded between the countries and regions of Europe did not alter profoundly between 1840 and 1914, when railways provided particular regions with greater opportunities to export products in which they enjoyed cost advantages (63).

For several economies (including Britain and Holland) well served by waterborne and road transport, relatively small gains flowed from the integration of national economies and from linking interior regions and natural resources to world markets. Even in Spain, the location of vineyards and mineral ore deposits near to the sea implied that railways were not a necessary condition for the rapid growth of wine and ore exports in the second half of the nineteenth century, although they did bring Spain's isolated agricultural regions into competition with each other (64). In Italy, despite the high hopes of post-unification governments, the trunk lines built before the 1880s did not promote a rapid expansion of inter-regional trade. Many regions remained too poor to trade, their capacity to absorb exports from other parts of Italy remained low and complementarity in production among Italian regions was not marked. Integration exacerbated dualism between North and South and encouraged imports (65). But rail links from the Upper Po Valley to the sea permitted the industrial triangle (Turin, Milan and Genoa) to develop (66).

For other economies the stimulus to production seems more marked. Thus Belgium's planners tried from the outset to build lines which would integrate the local economy into Western Europe (67). In Russia railroads increased regional specialisation and 'there was a major reorientation of resources towards those areas offering a higher level of productivity in the cultivation of rye and wheat'. Exports of Russian grain increased roughly four times between 1860-4 and 1875-9 and added about 4 per cent to national income (68).

Furthermore, they linked Donbas coking coal with Krivoi Rog iron ore deposits and created a new heavy industry complex in the Ukraine (69). Their forward linkage effects on the growth of Germany's coal and metallurgical industries were profound enough for Fremdling to label the railroad as 'the hero of Germany's industrial revolution' (70). In France their advent led (as Caron shows) to the spread of wine cultivation in Languedoc, and stock raising in the Charente (71).

Railways carried forward the reduction in transportation costs initiated by waterborne carriers and thereby widened markets, stimulated competition and provided access to natural resources and more efficient locations for productive activity. Their capacity to bring about reductions in costs per ton-kilometre were maximised in those parts of Europe where access to productive resources was prohibitively expensive by waterways and roads and where they successfully appropriated freight over medium hauls from road transportation. Only in special cases, such as the trade in meat or perishable farm produce, were railways a necessary condition for the extension of the market. Access to markets was, for most products, available by road and in many places by water.

Clearly it will not be possible to quantify the manifold effects of railways on Europe's regional and international trade. And associated with the spread of railways, there also occurred geographical shifts by enterprises within well-defined economic regions and such gains from the relocation of economic activity could, in aggregate, exceed the gains from longer distance trade. But they seem difficult to detect and impossible to measure, except as a by-product of business or industrial history. But railways invariably lowered freight costs per ton-kilometre and the cuts in transport margins were passed on in the form of lower prices. Such reductions seem to have been particularly important for fuels, construction materials and agricultural surpluses - meat and wine. Thus railways exercised their most profound effects on long-run growth by altering the structure of relative prices and the terms of trade between regions and sectors of the European economy. They appear to have had greater effects on primary production - agriculture and mining - than upon manufacturing industry. But their dynamic effects on capital formation, technical progress and the reallocation of resources which flowed from the reduction in transport mark-ups have not been measured. Nevertheless, Williamson's central question: what would have happened to economic development if price differentials had remained at their pre-railway levels, is the kind of counterfactual approach which will appeal to a traditional historiography which sees railways as a pervasive and extremely important element in the economic growth of

Western Europe from 1840-1914.

Not all regions benefited equally from railways. As always the gains depended on the elasticities of demand and supply or the capacity of a given economy to take advantage of the opportunities to trade afforded by railways. If consumers resident in a particular region responded to falling prices by importing heavily from another region, a substantial gain would occur in consumer surplus and their welfare. If a region's capacity to supply was elastic because resources were underemployed in the pre-rail situation or because inputs could be augmented at low cost, production could expand rapidly in response to the diffusion of cheaper transport. Unfortunately there were many parts of Europe (particularly in southern and eastern Europe) whose people remained too poor to take advantage of the opportunities for cheaper and more variegated consumption provided by railways. Their capacity to respond to the widening of markets continued to be restrained by unfavourable resource endowments and rising marginal costs of production.

For these regions the gains from railways before 1914 were positive but more limited than the gains which accrued to North Western Europe and the United States. For some parts of Europe (Southern Italy, Castille, Hungary and Romania) the benefits from trade creation may have been offset by the losses from trade diversion to other regions. Their workers had no option but to migrate in search of higher incomes (72).

Perhaps the contrasting economic histories of Germany and Spain from 1850-1914 will illustrate how railways could carry forward but not propel rapid industrialisation. In Germany the railroad linked backwards and forward to agriculture and to a multiplicity of industries (already in the process of sustained economic growth), while Spain lacked several pre-conditions for an accelerated rate of economic change. Spanish agricultural productivity was lower; its industrial base (particularly its metallurgical and engineering sectors) was less developed; regional specialisation seems less marked; and the complementary networks of roads and waterways were not in place. Spanish railways did shift the Spanish economy closer to its production possibility boundary, but their effects were less profound than they were in Germany (73). Thus in countries like Spain, Italy, the Austro-Hungarian Empire, Tsarist Russia and other parts of Eastern Europe railways accomplished a great deal by 1914. But despite all the euphoria which accompanied its introduction and the hopes placed on its supposed potential for both political and economic integration, the iron road was never a panacea for development.

NOTES AND REFERENCES

I would like to thank Stefano Fenoaltea for helpful comments on this chapter.

1. Relevant data is contained in B. Mitchell, *European Historical Statistics* (London, 1975) section G.
2. A. de Fouille, *La transformation des moyens de transports* (Paris, 1880) pp. 141-7 and L. G. McPherson, *Transportation in Europe* (London, 1910) pp. 69-78.
3. J. C. Toutain, *Les Transports en France* (Paris, 1967) p. 286.
4. J. De Vries, *Barges and Capitalism, Passenger Transportation in the Dutch Economy* (Wageningen, 1978) p. 236.
5. See pp. 34, 214-15 and Toutain, *Les Transports en France*, p. 298.
6. See pp. 121, 132, 152.
7. See pp. 40-2, 132-3; G. Hawke, *Railways and Economic Growth in England and Wales* (Oxford, 1970) Ch. XI; F. Caron, 'Recherches sur le Capital Des Voies de Communication en France au XIXe Siècle' in *Colloques Internationaux du CNRS*, no. 540, *L'Industrialization en Europe au XIXe Siècle* (Lyon, 1970) pp. 256-61; R. Fremdling, *Eisenbahnen und Deutsches Wirtschaftswachstum, 1840-1879* (Dortmund, 1975) pp. 35-51.
8. Toutain, *Les Transports en France*, p. 277.
9. Hawke, *Railways and Economic Growth*, p. 44.
10. Toutain, *Les Transports en France*, pp. 277-9.
11. See pp. 215 and 220.
12. Hawke, *Railways and Economic Growth*, pp. 179-80.
13. De Vries, *Barges and Capitalism*, pp. 245-8 and 359-61.
14. See pp. 42-5. 95-6 and 220; Toutain, *Les Transports en France*, pp. 277-8 and Hawke, *Railways and Economic Growth*, pp. 85-6, 173, 187-8.
15. Toutain, *Les Transports en France*, p. 283; and Fremdling, *Eisenbahnen und Deutsches Wirtschaftswachtum*, pp. 57-60.
16. See pp. 82-6.
17. L. Girard, 'Transport' in M. Postan and H. J. Habakkuk (eds), *Cambridge Economic History of Europe*, vol. VI, part 11 (Cambridge, 1966) and H. Parris, *Government and the Railways in Nineteenth Century Britain* (London, 1965).
18. Toutain, *Les Transports en France*, p. 244.
19. P. Bairoch, *Commerce extérieur et developpment économique de l'Europe au XIXe Siècle* (Paris, 1976) p. 32.
20. Toutain, *Les Transports en France*, p. 248.
21. See p. 221.

22. P. K. O'Brien, *The New Economic History of Railways* (London, 1977) chs. 2 and 3 and see pp. 184-8.
23. R. Fogel, 'Notes on the Social Savings Controversy', *Journal of Economic History*, XXXIX (1979) pp. 13-38.
24. J. G. Williamson, *Late Nineteenth Century American Development: A General Equilibrium History* (Cambridge, 1974) pp. 184-7.
25. See pp. 229, 232-3.
26. See pp. 154-7.
27. O'Brien, *New Economic History of Railways*, Fogel, 'Notes', and T. R. Gourvish, *Railways and the British Economy, 1830-1914* (London, 1980).
28. Hawke, *Railways and Economic Growth*, ch. XI; A. Fishlow, 'Productivity and Technological Change in the Railroad Sector' in National Bureau of Economic Research, *Studies in Income and Wealth*, 30 (New York, 1965) pp. 583-646; Fremdling, *Eisenbahnen und Deutsches Wirtschaftswachstum*, pp. 35-101; Caron, 'Recherches sur le Capital des Voies', pp. 237-61, and see pp. 34-5 and 40-2.
29. Fremdling, *Eisenbahnen und Deutsches Wirtschaftswachstum*, pp. 18-19 and 57.
30. Toutain, *Les Transports en France*, p. 278 and see p. 216.
31. R. Price, *The Economic Modernization of France, 1750-1880* (London, 1975) p. 16; Toutain, *Les Transports en France*, (New York, 1979) pp. 68-9, see pp. 139, 218 and 221.
32. See pp. 217-20. In France, the differential was roughly 2.4:1 in 1845-54 and widened to 4.6:1, 1905-13 - see Toutain, *Les Transports en France*, p. 279.
33. P. S. Bagwell, *The Transport Revolution from 1770* (London, 1974) pp. 145-9 and F. M. L. Thompson, 'Nineteenth Century Horse Sense' in *Economic History Review*, XXIX (1976) p. 78.
34. See pp. 152-4.
35. Unpublished paper delivered by A. Aslund to St Antony's Seminar in International Economic History, 1978 and Metzer, 'Railroads in Tsarist Russia', pp. 88-96.
36. See pp. 93-4.
37. See p. 220.
38. See pp. 150, 155-6.
39. Metzer, 'Railroads in Tsarist Russia', pp. 95-6.
40. See p. 191.
41. In France in 1830 80 per cent of the receipts for transport of merchandise accrued in the road system:by 1895-1904 the proportion had fallen to 41 per cent - see Toutain, *Les Transports en France*, p. 286; De Vries, *Barges and Capitalism*, pp. 223-30, and Fogel, 'Notes on the Social Savings Controversy', pp. 49-51 and see pp. 216-21.
42. Fogel, 'Notes on the Social Savings Controversy'; J. G. Williamson, 'Greasing the Wheels of Sputtering Export

42. *Continued*
Engines: Midwestern Grains and American Growth', *Explorations in Economic History,* 17 (1980) pp. 211-12 and see S. Fenoaltea, 'Railroads and Italian Industrial Growth, 1861-1913', *Exploration in Economic History,* IX (1972).
43. See pp. 60-72.
44. See p. 72.
45. Hawke, *Railways and Economic Growth,* pp. 215-30; Caron, 'Recherches sur le Capital Des Voies'. pp. 239-47, see pp. 45, 123-32, 212-13; J. N. Westwood, *A History of Russian Railways* (London, 1964) pp. 91-8 and R. Fremdling, 'Railroads and German Economic Growth: A Leading Sector Analysis', *Journal of Economic History,* XXXVII (1977) pp. 587-8.
46. F. Caron, 'French Railroad Investment, 1850-1914' in R. Cameron (ed), *Essays in French Economic History* (Homewood, 1970) pp. 315-40; R. Fogel, 'Railroads as an Analogy to the Space Effort', in P. Temin (ed), *New Economic History* (London, 1973) pp. 238-42; Fenoaltea, 'Railroads and Italian Industrial Growth', p. 336 but see his amended estimates, pp. 64-5. R. Tilly, 'Capital Formation in Germany in the Nineteenth Century' in P. Mathias and M. Postan (eds), *The Cambridge Economic History of Europe,* vol. VII (Cambridge, 1978) pp. 414-18.
47. See pp. 67, 158-9; V. K. Yatsunsky, 'The Industrial Revolution in Russia', in W. Blackwell (ed), *Russian Economic Development* (New York, 1974) p. 127, and J. Nadal, 'Spain, 1830-1914', in C. Cipolla (ed), *The Fontana Economic History of Europe,* vol. 4(2) (London, 1973) pp. 596-9.
48. See pp. 39, 123, 127, 130 and F. Crouzet, 'Essor declin et Renaissance de l'industrie Française des locomotives, 1838-1914', *Revue d'Histoire Economique et Sociale,* 55 (1977), pp. 121-52.
49. Nadal, 'Spain', pp. 596-9 and 551-2; F. J. Coppa, 'The Italian Tariff and the Conflict Between Agriculture and Industry. The Commercial Policy of Liberal Italy, 1860-1922', *Journal of Economic History,* XXIV (1964) pp. 742-6; L. Cafagna, 'The Industrial Revolution in Italy, 1830-1914', in C. Cipolla (ed), *Fontana Economic History of Europe,* vol. 4(1) (London, 1973) pp. 286-92; M. Falkus, *The Industrialization of Russia, 1700-1914* (London, 1972) pp. 56-7.
50. Crouzet, 'Essor declin et renaissance', pp. 121-52; Caron, *Economic History of Modern France,* pp. 155-9 and see pp. 123-30 and 209-12.
51. O. Crisp, 'Labour and Industrialization in Russia', in P. Mathias and M. Postan (eds), *Cambridge Economic History of Europe,* vol. VII(2) (Cambridge, 1978).

52. See p. 64 and B. Mitchell, 'Statistical Appendix, T.4 to C. Cipolla (ed), *Fontana Economic History of Europe*, vol. 4(2) (London, 1973).
53. See p. 57.
54. See pp. 123-4.
55. Fenoaltea, 'Railroads and Italian Industry Growth', and see p. 72 and Williamson, *Late Nineteenth Century American Development*, ch. 9.
56. Hawke, *Railways and Economic Growth*, p. 376 and Fishlow, pp. 103-16.
57. See pp. 34-5 and Fenoaltea, 'Railroads and Italian Industrial Growth', pp. 338-9.
58. See p. 123 and Fremdling, *Eisenbahnen und Deutsches Wirtschaftswachstum*, pp. 151-8.
59. Williamson, 'Greasing the Wheels', pp. 203-9; Fishlow, pp. 117-18 and pp. 302-3.
60. See pp. 164 and 205 and Fenoaltea, 'Railroads and Italian Industrial Growth', p. 327, and O. Crisp, *Studies in the Russian Economy Before 1914* (London, 1976) chs. 7 and 8.
61. See pp. 28-9 and M. C. Reed, *Investment in Railways in Britain, 1820-40* (London, 1975) and G. R. Hawke and M. C. Reed, 'Railway Capital in the United Kingdom in the Nineteenth Century', *Economic History Review*, XXII (1969) pp. 269-86.
62. But see R. Dumke, 'The Political Economy of German Unification: Tariffs Trade and Politics in the Zollverein Era' (unpublished PhD, University of Wisconsin, 1976).
63. Bairoch, *Commerce exterieur*, pp. 81-7.
64. See pp. 165-66.
65. V. Zamagni, 'Railways Building and Market Integration in Post Unification Italy', unpublished paper delivered to RENFE Conference on *Railways and Market Integration* (Madrid, 1980).
66. See pp. 95-6.
67. See p.206.
68. J. Metzer, 'Railroad Development and Market Integration: The Case of Tsarist Russia', *Journal of Economic History*, XXXIV (1974) pp. 529-49 and W. Kelly, 'Railroad Development and Market Integration in Tsarist Russia: Evidence on Oil Products and Grain', *Journal of Economic History*, XXXVI (1976) pp. 908-16.
69. C. White, 'The Concept of Social Savings in Theory and Practice', *Economic History Review*, XXIX (1976) pp. 88-9.
70. See p. 137.
71. See pp. 45-6.
72. Zamagni, 'Railways Building and Market Integration', and

72. *Continued*
V. Zamagni, *Industrializzazione Esquilibri Regionali in Italia* (1978), chs. 2 and 5.
73. See pp. 148-9, 164-6.

W. N. Parker

The pre-history of the nineteenth century

THE FORMS OF LATE MEDIEVAL INDUSTRY[1]

Industrial activity in Europe in the late fifteenth century fell typically into five forms. Two of these were destined to decline over the following several centuries; one was to continue a vigorous life over the whole period covered in this essay, then virtually to disappear; and two, under

Originally published in Peter Burke (ed.), *The New Cambridge Modern History,* XIII, Compan. Vol. (Cambridge: Cambridge University Press, 1979), 43–79.

[1]Recent surveys have made this body of experience somewhat more accessible to English-language readers. On the subjects treated in this and the following section, see: Fernand Braudel, *Capitalism and Material Life* 1400–1800 (New York, Harper & Row, 1973); *The Cambridge Economic History of Europe* (London, Cambridge University Press, 1952–67); Jan De Vries, *The Economy of Europe in an Age of Crisis 1600–1750* (London, Cambridge University Press, 1976); Hermann Kellenbenz, *The Rise of the European Economy* (London, Weidenfeld & Nicolson, 1976); Gino Luzzatto, *An Economic History of Italy from the Fall of the Roman Empire to the Beginning of the Sixteenth Century* (London, Routledge & Kegan Paul, 1961); H. A. Miskimin, Jr., *The Economy of Early Renaissance Europe 1300–1460* (Englewood Cliffs, New Jersey, Prentice-Hall, Inc., 1969); Domenico Sella, 'European Industries 1500–1700' in C. M. Cipolla (ed.), *The Fontana Economic History of Europe,* vol. 2 (London, Collins/Fontana Books, 1974) and Sylvia Thrupp, 'Medieval Industry 1000–1500' in volume one of the same series (1972).

Among the many articles, see: Max Barkhausen, 'Government Control and Free Enterprise in Western Germany and the Low Countries in the Eighteenth Century' in Peter Earle (ed.), *Essays in European Economic History 1500–1800* (London, Oxford University Press, 1974); Eleanora Carus-Wilson, 'The Woollen Industry' in M. M. Postan and E. E. Rich (eds.), *The Cambridge Economic History of Europe,* vol. II (London, Cambridge University Press, 1952); Hermann Kellenbenz, 'Rural Industries in the West from the End of the Middle Ages to the Eighteenth Century' in Peter Earle (ed.), *Essays in European Economic History 1500–1800;* Herbert Kisch, 'The Impact of the French Revolution on the Lower Rhine Textile Districts – Some Comments on Economic Development and Social Change', *Economic History Review,* 2nd ser., 15 (1962), 304–27, and 'Textile Industries in The Rhineland and Silesia, A Comparative Study', *Journal of Economic History,* 19 (1959), 541–69; Marian Malowist, 'The Economic and Social Development of the Baltic Countries from the Fifteenth to the Seventeenth Century', *Economic History Review,* 2nd ser., 12 (1959), 177–89; Joan Thirsk, 'Industries in the Countryside' in F. J. Fisher pressures from changes in technology, were to blend together to create the industrial technique and organisation, larger-scale and continuously dynamic, that we recognise as characteristically modern.

The *village industry,* descended from the specialised crafts on manorial estates, was perhaps the most widespread of these forms. The serf status of the artisan, continued or restored in eastern Europe, had been permanently transmuted in the West to that of free worker owning his tools and materials. But markets were local, pay was often made in kind, and the artisan, particularly if he held a bit of land from a lord or one of his subtenants, was effectively immobilized. The shoemaker, the smith, the carpenter, the thatcher, the mason, the miller, the butcher, the baker, the weaver – all were distributed in local markets over the countryside, drawing upon the locality for most materials and serving the households of village and rural families. Their work was supplemented by the industry of itinerant craftsmen who transported their capital – i.e. their skills and a few tools – from place to place, eating their way through the countryside, sometimes in the training years of an urban apprenticeship, sometimes in a permanently gypsy-like existence. Below the level of village industry, the primitive *industry of peasant households* for their own or local consumption continued in many more remote areas. Except for the basic tasks of food preparation, it is difficult to find in fifteenth-century Europe, or thereafter, examples of the degree of self-sufficiency in a rural household that characterised the extreme conditions of the American frontier. The village form of social organisations was designed, one might almost suppose, to avoid it, and to afford to an agriculture of low productivity the means to release a few specialised workers for industrial tasks.[2]

(ed.), *Essays in the Economic and Social History of Tudor and Stuart England* (London, Cambridge University Press, 1961); Herman Van Der Wee, 'The Structural Changes and Specialization in the Industry of the Southern Netherlands, 1100–1600', *Economic History Review,* 2nd ser., 28 (1975), 203–21; H. Van Werveke, 'Industrial Growth in the Middle Ages: The Cloth Industry of Flanders', *Economic History Review,* 2nd ser., 6 (1953–4), 237–54.

For older literature, see: Karl Bücher, 'Gewerbe' in the *Handwörterbuch der Stattswissenschuften,* vol. 4 (Jena, G. Fischer, 1892) and *Industrial Evolution* (1901; reprinted 1967; New York, Burt Franklin); J. U. Nef, 'Industrial Enterprise at the Time of the Reformation, c. 1515–c. 1540' and 'Mining and Metallurgy in Medieval Society' in *The Conquest of the Material World* (Chicago, University of Chicago Press, 1964); George Unwin, *Industrial Organization in the Sixteenth and Seventeenth Centuries* (1904; reprinted 1965; New York, Augustus M. Kelley); Max Weber, *General Economic History* translated by Frank Knight (New York, Greenberg, 1927).

[2]The standard references on the history of Renaissance technology are the rather widely known and compendious volumes: Maurice Daumas (ed.), *A History of Technology and Invention,* vol. 2 (New York, New Crown Pub-

The pre-history of the nineteenth century

An immense gap in skill and organisational complexity existed between village and peasant industry and that of the *workshops of urban artisans*. In north Italian cities and the Flemish towns they are as well known to us through the history of the decorative arts, including at their highest the masterpieces of painting and sculpture of the Italian and north European Renaissance, as through the history of useful industry. The form of organisation was much the same in the fine and the practical arts and had not much changed since the flourishing of the craft guilds in the twelfth to fourteenth centuries. A master workman trained journeymen and apprentices, the latter bound to him, almost as family members, for a period of years. The master in turn was controlled to a degree through the guild, which set prices, terms of apprenticeship and standards of quality. The master workman had his own customers, or dealt with a merchant who also brought in supplies. The system, like the village agriculture of the period, remained a mixture of group control, individual initiative, and private property. The resurgence of princely authority since the late Middle Ages had destroyed some of the political power of the guilds but in many cities in 1500 they still formed an important component in town government.

A fourth industrial form present at the outset of the expansive period of European capitalism was *Montanindustrie*, mining, smelting, charcoal burning and quarrying, located with reference to the sources of supplies of the natural raw material. Since these were deep in the mountains and forests, the industries exploiting them tended to be part of landed estates, with labourers closer still to a serf-like status and controls altogether less capitalistic than in the village or urban workshops.

lishers, 1969); T. K. Derry and T.I. Williams, *A Short History of Technology* (London, Oxford University Press, 1961); Melvin Kranzberg and C. W. Pursell, Jr (eds.), *Technology in Western Civilization*, vol. I (New York, Oxford University Press, 1967); Charles Singer *et al.* (eds.), *A History of Technology*, vol. 3 (London, Oxford University Press, 1957).

My own knowledge owes most to A. P. Usher's classic treatment, *A History of Mechanical Invention* (revised edn, New York, McGraw-Hill Book Company, Inc., 1954) and to the books and articles of J. U. Nef, especially *The Rise of the British Coal Industry* (London, G. Routledge & Sons, Ltd., 1932) and his essays reprinted as *The Conquest of the Material World* (Chicago, University of Chicago Press, 1964); also to C. M. Cipolla's interesting little books, *Clocks and Culture 1300–1700* (New York, Walker and Company, 1967) and *Guns and Sails in the Early Phase of European Expansion 1400–1700* (London, Collins, 1965) and Samuel Lilley, *Men, Machines and History* (London, Lawrence & Wishart, 1965) as well as his chapter 'Technological Progress and the Industrial Revolution 1700–1914' in C. M. Cipolla (ed.), *The Fontana Economic History of Europe*, vol. 3 (London, Collins/Fontana, 1973) and Hermann Kellenbenz's chapter 'Technology in the Age of the Scientific Revolution 1500–1700' in volume two of the same series.

Europe's industrialization: the pre-history

The iron industry, often considered in this category, was in fact only partly so. Small iron deposits in shallow diggings were exploited at many scattered locations. Iron production was located with reference to ore and charcoal supplies and, with the fifteenth- and sixteenth-century development of the blast furnaces and rolling mill, to waterpower sources as well. But the further working of bar iron occurred at forges near market locations.

Finally, *merchant-organised networks combining rural and urban labour* were widely employed. In the medieval wool trade a famous division of labour had existed between England, which grew the wool, Flanders, which spun and wove it, and Italy, which dyed and finished it. The growth of the Italian woollen industry in the fourteenth century had displaced this trade by putting out materials within the Italian countryside. But in East Anglia and the west country in England, and in patches on the Continent – in Flanders, Switzerland, parts of northern France – forms of a putting out system had begun to flourish wherever rural labour could be put to use or where waterpower in the fulling operation had drawn that part of the finishing trades to the countryside.

In 1500 the forms of industrial organisation in western Europe then were the following: (a) village and local specialised industry; (b) peasant industry for the household; (c) urban artisan industry; (d) materials-oriented industries in the countryside; (e) merchant-organised systems, combining rural and urban labour. All these forms had been present in the thirteenth century, and all persisted in one corner of Europe or another up through the nineteenth century. The changes of the early and middle modern period, i.e. from 1500 to 1800, which are the subject of this essay, occurred steadily throughout these 350 years, in response to a number of economic and technical factors, in particular as a result of the interaction between market growth and technical change. Before the mid-eighteenth century much the major causative factor, insofar as it is possible to weigh such things, was, as Adam Smith discerned, the steady growth of markets – the increase in the volumes of industrial goods which could be sold.

THE GROWTH OF THE MARKET

To analyse the reasons for the market growth would lead the discussion far into the total economic, social and political history of the early modern period. Evidently a mass of self-reinforcing expansionary processes lay implicit in the European environment and social system of the fifteenth century. They have been only very incompletely laid bare. Population growth, resuming its upward course after the catastrophes of the fourteenth century, must have expanded the margin of cultivation, and in so doing have increased the absolute surplus available to support a non-agricultural workforce. If then economies of scale were present in industry taken as a whole, a rise in industrial productivity would ensue and with it a rise in the market for both manufactures and agricultural products. Or again the growth of the state – the notable feature of Western political history in the sixteenth and seventeenth centuries – centralised and magnified demand for certain specific industrial products – for military equipment, for ships, for coinage, for the constructions and luxury consumption of princely courts. The stream of landed revenues must have been in part diverted to the hands of those who demanded goods with a higher skill and materials component and a lower component of sheer labour services. The growth of taxes, with royal imposts piled on top of feudal dues, must have increased the slice taken from the peasantry whose localised demand patterns were less favourable to concentrated industrial activities. In this same category of explanation must be placed the famous sixteenth-century inflation – perhaps four-fold over the century and affecting agricultural goods more strongly than manufactures. Whatever the distortions produced in the distribution of income, it would not be surprising if a price rise in the presence of some reserves of rural underemployment should have stimulated total demand. It would be interesting, too, to speculate on whether larger supplies of the precious metals and the growth in credit instruments and forms of debt did not themselves extend the market simply by facilitating trade, encouraging the conversion of barter transactions to monetary ones and permitting the accomplishment of trading transactions which under a scarcity of money would not have been consummated at all.

The growth of trade itself may be looked on as an exogenous prime mover insofar as improvements in navigation, ship construction, and the increase of geographical knowledge were involved. Here within the growth of trade itself occurred a reciprocal process, moving from its expansion to the knowledge, growth, and technical changes which made further trade possible. Beyond that, the distributional shifts and social and organisational changes accompanying the trade expansion both affected the growth of the market for industrial goods and the changes in the organisation forms within industry itself; this will be shown in more detail below. Directly, one can attribute major economic effects before the eighteenth century not to the overseas expansion, but to the growth of trade within Europe, the Dutch–Baltic trade, and the trade between the North Sea and the Mediterranean. Yet the overseas trade – round Africa and to the Americas – must have acted as an important exogenous stimulus, whose effects were multiplied within Europe itself. Closely related were the social and intellectual changes which, intertwined with the contemporary religious and political change, made western Europe in the sixteenth century a seedbed of individualistic mercantile and capitalistic industrial enterprise.

The market growth then was accompanied by changes in the shape of demand, by the development of forms of business and market organisation, and by a spirit of enterprise which, taken within its institutional forms, we call capitalistic. It should be emphasised that the effects of market expansion were felt in the sixteenth to eighteenth centuries not only in the mercantile sector, but in industry as well. Village and peasant industry for immediate consumption probably did not flourish, although it maintained its share of local markets until the displacements produced by the technical changes of the eighteenth and nineteenth centuries. But urban workshops, mines, and smelting works prospered, the latter benefiting from some notable technical improvements. To them were added, particularly in the France of Colbert, the royal manufactories, the mints, arsenals, potteries, and textile factories under royal sponsorship and finance. Growth of a state sector was accompanied by the sponsorship of a technical discovery and of science, described in the following section.

Among the industrial forms, it was the merchant-organised and merchant-dominated rural industry which enjoyed the greatest expansion. Where trade and mercantile influence was strongest – in the Midlands and the north of England and along the great river of industry that ran from the Low Countries up the Rhine, across south Germany and over the north Italian plain – the activity was most striking. The actual penetration of the countryside, the use of surplus rural labour, the complex movements of materials and foods had gone far beyond the mere absorption of seasonally idle agricultural workers, or of women of farm households. A rural industrial labour force was present in these locations engaged in full-time industrial activity for the market. Tasks were divided, small pieces of capital intruded themselves at each point. The demand patterns, wealth patterns, and especially the demographic behaviour of this labour force were radically different from those of a peasantry or of industrial workers in cities. At the same time, an alternative system of industrial organisation had not died out. Urban workshops, mills and mines, royal manufactories had increased in number and benefited from the technical changes of the sixteenth century more perhaps than the cottage industry could.

THE PATH OF TECHNICAL CHANGE

Modern industrial society takes its origin in the eighteenth century at the intersection of two historical processes which, though never com-

The pre-history of the nineteenth century

pletely separate from one another, had developed during the Middle Ages and Renaissance in relative independence. One of these was the organisational development identifiable as early capitalistic enterprise. The other was the complex and uncertain process of technical changes.

That these two features of industrial history had existed in isolation in earlier periods is undeniable. Trade and production for private profit occurred in ancient Greece, in South Asia, and in medieval Europe, together with the introduction and diffusion of money as a means of payment. Technical changes were not notable in mercantile capitalism of this sort; indeed it is not clear that the mercantile mentality with its quick calculations and short time horizons is best suited to understanding and fostering the uncertain, obscure, and capital- and skill-intensive activities by which production processes are improved. On the other hand, technical changes had appeared in societies – for example in China – where the drive for maximum profit or the pressure of market failure was relatively remote. The very slowness of technical change over mankind's history, its spottiness, the lack of ready diffusion of its results, may be attributable to the relative isolation of peasant producers, royal households and craft workshops from the force of competition on capitalistically organised markets. So long as merchants dealt in the natural surplus of a region's agriculture, or in goods produced in the local monopolies of the countryside or the guild-dominated city, capitalist competition did not systematically penetrate the structure of production. Technology remained largely a matter of the transmission of the considerable body of hand skills, the arts of industry and agriculture, acquired painfully over thousands of years of industrial history and held tenuously in the brains and trained muscles of the living generation of craftsmen or in the pages of a very few hand-copied texts and treatises. In technology's long and uncertain development before 1800, new ideas appear rarely and when they do, it is as if in the thought processes of an absent-minded man. They do not diffuse readily over space, nor are they followed up in all their refinements and implications even at their points of origin. There is a lack of concentration in the history, a lack of cumulative effect in the development.

Nevertheless the industry of the sixteenth century benefited from an accretion of inventions that had occurred in Europe during the preceding 500 years. Among power sources, to be sure, no striking innovation occurred. Only in the development of firearms was the expansive power of gases harnessed to any use. Wind and water remained, and were to remain till the late eighteenth century, the only inanimate prime movers with appreciable industrial use. Except for water in mill ponds and dams, they could not be stored; hence industrial operations beyond the strength of men or beasts were as dependent on the variability of natural forces as was agriculture. In the face of this restriction, the main development in the sixteenth and seventeenth centuries was the diffusion, development and generalisation in a variety of uses, of the water wheel. In mining, in iron working, in fulling and other operations requiring a stamping and hammering motion, the water wheel hitched to many ingenious gearing devices came into much wider use.

Connected to this development were several improvements in the other components of an integrated technology: the provision of raw materials and the transmission of power. Materials supplies were increased in the sixteenth century by improvements in mining, in paper-making, and through the development of the blast furnace, which spread probably from the neighborhood of Liège up the Rhine to the south-east, and also north across the channel. The largest branch of industry – textiles – remained dependent, however, on traditional raw materials: wool from England and Spain, flax from the North Sea coast, and a little cotton from Egypt and the Middle East. It was the extension of trade within Europe and the geographical exploration overseas rather than technical change that widened European industry's resource base. Precious metals from the New World, and increased supplies of Baltic timber and Swedish bar iron were important accessions.

On the whole industry improved most through the further development of mechanisms for the transmission of power. Here the structure of the technology was, in a sense, best able to yield to the pressures and incentives of a growing demand. The lathe, that marvellous late medieval tool, was developed, improved and adapted to many uses. The products of the smithy, forge and machine shop were not the complex forms of machinery for further production known to the late eighteenth century, but direct consumer goods: firearms, clocks and watches, scientific instruments, furniture, hardware of all sorts, crude agricultural implements. No invention in the operations of the textile industry in its many branches approached that late medieval invention, the spinning-wheel, in productivity-raising effect. Where inventions occurred, for example, the stocking frame and the ribbon loom, they had a striking, but rather localised impact. Diffusion was slow, and the basic operations of cloth-making remained unaffected. Most characteristic of the sixteenth and seventeenth centuries was the strenuous activity of the industries involving the largely unmechanised assembly of parts and materials – construction and ship-building. Here organisation was important; although designs, styles and materials altered, the crafts and their tools remained much as the Middle Ages had left them.

In the three centuries prior to 1750, then, some changes occurred in the three branches of technology: power generation, power transmission and materials production. Particularly between about 1450 and

1600, a mini-revolution may be identified in the application of waterpower in mining and iron-making, the developments in smelting, and in the invention of ingenious mechanisms, especially in branches of production specifically stimulated by a spreading luxury demand – firearms, printing, clocks. Many of the improvements originated in south Germany, and the diffusion of inventions was fairly rapid after the invention of printing and where the items themselves moved in trade. In the imaginative notebooks of Leonardo da Vinci sketches of these Renaissance inventions went far beyond actual practice, but retained that scattered quality characteristic of all pre-modern technical change. They depended in no respect upon the introduction of any drastically new scientific or engineering principle, such as was to characterise invention in the late eighteenth and early nineteenth centuries, and they had little interaction with one another. No chains of rapid inventive progress were forged to pull productivity along in one industry after another, with the steady upward movement that became the mark of the Industrial Revolution and the industrial history that followed it. The productivity growth may have contributed to the relatively less rapidly rising prices of industrial goods in the sixteenth-century inflation but the growth of demand appeared still largely dependent on the factors mentioned above – population movements, political change, geographical discovery, the increase in trade. Technology was learning, in a sense, to respond to market incentives, but it could not yet lead the way to continuous market growth.

THE GENERATING INSTITUTIONS OF SCIENTIFIC AND TECHNICAL KNOWLEDGE[3]

The historian, at least when he works on a period where data are scarce, is always part-novelist, employing a narrative rhetoric in which the tone

[3]Some useful references from the large literature in this field include the following: J. D. Bernal, *Science in History*, vol. 2 (London, Penguin Books, 1969); A. F. Burstall, *A History of Mechanical Engineering* (Cambridge, Massachusetts, M.I.T. Press, 1965); E. A. Burtt, *The Metaphysical Foundations of Modern Physical Science* (New York, Doubleday, 1955); Herbert Butterfield, *The Origins of Modern Science 1300–1800* (London, G. Bell and Sons Ltd., 1950); G. N. Clark, *Science and Social Welfare in the Age of Newton* (second edn, London, Oxford University Press, 1949) and *The Seventeenth Century* (New York, Oxford University Press, 1961); A. C. Crombie, *Mechanical and Early Modern Science*, 2 vols. (New York, Doubleday, 1959); A. R. Hall, *The Scientific Revolution, 1500–1800* (Boston, Beacon Press, 1956) and 'Scientific Method and the Progress of Techniques' in E. E. Rich and C. H. Wilson (eds.), *The Cambridge Economic History of Europe*, vol. IV (London, Cambridge University Press, 1967); Peter Mathias, "Who Unbound Prometheus? Science

and emphases of the discussion and the arrangement of its parts contribute to its interest and its verisimilitude. At this point in this essay, it would be appropriate and straightforward to move directly into the English Industrial Revolution. When we do that, in the following section, it will become apparent how readily that central development follows upon the general market growth, the articulated economic institutions and the rather diffused technological change of the sixteenth and seventeenth centuries. For a treatment, which encompasses the nineteenth century, even in part, on the Continent, however, it is necessary to consider another early modern development, whose connection with the inventions of the fifty years following 1750 is thought by most scholars to have been tenuous, but whose underlying importance for the path of modern industrial development as a whole can hardly be denied. This development is, of course, the growth of fundamental science, with a particular body of social institutions to carry it on, and a particular mentality, a way of looking at the world that we now recognise as 'modern'.

Scientific thought is akin to the rising capitalism of the Renaissance in several respects. Both are materialistic philosophies *of this world*, and both – at least in their European form – were conceived as activities of individuals rather than of social or corporate entities. Both began to grow in European society before the Protestant Reformation and were indeed part of the social and intellectual ferment out of which the Renaissance, the Reformation, and the modern national state arose. Both involved a trust in tangible sense data and both were rebellious against authority, especially when it interfered with the individual pursuit of gain, or of scientific truth. The medieval Catholic world had furnished not only a relation of man to God and His Church, but also a sense of the social whole. It was a world in which ideally every part and

and Technical Change, 1600–1800' in Peter Mathias (ed.), *Science and Society 1600–1900* (London, Cambridge University Press, 1972); R. K. Merton, *Science, Technology and Society in Seventeenth Century England* (New York, Harper Torchbooks, 1970); René Talon, *Reason and Chance in Scientific Discovery* (New York, Philosophical Library Inc., 1957); A. N. Whitehead, *Science and the Modern World* (New York, Pelican Mentor Books, 1948); Edgar Zilsel, 'The Sociological Roots of Science', *American Journal of Sociology*, 47 (1942), 544–62.

On the ancient argument over Protestantism and Capitalism, an article by Herbert Lüthy, 'Once Again: Calvinism and Capitalism', *Encounter*, 22 (1964), 26–38, contains a new point of view which I have adopted here. A recent thoughtful statement of a different, but by no means contradictory view, is made by Albert O. Hirschman, *The Passion and the Interests* (Princeton, New Jersey, Princeton University Press, 1977).

person depended upon every other, the whole bound together by a common belief in another world, and by a common magic – the magic of the church and its sacraments, as much as in any primitive society studied by anthropologists in recent times. Against this social sense, with its supernatural sanctions, capitalism put the pursuit of individual gain, without regard to just prices, usury restrictions, or any ultra-mundane devotion. Science, though long retaining a magical and religious aura, depended in Galileo or Bacon on an individual mind's search for truth by observation and experiment. The Protestant Reformation, expressing a similar individualism in the search for the soul's salvation, achieved a rapid, religious symbiosis with both capitalism and science which Catholicism never could attain.

Yet neither capitalism nor science could avoid or dispense with the institutionalisation of their thought and behaviour patterns, the regularisation of the norms by which both money-making and truth-finding might be legitimately achieved and success tested and identified, and the development of devices for communicating their culture to others and to successive generations. These tasks were vested only partially in the state; private business activities, private agreements and codes of behaviour, private meetings and correspondence formed the basic stratum of a developing modern culture in the economic and intellectual life of northern Europe in the seventeenth and eighteenth centuries. The small peer groups – companies of merchants, societies of amateur scientists, corresponding members of university faculties – began to form a significant social class, an aristocracy of money, enterprise or intellect within which the rewards and sanctions exceeded anything that a king or his courts could have imposed.

The free European market both in goods and ideas developed, however, in the presence of – one might almost say under the very nose of – a state apparatus which in the sixteenth and seventeenth centuries did not remotely resemble that of the nineteenth century when capitalism and science had become dominant. The relation to this state – the royal and centralist state of sixteenth-century England and seventeenth-century France – was ambivalent and, like most ambivalent relationships, stormy. Merchants and manufacturers depended on the growing power of the central government for many things: first, for special privileges, grants of monopoly, contracts and trading rights; second, for the establishment of a currency system and the chartering and protection of financial institutions; third, for the legitimising of commercial contracts, the protection of property, and to a considerable degree the control of the labour force. On the other hand, their interests were not synonymous with those of the state, and the presence both of the landed interest and the monarch himself ensured a conflict. Clearly for commercial expansion, the optimal arrangement was that experienced in the Dutch Republic after its successful revolt against the Spanish emperor and in England after the 'Glorious Revolution' of 1688: i.e. the substantial hegemony of the mercantile interest in the conduct of government. In France, the policies of Richelieu, Mazarin, and Colbert directed toward fostering trade and industry were a rather weak substitute for the stimulus given to trade by the provision of a greater degree of bourgeois freedom in the Protestant lands.[4]

However, for the development of science, as distinguished from practical improvements in technology, it is not clear that the dominance of society by the commercial classes was the most favourable arrangement. The role of scientist, even in the amateur or non-institutionalised science of the Renaissance, is closer to that of the theologian or scholar-priest than to that of the inventor of practical technology. He is the elaborator of the true view of the world, employing a method involving both reasoning and appeals to sense data, to give men an understanding of where they are in relation to the universe, and what is the ultimate constitution of matter and material forces. The sponsorship of such an institution devolved naturally upon those who in previous ages had sponsored the church and ecclesiastical foundations and activities. Included among these foundations were the universities, which in the sixteenth and seventeenth centuries were beginning to change from the monkish groups of the Middle Ages engaged in theological disputation into the general purpose institutions of knowledge we know today.

Science then, of the more or less 'pure' variety – the natural philosophy which investigated astronomy, mechanics, and even chemistry – found the growing interest in its methods and its findings institutionalised and sponsored in several directions: first, in small private groups and societies of interested men, rich enough to pursue such a hobby; second, under specific sponsorship and financial support by noblemen and monarchs and even by wealthy bankers or merchants, as part of a general sponsorship of the arts; third, in the universities, as separately endowed foundations under royal or ecclesiastical patronage. In the Protestant states, private groups with some royal sponsorship were largely responsible for the growing body of experimentation and research; in southern Europe, science had to live under the watchful eye of the church and the Inquisition; in France, a characteristic mixture of Catholic and Protestant forms, in this as in many other things, offered perhaps the most favour-

[4]Nothing has surpassed the treatment of French and other mercantilisms by C. H. Wilson in 'Trade, Society and the State' in E. E. Rich and C. H. Wilson (eds.), *The Cambridge Economic History of Europe*, vol. IV (London, Cambridge University Press, 1967).

able climate for the development. The Inquisition was absent; sponsorship by the crown, eager to exhibit itself as the source of all knowledge and all light, was generous, yet a spirit of free rationalism was not smothered. By the eighteenth century, these institutions had matured into a strong network of intercommunicating groups. Even in Prussia, Russia, and Austria, the monarchs of the Enlightenment, affected by a culture that had its home and origin in France, founded schools and academies. At length that Black Prince of the Enlightenment, the Emperor Napoleon I, in France itself, established a legal, educational, and professional structure for the country which crystallised and routinised the seventeenth- and eighteenth-century practice.

From Galileo to Darwin, then, European science participated in, and even led, the general development of a triumphant and glorious secular culture, its rationalism animated by a glowing faith in reason, its materialism made endurable by a strongly felt esthetic, its potentially corroding individualism checked and channelled by an emergent nationalism, by strong state power, and the sense of participation internationally in a developing bourgeois culture, replacing the universal aristocratic and ecclesiastical culture of Catholic Christianity. In this cultural climate the Protestant sects – Lutheran, Calvinist, Puritan, Baptist, and even the national churches that had broken with Rome – flourished. But the back of ecclesiastical domination of thought and science, and the fine arts had been broken by Luther's revolt and Henry VIII's bullying, and by the wars of religion in the Netherlands, France, and the German states. Theology was no longer the queen of the sciences. Catholicism was no longer Europe's state religion. Despite the emphatic personal ethics of its sixteenth- and seventeenth-century founders, and the persistence of such an ethic far into the nineteenth century, Protestantism was in fact a much weaker form of social control than Catholicism had been. By the nineteenth century, the scientist, the political economist, the politician, and the businessman had replaced the courtier and the priest.

From the viewpoint of industrial history, it is not clear that European science until the middle of the nineteenth century was of much practical value. Industrial technology, and agricultural technology too, developed by 'tinkering', i.e., by rather random experimentation aimed at some useful object. The details of the process are considered in the next section, where it becomes evident that the accumulation of various series of such efforts, produced at length the climax of interaction that we call the Industrial Revolution. The scientific investigation of nature proceeded along somewhat different lines. Yet parallelisms or interconnections between technological and scientific progress may be observed. For one thing, it seems quite apparent that much the same attitudes of mind motivated and guided both processes. A rational faith in the orderliness and predictability of physical processes and the stability of physical materials was combined with a refusal to accept any evidences of this faith except those offered by direct observation and material demonstration – a rejection of authority and of history, of all that could not be personally and individually seen, communicated, and made available to be confirmed by others. The difference between scientists and practical men before the nineteenth century lay in the scientists' effort to generalise their results with the instrument of mathematics, so as to produce 'laws' of nature. Where technological change stopped with the development of a useful device, process, or material, going on only to employ it in further uses or in other inventions, the effort of scientists was to produce a general statement which would state the enduring relationship, preferably with a formula showing the magnitudes of the quantities and effects involved.

Given the similarities in the basic animating attitudes, it is no accident that both 'pure' science and applied technology experienced a lift in the intellectual climate of the centuries in which protestantism and capitalism grew, and a culture of rational humanism spread out from Italy over western and northern Europe. Nor is it an accident that the branch of science and technology which was first to yield up its secrets to the curiosity and contrivances of men in these centuries was that in which natural forces and materials are most ostentatiously displayed to the naked eye: the science and art of mechanics. By the eighteenth century, both the science and the art were well advanced, and one can then see forming between the two, the profession so important for nineteenth-century development, the mechanical engineer, acquainted with scientific principles, possessed of an adequate knowledge of mathematics, concerned with the exact and quantitative statement of a phenomenon or relationship, and interested in relating this knowledge and technique to the improvement of the useful arts, of machinery construction, bridge-building, road-building, mining, navigation. The development of such a profession requires that both the science and the technology in a field be at a certain point of development, each having arrived there separately to a degree and by empirical methods, a point which makes it possible to relate theory and practice in a stream of systematic improvements. But to say that this relationship could occur, first in mechanics, then in hydraulics (a branch of mechanics), then – beginning in the mid-nineteenth century in chemistry, finally in electricity, and the life sciences – is not to argue of either scientific progress or of technological change that one was the cause of the other. Bits of interconnection can of course be found. The devices and materials developed in industry were available for scientific experimentation, and the demands of scientists had a stimulating effect on the development of measuring, time-keeping, and other instrumental equipment. But relative to industrial markets generally, the demands of scientists were but one of many luxury demands of individuals and the state – like the demand for fine tapestries, firearms, china – which a developing body of industrial crafts could serve. By and large, in the seventeenth and eighteenth centuries science and technology grew up together like twins in a family out of a common culture which had deep-hidden social and socio-psychological, and ideological, origins.

THE INDUSTRIAL REVOLUTION: TECHNOLOGICAL ASPECTS

The brilliant mercantile expansion of northern Europe in the seventeenth and eighteenth centuries was accompanied by a measure of technical change. Europe was here involved in a complex social process in which three aspects were of predominant importance: mentality, scale and feedbacks. Of the mentality we have already spoken: an inquisitiveness about nature and a greedy desire to improve on her workings for practical ends. The scale depended partly on the diffusion of these attitudes and the links of communication between inventors and producers at points as distant as Italy and England, Sweden and Spain. With the economic and industrial awakening of northern Europe after 1550, a sharp increase had occurred in the area and in the population over which trade and the exchange of ideas and devices of technology were diffused. The growth in intra-European trade, plus the small but marginally very significant links with the nascent overseas empires in the Indies, the African coast, and the Americas, meant a growth in market size which had many economic and productivity-raising effects. These depended largely on the spreading of fixed costs in tangible and intangible social capital of which the new nation states could avail themselves. Rivers, harbours, shipyards, dock facilities became more crowded and more fully used. Knowledge of ship construction, the shipping lanes, and navigation became more widely spread and shared. As shipping routes became more complex, with numerous burgeoning ports of call, waste space and empty return hauls diminished.[5] The states grew strong enough to war on each other, which was a waste, but also to suppress internal tolls, bandits and marauders on land and sea, and to establish that chief public good of the modern state, internal peace and order, so as to permit the easy development of commercial practices, and laws of property and contract, enforced by the king's courts.[6] The growth of navies made possible the convoying of

[5]D. C. North, 'Sources of Productivity Change in Ocean Shipping', *Journal of Political Economy*, 76 (1968), 45–69.
[6]D. C. North and R. P. Thomas, *The Rise of the Western World* (London,

unarmed merchant ships, and the use for cargo of the space previously taken up by guns and fighting men. As such extensions of scale lowered production and transport costs, incomes rose and the effective scale of the market was further increased. This purely economic feedback, dependent upon the phenomenon of decreasing social costs, was not, however, the most important effect of the expansion in the European economy and its interconnected commercial and industrial culture. To understand the crucial effect on the process of technological change, it is necessary to consider the process in a little more detail.[7]

Technological change, at least before the age of the research laboratory and a developed engineering and scientific base, depended upon the unorganised ideas and obsessions of individual inventors. These men came from various occupations, and often were themselves tool-users, conversant with some branch of production and observant of means of improving it. They became characteristically seized with a specific problem – the mechanising of stocking knitting, the casting of large pots, the working out of a control mechanism for a clock, the smelting of iron ore with coal – which they pursued with single-minded intent. Whether they solved it or not was partly a matter of luck, but it was also dependent on the ideas, materials, instruments and auxiliary devices with which their minds and their workshops were furnished. Given that an inventor was seized by a problem, his efforts to solve it were carried

Cambridge University Press, 1973); W. S. Holdsworth, *A History of English Law*, vol. 8, 10, 11 (7th edn revised, London, Methuen & Co. Ltd., 1973, 1966, 1973).

[7]See references cited in note 2, pp. 16–17. A good bibliography on all the subjects in this and the following section is given in David Landes' contribution to *The Cambridge Economic History of Europe*, 'Technological Change and Development in Western Europe, 1750–1914', in vol. VI, part 2, H. J. Habakkuk and M. M. Postan (eds.) (London, Cambridge University Press, 1965), extended and published separately, without bibliography, as *The Unbound Prometheus* (London, Cambridge University Press, 1969). The role of science and technology has been re-examined by A. E. Musson and Eric Robinson, *Science and Technology in the Industrial Revolution* (Manchester, Manchester University Press, 1969) and D. S. L. Cardwell, *Turning Points in Western Technology* (New York, Science History Publications, 1972) and *The Organisation of Science in England* (revised edn, London, Heinemann, 1972). The coal technology and its effects on industrial skills and locations have been the subject of researches by J. R. Harris and his students. See J. R. Harris, 'Skills, Coal and British Industry in the Eighteenth Century', *History*, 61 (1976), 167–82; Jennifer Tann, 'Fuel Saving in the Process Industries during the Industrial Revolution', *Business History*, 15 (1973), 149–59. Two other recent treatments with new materials are R. L. Hills, *Power in the Industrial Revolution* (Manchester, Manchester University Press, 1970) and C. K. Hyde, *Technological Change and the British Iron Industry, 1700–1870* (Princeton, New Jersey, Princeton University Press, 1977).

out on a stage which was set by all that he knew, and all the equipment and materials available at the time. At some point, in a successful invention, a moment arrived when, after weeks or years of conscious and unconscious concern with the solution, the combination of elements in the environment was hit upon, by accident, even in a dream, or sometimes with conscious design, which provided a feasible, economical solution. Following this, a period, called by Usher the period of 'critical revision', occurred in which the invention was refined, ancillary improvements were introduced, a model was made, a patent acquired and production begun. Even here, and in the early stages of the production process, many small inventions, small novelties were introduced. The whole was an act of creation perfectly analogous to the creation of a pure work of art or of the intellect.

Now the function of the environment in this process was to provide the inventor both with the problem and the means and stimulus for its solution. But the availability of such ideas and materials depended much upon how wide was the inventor's world and vision. The growth in the scale of the economy, and of western European industrial society, was a widening of this world. To provide in England, Swedish iron, or ideas published in German books, was to offer to English inventors materials and information which greatly facilitated their efforts. The function of scale expansion then, apart from spreading the fixed costs of equipment, transport, public goods, and knowledge, was to increase the communication of such ideas and the availability of such materials and equipment. Across western Europe, in all the industrial areas, a race of inventors appeared in the wake of the industrial and commercial expansion. But it required a particular intensity of economic life, a rather strong concentration of industrial opportunity, a rather close-knit nexus of communication to produce a flowering of inventive activity. The combination of attitudes, ideas, knowledge, ambitions, commercial opportunity and the protective institutions of property and patent rights – all this was required to coax out invention and then, once it was begun, to give it its head.

Hence it was that 'the' Industrial Revolution occurred not in the old industrial areas of the continent, in royal factories, or towns dominated by princes or the remains of guilds – but in England where trade expansion, capitalist institutions, a pragmatic view of the world, and a social structure that gave common tradesmen and mechanics appreciable freedom, were all simultaneously present and on an adequate scale. The Netherlands, a closely analogous case in many respects, could not furnish so large an internal market or so broad a base in industry and industrial resources. England, on the other hand, which had shared but not dominated the European expansion in scientific knowledge, main-

Europe's industrialization: the pre-history

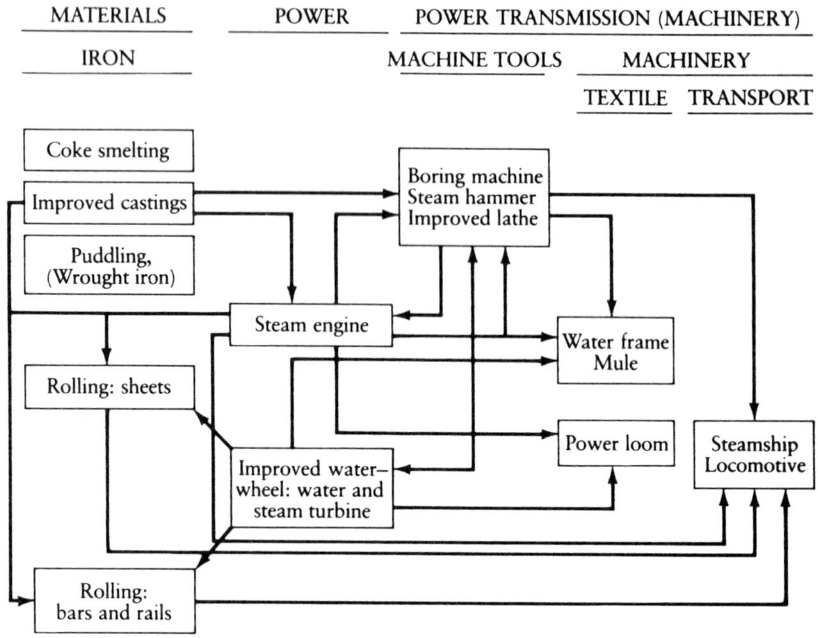

Fig. 1. The Industrial Revolution: sequence and technical interconnections, 1750–1850.

taining her connection with the intellectual sources of invention, offered a commercial climate in which the activity could flourish.

Given this locus for inventions, it is of interest to examine their specific interconnections. Figure 1 divides inventions into three groups: the production of materials, the generation of power from inanimate sources, and the transmission of power – its 'harnessing' to move the right part, at the right speed in the right direction. These branches of technology in the eighteenth century existed in what may seem by some retrospective standard to have been uneven states of development. The third, being mainly the province of mechanical technology, was the most advanced. Using parts made of wood or of iron, employing the power of animals (including men), of gravity, generally with the medium of falling or moving water, or of the wind, a body of equipment had developed since antiquity, which under the inducements of the sixteenth and seventeenth centuries had acquired considerable sophistication. Until the middle of the nineteenth century, it was these mechanisms which showed the greatest development. Most of the industrial revolution was the extension of machinery – now built of iron – into new types of operations and new branches of manufacture and agriculture. Underlying the whole development of the mechanical side was the elaboration of machine tools and the development of a profession of mechanical engineering whose calculations could improve design and efficiency. The mechanical inventions of the period 1760–1850 make a very long list indeed, and developed partly out of one another, employing parts, devices, or ideas to produce a widening array of products.

Alongside these inventions, the other two branches of technology have only two major changes to offer: in materials production, the improvement and cheapening of iron through smelting, refining, and further working; and in power generation, the development of the steam engine, itself largely a mechanical invention except for the novel power source employed. The important improvements in the water wheel came largely as the result of careful application of principles of mechanical and hydraulic engineering to an old device. It may be thought a convenient formulation to call these 'enabling inventions,' which occurred at just the time and to just the degree to permit the mechanical industrial revolution to flourish.

It would take a long book to begin to specify all the interconnections among the mechanical inventions. The story of Watt's dependence on Wilkinson's boring machine to give him cylinders machined to a fine enough tolerance is but one example. Through the mass of technical interrelations an evolving body of machine technology was initiated. Its creation required a number of economic, social and intellectual conditions, notably: (1) a moderately large and well-organised market sufficient to make invention profitable; (2) within that market, some bottlenecks focussing attention on specific points in the technical processes of manufacture, offering rewards for the solution to specific technical problems; (3) a body of skilled mechanics, tool-users, and ingenious practical men who kept in touch with one another and with manufacturers; (4) means of finance at least at the very modest levels required in this sort of research and development; (5) an absence of interference or excessive direction or planning from a central authority, whether the state or money market. In these circumstances, invention could become a kind of folk activity, done repeatedly, on a very small scale, by very many different operators. An attack of this kind on the secrets of nature is sometimes called a 'shotgun approach'–the firing of many small missiles at the target. Obviously such a process entails great individual loss, many bankruptcies and much suffering. In the absence of full knowledge of the underlying science, no other way of guiding invention was possible.

Given this social technique of invention, it was certainly no accident that mechanical invention flourished while subtler forms and problems of technology–in chemistry or physics, animal breeding or the plant sciences for example–lay untouched. Most of its basic knowledge goes back to Aristotle and Archimedes and the development of mechanisms for specific tasks was not a matter of research but of contrivance. Ingenuity and imagination were required, but the instrument of observation was the naked eye and, unlike the case in agricultural sciences, the success or failure of an experiment could be immediately known. No historian of technology has yet arrayed discoveries in various branches in order of their inherent difficulty, or in relation to the capital equipment in instruments, scientific knowledge, and experience required to bring them to light. Yet such an investigation must underlie the understanding of why modern technology developed in the forms and sequences that in fact it did.

The fact that mechanical inventions came first, employing iron and moved by waterpower or by the steam engine using wood or coal is, however, a fact of vast significance for modern industrial history. Its significance for the development of the organisational form of modern industry – the factory – built around a central power source is considered in the next section. Equally important was the way it reinforced, almost as if by design, the commercial supremacy of England for the hundred years after 1770. The technology of coal and iron was, by modern standards, a crude and simple technology which favoured, locationally speaking, areas well-endowed with those minerals. Such accidents happen in economic history where the effects of resources, techniques, and commercial culture are so closely intertwined. How different might the picture of industrial location in the nineteenth century have been had not coal, but oil, been the first prime industrial fuel, and water power or water vapour figured less prominently! The large coal and ore resources conveniently located near water transport in Great Britain set the seal, as it were, on the predominance that Britain's commercial expansion had already initiated. Other coal and iron areas at that period of history had generally one raw material or the other, but not both, or were themselves remote from intense commercial activity, and British capital was not yet moving out to such locations with the vigour and abundance that appeared in its worldwide expansion after 1850. Other strongly commercial areas, the Netherlands or later New England, did not overlay, in close contact, the extensive subsoil deposits that could create a local mining industry. In Britain, both parts of the puzzle fell into place and a complex industrial development suffered no check from any direction. But while all this is true, it was also the case that Britain really held no secrets in the minds and equipment of her inventors which others could not readily copy. If the Industrial Revolution was built on a combination of aggressiveness and physical advantage, there might come a time when others who had or could acquire similar endowment could adopt the attitudes that produced so exciting a result. The British development was like the growth of a tropical forest in a favourable climate and on the soil of a shallow technology which proved exceedingly fertile for a short space of historical time. As technology advanced in step with a developing body of science and scientific techniques of gaining more knowledge, its roots went deeper, its variety became richer, and its applications in the world's areas came to depend less on climate, harbours and mineral wealth at home, and more upon the adaptability to local conditions that a profound knowledge of the ways of nature would permit.

As the Pandora's box of modern technical knowledge was opened, the creatures which lay on top swarmed out first. And the situation was made the more complicated by the fact of not one, but two Pandora's boxes – one labeled Technology and one labeled Science. It is not fair to English science to suppose that because inventions poured out of the Technology box in such profusion, the Science box lay unopened. But in fact for British technology the commercial opportunities in the late eighteenth and early nineteenth centuries were so rich that the mass of inventions of immediate practicality darkened the sun. Like rich lumps of ore lying on the surface of a deep deposit, the mechanical inventions, involving simply the physical relations of one piece of visible matter to another, lay open, shiny and attractive, available at the cost of a little ingenuity and cleverness. What lay below the external appearance of matter, what forces bound it together, even giving it life, could not be known and worked for the practical power it gave over nature until the Science box had been opened for several generations. A science of mechanics could be developed in the eighteenth century, and a mechanical conception of nature, originating in the study of celestial mechanics, could take hold of men's imaginations at the same time that mechanical technology could proliferate in practical uses. As the profession of mechanical engineer developed, the science could even be of use in the art of invention, particularly in perfecting its details and increasing its efficiency. And purely empirical research could go a long way towards harnessing even forces as remote and dangerous as electricity or the chemical processes of metallurgy or the explosive power of gunpowder or combustible vapours. But as the mechanical technology presupposed a certain level of familiarity with mechanics, these later technological developments came about in the presence of an advancing science already looking into their essential nature and causes. The growth of that sort of knowledge in turn required experience with, and improvements on, the methods, instruments, and language of science itself.

THE INDUSTRIAL REVOLUTION: ORGANISATIONAL ASPECTS

The path of mechanisation and its attendant industrial change through the industries and regions of western Europe is sketched briefly in the final section of this essay. Before discussing the spread of technology, however, it is important to specify more exactly the old and the novel elements in the socio-economic organisation which surrounded, permitted, and even stimulated it, and ultimately were so strongly shaped by it. Changes in the status and condition of the working class were the most sensational social effects of the new industrial forms and techniques. A subtle analogy exists, however, between the balance of liberty and authority in labour organisation and similar balances in capital markets and in the social and intellectual history more generally.

In all respects English society appears to have been freer, more fluid, closer to an ideal market economy, even in the eighteenth century, than were the kingdoms of continental Europe. Much erosion of medieval and even of mercantilist restrictions and conceptions had occurred with the development of trade and the working class. The almost purely capitalistic form on which English trade was conducted and financed promoted the diffusion of a money economy and market relationships and motions from international commodity markets into the markets for capital, technology, and labour. The wide extension of the putting-out system – the creation of mercantile enterprise – gave entrée for capital into industrial processes, but it was not wholly responsible for the revolutionary result. Putting-out existed on the Continent, too, without leading readily into a dynamic process of technical change. But the fact that some mobility of capital, ideas and labour was good does not imply that more would have been better. English society of the late eighteenth century must have contained a balance of mobility and rigidity, of fluidity and 'lumpiness', of freedom and authority very nearly optimal for economic growth in that stage of technology and markets. To make this point more explicitly, we must consider the function played by monopoly and immobilities in productive factors in three areas: the generation of technology, the accumulation of capital, and the organisation of the labour force.

Technological change.[8] In a competitive economy organised by small-scale producers, the generation of new techniques is both stimulated and directed by economic circumstance. If a new product or new process can be devised and temporarily or permanently monopolised by a producer, monopoly profit can be derived from it. The incentive to innovate should be high, particularly if the market is wide, but not so wide as to make its monopolisation wholly beyond the reach of any

[8]Some of the extensive literature on induced innovation and related topics is examined, and many important insights given, in Nathan Rosenberg's collected essays, *Perspectives on Technology* (London, Cambridge University Press, 1976) especially chapters 4–6.

single producer who can obtain a modest cost or sales advantage by an invention. Or in a competitive industry producers may be threatened with the narrowing of profit margins, or even with bankruptcy by sudden external developments. Competing products suddenly coming in from abroad or from regions opened up by transport improvements, a tightening of the labour market, or rises in raw materials costs, may produce this threat. Producers then may feel pressure to seek for innovations, to shift out of production, to find other sources of labour or supplies and to apply any other available profit-maintaining or cost-reducing devices.

Innovation induced in these ways is part of competitive producers' efforts to escape from the pressure of competition on profits by whatever means are at hand. The effectiveness of the incentive was recognised in England and on the Continent in the eighteenth century by the development of patent laws, themselves a device drawn from the arsenal of state mercantilist policy. Patents, by giving a monopoly, or property right, in an invention, need not have a total social effect of stimulating the competitive search for innovation or efficiency. The effects of royal grants of monopoly in the seventeenth century differed little from the effects of the guild restrictions of medieval industry. But given for a limited period in an atmosphere of competition, patents may offer the promise or the illusion of encouraging invention without perpetuating monopoly. On the whole patents appear to have added a bit to the incentive to innovate without much encumbering the spread of knowledge. Furthermore, as an unintended side-effect, patenting created a new product: technology, which could be licensed or sold. It helped to professionalise the trades of inventor and engineer and to separate invention from production, creating a specialised activity with its own organisation and rewards. This aspect of the development did not become prominent until late in the nineteenth century.

Now the special historical conditions under which patents and property rights in knowledge could have an incentive effect on the creation of new technology appear to have obtained in the hundred years from 1750 to 1850 in Great Britain. Most important in the result was the relatively small scale of production. An industrial production derived not from the monopolies of craft shops but in large measure from the mercantile organisation of the putting-out system was animated less by the 'instinct of workmanship' than by the drive for profits. Even in a monopoly, if a profit-maximising spirit is preserved, the possibility of adding to profits acts as an inducement to invention. But the atmosphere was one of a world not of monopoly but of small competitive ventures, each seeking to capture a bit of profit, as in a trading venture, and needing some special place in the production structure to do so.

Europe's industrialization: the pre-history

Capital was fairly mobile within the trades and competition was keen. How much more then must the drive for profits have been active where profits were continually threatened, where competition or external developments brought the face of bankruptcy and ruin up close, pressed stark against every producer's window! Not monopoly, not pure competition, but the uneasy monopoly in a competitive world offered the economic – as distinct from a sheer intellectual – stimulus to invention.[9]

A rush to invention as a means to gain or to maintain profits in a competitive economy suggests that invention is in some sense easier to come by than the rearrangement of capital and productive factors through markets along different lines. Indeed this is what is meant when it is said that invention is spurred on by the fear of loss along a given line of production. Such loss can be felt only if productive factors are immobile, if sunk capital or acquired skills cannot be converted or liquidated and shifted to other use, if the supply of a factor to an individual producer is not perfectly elastic. In order for bottlenecks in production to arise, creating quasi-rents for some and threatening ruin for others, a degree of factor immobility is necessary, and it must appear to producers more nearly possible to break such bottlenecks cheaply, not by shifting factor proportions, or moving out of the industry, but by focussing inventive talent on some portion of the production process. Inventors must be, in a sense, more mobile than capital or labour, if economic circumstances as such are to produce a spur to invention. But in fact it is just this generalisation of mechanical and inventive talent, this flexibility and mobility that appear to have existed in eighteenth-century England and nineteenth-century USA – perhaps even more strongly in the latter where a 'jack of all trades' tradition had been stimulated by the conditions of the frontier.

The scale on which inventive activity occurs and the openness of lines of communication among inventors and between them and producers is also an important feature. Large-scale organisation, with good communications, makes for specialised activity which narrows an inventor's focus, but a large economy also provides a mass of industry over which inventive talent can work. Were the market to reign supreme, without other means of communication or other incentive to inventors, specialisation might become too narrow and invention too closely directed to specific ends. The inventor must see beyond the end of his nose, or of a profit and loss account. Were the market not to exist at all, an inventive culture might intercommunicate, but the spurs and the signals to move

[9]F. D. Prager, 'A History of Intellectual Property from 1545 to 1787', *Journal of the Patent Office Society*, 26 (1944), 711–60. See also the remarks in D. C. North and R. P. Thomas, *The Rise of the Western World* (London, Cambridge University Press, 1973), pp. 152–6.

and direct useful activity would be absent. Here, as at so many other points, an appropriate mix of market and non-market organisation and motivation appears to have been required for economic and industrial growth.

The incentive effect, as just described, is supplemented in modern technical change by another effect of economic circumstances, called by Rosenberg, 'focussing' – the steering of an inventor, or a group of inventors in an industry, toward a specific problem. In what has been said above there is no clear reason why the economy should cause invention to focus on one problem rather than another. Invention may be induced in response to external circumstance but in general not located or focussed by changed economic conditions. If labour grows expensive, there is no particular reason for producers looking for labour-saving techniques, rather than some other sources of savings. The structure of production and technology should adjust where it is weakest, i.e. where factors are most mobile or where technical changes are closest to the horizon. However, the small-scale competitive firms of eighteenth-century England utilised rigid and rather narrow techniques. If costs in a process rose, the option was not open to reduce costs in other firms or branches of production; the process itself had to be adjusted or the trade abandoned by the producer. In coke smelting, for instance, a rise in the price of charcoal might in theory have caused capital to move elsewhere. But in the smelting firms, it may plausibly also have focussed inventive activity on efforts to find a substitute fuel. A rise in the price of yarn, derived from increased demands of weavers using the flying shuttle in the eighteenth century, or by suspension of English yarn imports on the Continent in the Napoleonic wars, might have been compensated by some shifts of workers into spinning. But it would also understandably focus inventive activity on improving the productivity of the spinners already at work. And if, as it happened, that was just a place where invention was practicable, where the structure of the existing technology was relatively easier to change, then spinning inventions would be the result.

The mechanical technology of the eighteenth century appears to have been particularly susceptible to such economic inducements and focussing. The situation in industry indeed was not as dissimilar to that in agriculture as has sometimes been supposed. In both sectors the skills of workers trained in the specific trade since youth had formed the basis of the technology. In both sectors, the excitement of market growth, stirring the imagination of materialistic and money-minded men, encouraged efforts to improve existing techniques. In both cases under such pressures the body of existing technology proved itself capable of extensive improvement without undergoing striking change in fundamental principles. In industry the improvement was accompanied by the perfecting of a science of mechanics (which could later be applied to agriculture as well); both art and science were capable of being improved by the unaided power of the mind and the eye, by observation, thought, and contrivance, and the technology could move on within half a century to more complex realms.

Organisation of capital and labour.[10] Strictly speaking, a 'market for technology' did not exist before very recent decades in industrial history. In the eighteenth century there was a market for products, and not a very perfect market at that. But where it was active, thanks to mercantile activities, it made its influence and excitement felt backwards into the structure of production and even of technology. In doing so, the mixture of market and non-market elements, of old and new motives, of pockets of security and profit in the blowing wind of competition created an excellent environment for rapid, simple technological change. As the techniques were derived from production experience, so did they remain closely entwined with industrial activity, providing a growing inventory of tools and ideas for further development.

The situation was not greatly different with respect to the accumulation of capital. Capital was thought of as either very short- or very long-lived. The national wealth, when it was thought of at all, was viewed either in the mercantilist notion of a stock of precious metal, or in Adam Smith's enlightened view, as land improvements and public buildings, and the energies and skills of the population. Private capital was a merchant's stock in trade or a manufacturer's inventory of materials and goods in process or a fund of money to be turned over in trade or through wage payments in the somewhat more lengthy processes of manufacture. As in the case of technology, there was not so much a market for capital as a responsiveness on the part of capital to market

[10]The *loci classici* for a discussion of these subjects are the famous chapters in Adam Smith, *The Wealth of Nations,* book 1, chapters 1–3, and Karl Marx, *Capital,* vol. I, chapters 13–15. An interesting, and provoking, recent addition to the discussion is by S. A. Marglin, 'What Do Bosses Do? The Origins and Functions of Hierarchy in Capitalist Production', *Review of Radical Political Economics,* 6 (1974), 60–112. On the supply of capital to industry in the Industrial Revolution, see the essays collected by Francois Crouzet, with a valuable editor's introduction: *Capital in the Industrial Revolution* (London, Methuen & Co. Ltd., 1972). Management methods are discussed in Sidney Pollard, *The Genesis of Modern Management* (Cambridge, Massachusetts, Harvard University Press, 1965) and sociological aspects of labour force organisation in Neil Smelser, *Social Change in the Industrial Revolution* (Chicago, University of Chicago Press, 1959) and the work of E. P. Thompson, *The Making of the English Working Class* (New York, Random House, 1963).

facts and opportunities. The liquidity which a primitive capital market afforded, and the communication of knowledge of investment opportunities encouraged saving and directed capital into profitable uses. But most of the industrial capital in England appears to have derived from industrial operations; indeed in early firms the distinction between capital and income, or capital expenditures and current expenses, was not made at all clearly. Such a situation was conducive to industrial saving, and to reinvestment at points where a hard-headed entrepreneur could see prospects of a sizeable gain. Money and credit, and even mercantile banks, were essential to keep the system running, but one may wonder whether a sensitive market for industrial capital would have improved the rate of accumulation or the direction of its investment. As it was, with each firm depending largely on its own resources the incentive to save was high, and the investment was made by those with best knowledge of the opportunity. It is doubtful whether the collection of brokers and stock-jobbers accompanying the capital markets of the 1870s and 1880s would have either induced more saving or known how to advise investors more wisely than the entrepreneur-savers of the Industrial Revolution were able to do for themselves.

If the organisation of capital and enterprise in the Industrial Revolution involved no separation of ownership and control, the reverse may be said for the organisation of labour. I do not refer here to the notorious separation of the worker from his 'means of production', the tiny physical capital of which he had availed himself under the putting-out or craft shop systems of organisation. Within both those organisational forms, a hierarchical organisation prevailed, based on the hierarchical organisation of the family. In a shop the master controlled the labour of his family and, for the length of their terms, that of journeymen and apprentices. Within such tiny political units, each man had his duties and station, and no doubt customary rights and obligations assumed the form of informal law, just as had occurred on manorial estates on a larger scale. Similarly also in rural domestic industry, a man commanded the labour of his wife and children, subject to all the sanctions which through proximity and affection they could inflict upon him. Nor is it clear that the position of women was distinctly inferior to that of men in the peasant or peasant-industrial household. What distinguished the fully developed putting-out system from these was the intimate and customary tie of the head of the family or para-familial labour unit to the furnisher of the materials and the market. To the degree that the merchant employer held a regional monopoly over 'his' cottage workers, the latter were subject to close bargaining whether they 'owned' their tools or not. Materials could be allotted and piece rates fixed to keep labour's share at a minimum, and the difficulty of organising a group of workers established that preponderance of bargaining power on the side of the merchant employer, which was to become so marked in the developed factory and wage labour system.

In these conditions, the establishment of a few factories, i.e. collections of workers or worker family-groups under one roof and subject to one direction, may have made capitalist control of a labour force easier in the short run. But ultimately, as Marx first pointed out, factory organisation, though facilitating longer hours, stricter discipline and more careful supervision of time and materials, in the end removed the capitalist's great advantage over his workers: the difficulty they had in communicating with one another when they worked at dispersed locations. Instead of coming together at church or tavern on social occasions only, under the eye of priest or constable, the workers in a factory were thrown together in daily, intimate, professional contact. It is little wonder that workingmen's associations, which had led to a marginal existence among apprentices or rural workers, pushed beyond the limits of endurance, began to grow into regular and continuing bodies as factory organisation extended its scope. The titular loss of independence through loss of ownership of a few tools was finally more than balanced by the gain in easy access to the means of group solidarity, though this did not necessarily help the workmen immediately affected.

In those trades where the advantage of a factory organisation was confirmed by the grouping of complex geared and belt-driven machinery around a single power source, it was almost inevitable that the organisation would be carried out by a capitalist grouping workers at just the right points in the machine process, supplementing the failings of the machine by manual skills and human muscle power. The system indeed was preserved anachronistically within some early spinning factories, and systems of piece rate and time rate competed with each other for dominance. 'The market' appeared in the buying and selling of labour power as of any other commodity, but it is equally important to note the points into which the market did not penetrate. Labour was sold, but not by the minute on a perfect market, subject to close calculation and instantaneous renegotiation. The workers could not hire capital in the same sense that capital could hire workers; rather the hierarchical form of family organisation was preserved and transferred here, and without many of the protections that customary law in a small political unit gave to family and workshop. Workers sold themselves for a space of time, in a form of daily slavery, agreeing to do each day what job the employer required in exchange for a wage. Such an organisation is not truly a market organisation in the same sense as foreign exchange markets, for example, where supplies are offered and withdrawn at a slight titillation of the price, and negotiators change sides from supplier to demander as quickly as thought passes through the mind. The capitalist firm was a political organisation in its internal structure, bound by markets at either end.

The question is: how well was this form, derived so naturally from the historical circumstances in which the new technology was created, adapted to those techniques? It is not possible to argue that the techniques themselves were created and adapted to the form without ignoring the immanent constraints of the technology. Are we not rather in the presence again of one of those historical coincidences without which the surprising growth of modern industry could never have occurred? Remarkable developments require remarkable explanations and England's sudden industrial development in textiles and iron was a wholly unexpected and, in the long run, unsustainable and extraordinary event. Why then should it appear or yield to some deterministic and holistic explanation?

Whatever the explanation, it is possible to argue in retrospect that the adaptation was a good one for establishing machine industry and for insuring its spread. The factory built around a central power source – a water wheel or large steam engine – possessed significant technical economies of scale, particularly since the steam engine was never as successful as was later the electric motor in adapting to a wide range of capacities. The scale economies made it profitable to keep machinery in continuous operation – particularly water wheels during seasons when water power was strong. (In many cases the steam engine was first used to supplement water power in dry seasons.) It meant also attaching as much machinery as possible to the wheel or engine once its fixed installation was accomplished, and so of producing a large volume of output. These economies of scale were closely allied with economy of continuous operation, particularly evident to private owners of capital who thought in terms of turning over their stock. Whether working with his own or borrowed money, it paid a manufacturer not to let his equipment stand idle. But continuous operation had another aspect. While the machinery was attached to the power source, it all moved together. Belts or specific machinery might be temporarily detached, but on the whole workers had to operate with the machine and to time their motions to its requirements. This was the principal advantage that a wage system, with an authoritarian organisation, held over a system in which labourers might sell not their labour but their product as it passed from one form to another, or might renegotiate their labour contract minute by minute, or hour by hour. The resemblance of factory organisation to that of an army, in which each member had to respond quickly and on command, without knowing all the reasons why, did not escape notice. And, as in an army, the operation of the market principle within the organisation was rendered impossible, or at least unnecessary, by the fact that a few men – the managers – planned the whole operation and had no need of any inputs from below.

It remains to ask why, if slavery on the job was the technically most efficient status for the labourer, the freedom of the labourer off the job appears to have been wedded so closely to nineteenth-century industrialism. In Russia and in the American South, serf- or slave-operated factories were not unknown. Their limited success makes them the exceptions that prove the rule. The answer – that it is cheaper for an individual capitalist to avoid responsibility for workers off the job – is not wholly satisfactory since it confounds the individual with the social perspective. Unemployed or ill workers had to be supported by someone, and ultimately the capitalists would have to pay a sizeable share of the cost. The advantage of freedom over slavery for the workers in nineteenth-century growth appears to have lain in the value of mobility in an economy where new industries, firms, locations, and tasks were appearing every week. Where entrepreneurs were locked in to their already adopted technique and where capital was not nearly as liquid as it was later to become, ownership of workers by masters, or even binding patriarchal relations deprived an economy of an important element of flexibility. And, too, wherever freedom won out over slavery the drive to individual self-advancement could remain for workers as for entrepreneurs a strong engine of economic growth.

In England, the peculiar mixture of markets and authority could prevail because state controls had been relaxed to a greater degree than on the Continent. Here again the vestiges of an earlier, more authoritarian organisation remained, enough to preserve the regularity of life and of expectations. It is not that the state was weak; there was no question of who retained the monopoly of force. The central authority of the Tudor state, indeed of the Norman kingship, was retained. The writ of the king and the king's courts ran into every county, and the lords lieutenant and the state church kept watch on the vagaries of local and town government. Below that and well integrated into it, a social structure of classes sustained by a class deference derived from feudalism persisted in England as elsewhere in western Europe. The beginnings of a democratisation, or even a thorough embourgeoisement of this society were still a hundred years in the future, and a full modernisation and proletarianisation still further off. The difference between England and the Continent lay not so much in the balance of central and local authority as in who controlled that authority. Here thanks to the intrusion of a strong mercantile interest in the struggle between king and landed aristocracy, the power of the Renaissance and seventeenth-century monarch, so strong in France and on a miniature scale in the German states, had been balanced by the wealth of London and the Puritan independence of large areas of the countryside. The Revolution of 1688 was to set the seal on a political compromise which lasted till the Reform Act of 1832 – rather less a compromise than a combination of the heavy weight of a Protestant landed interest and a Protestant merchant and banking aristocracy to establish a limited monarchy. With things so arranged on top, and with enough flexibility to adjust squabbles within the ascendant Whig and Tory aristocracy, a government of incredible strength and toughness evolved, able to survive and surmount financial instability and crises, massive fraud and corruption, wars on the Continent and overseas, the loss of one empire and the development of another, and finally to organise the conservative forces of the Continent to the defeat of Napoleon. Little wonder that such a structure could also almost unconsciously control and channel the forces of a rising industrial development. No doubt its congeries of policies and laws, a fantastic scrap basket of bits left over from feudal and medieval restrictions, mercantilist encouragements, and responses to the pressures of particular situations and interests, produced a less than ideal or optimal effect on the development. One thing was clear; it was fully capable of providing stable support for the evolving body of commercial law, for the 'rights' of the individual property-holder, and for domination over the labouring poor. It was not a 'tool' of the propertied classes in any conscious, planned, or conspiratorial sense; had it been so, its evolution into full parliamentary democracy and twentieth-century socialist industrial organisation would have been far less steady and more bloody. But in the eighteenth century and the early nineteenth, it produced just the balance of authority and free markets that an early and unsteady capitalistic industrial organisation required to take its first steps toward maturity, strength, and dominance. On the Continent, developments were otherwise, as the following section will indicate.

THE SPREAD OF MODERN INDUSTRY[11]

The spread of mechanical techniques through industries and through geographical regions is a pair of processes with many common charac-

[11] W. O. Henderson, *Britain and Industrial Europe 1770–1870* (2nd edn, Leicester, Leicester University Press, 1965) traces some of the direct lines of connection from Britain to the Continent. Rondo Cameron's original and well-researched book, *France and the Economic Development of Europe 1800–1914* (Princeton, New Jersey, Princeton University Press, 1961) and the monumental work of Maurice Lévy-Leboyer, *Les Banques Européennes et l'Industrialisation Internationale* (Paris, Presses Universitaires de France, 1964) treat the Continental industrialisation, as does the earlier work of A. L. Dunham, *The Industrial Revolution in France 1815–1848* (New York, Exposition Press, 1955) and the

teristics. The mechanisation of a new industry required adaptation of the power and inventions to harness power to specific operations. The spread of a given machine technique, e.g. mechanical spinning, from one region required interested entrepreneurs, favourable factor cost conditions, suitable government policy, and supplies of capital and workers adaptable to the machine process. It too required minor inventions to adapt the equipment to specific raw materials, markets, climate, and labour force. The timing, speed, and form of the diffusion in both cases depended on special technical, economic or sociological conditions of the industry or region in question. In the competitive economy of northern Europe, the spread of mechanisation among industries was in one sense the more fundamental sort of diffusion since the minimum physical cost locations of new industries were determined by cost characteristics of the new technique. Given these characteristics, the ultimate dispersion of an industry was largely a matter of economic geography. Social and economic differences – the availability of capital, the training and immobility of the local labour force, the policy of the region's government – determined the lags, however, before the 'natural' economic factors and the force of competition took effect. This was especially true where the natural locational advantage of one region over another was rather small. In textiles, for example, differences in the location of enterprise, capital, cheap labour, and close relations with overseas outlets could set the advantage.

Diffusion among industries. In an earlier section, indication was given of the path of the Industrial Revolution among industrial processes.

essays by Jan Craeybeckx and Claude Fohlen in Rondo Cameron (ed.), *Essays in French Economic History* (Homewood, Illinois, Richard D. Irwin, Inc., 1970). To the classic textbook of J. H. Clapham, *The Economic Development of France and Germany 1815–1914* (4th edn, London, Cambridge University Press, 1963) and David Landes' treatment of the western European area as a whole (see *The Cambridge Economic History of Europe*, vol. VI, part 2, H. J. Habbakuk and M. M. Postan (eds.) (London, Cambridge University Press, 1965) and *The Unbound Prometheus* (London, Cambridge University Press, 1969) have now been added very good chapters in A. S. Milward and S. B. Saul, *The Economic Development of Conti-Europe 1780–1870* (London, George Allen & Unwin Ltd., 1973) and in C. M. Cipolla (ed.), *The Fontana Economic History of Europe*, vol. IV, especially chapter 1, 'France 1700–1914' by Claude Fohlen, and 2, 'Germany 1700–1914' by Knut Borchardt. The conference volume of the International Economic Association, edited by W. W. Rostow, *The Economics of Take-off into Sustained Growth* (New York, St Martin's Press, 1965) also contains a number of valuable articles on early development in various countries as well as an evaluation of the 'take-off' hypothesis of W. W. Rostow.

The pre-history of the nineteenth century

One may observe certain clusterings in the history of its progress. One was the spread of the power process from cotton-spinning through other branches of the textile industry – to wool and linen, and after a delay to mechanised weaving. Hand processes in the garment industry – e.g. shoemaking and sewing – presented specific technical problems which yielded, one after another between 1800 and 1850, to British and American ingenuity and enterprise. All this development was brought on partly by the force of economic pressures within the structure of textile and clothing production which focussed inventive activity, and partly by the rise in the level of technical opportunity for solving problems through improvements in materials, machine tools and control mechanisms. The history of the development of the sewing machine may be cited as a notable example of the eco-technic process at work.

Beyond the 'light' industries – textiles, boots and shoes, machine tools, and farm machinery – the latter developing more rapidly under favourable market and terrain conditions in North America after 1850 – there lay the engineering problems of heavier equipment in transport and power generation. Continuous improvements in the steam engine increased its efficiency and extended the range of capacities and pressures generated and contained. Such improvements occurred at all the locations where engines were used and produced, in England, Wales, Belgium, Germany – even on the banks of the Ohio in America. The adaptation of steam to water navigation is a classic story in the history of invention, and from Trevithick to Stephenson, the development of locomotives, braking mechanisms, and all the vast array of railroad inventions created the mid-century transformation of land transportation. In stationary engines, the first half of the nineteenth century saw the development of the water and steam turbine in France and England through the inventions of Fourneyron and the thorough investigation of the science of thermodynamics by Carnot.[12]

We have seen that mechanical inventions, as they spread to various industries and locations in northern Europe and North America, presupposed a large interconnected industrial region. This was required to ensure both adequate market size and the mass of intercommunicating inventive activity necessary to keep economic expansion and technical change in motion. In the eighteenth century, it appears that central England itself was a large enough area. A striking fact about the Industrial Revolution is the speed with which improvement extended from iron and machinery production to the manufacture of machine tools. Machine tools lie deep in the production processes of modern industry.

[12] D. S. L. Cardwell, *From Watt to Clausius* (Ithaca, New York, Cornell University Press, 1971).

Europe's industrialization: the pre-history

To make it worthwhile to devote efforts to improve the machines which make machines, a large market for machinery is essential. The increase in the productivity of machinery and in its durability when iron was used worked in exactly the reverse direction. A market for machinery of unusual scale must have been present to give the impetus. This is true in early nineteenth-century north-eastern United States, where the level of income, its distribution, the protection afforded by distance from English competition and (some allege) the scarcity of labour relative to capital, and of both relative to the ambitions of the population, helped also to allow an industry of specialised machine tool manufacture to grow quickly out of the machine shops of the textile mills. Once developed, the machine tool industry, employing water or steam power, improved and cheapened iron, and better and more closely machined parts in its own equipment could lift machine production out of the workshop of the mechanic and make it too a factory industry. The cheapening of machinery, rather than a fall in the rate of interest, has been largely responsible for the greater physical capital intensity of modern processes.

In England and Belgium, the close link of machinery production to the local iron industry cannot fail to be observed. With the possibilities of steam-powered machinery, an engineering industry was growing up around iron works, and the massive fuel requirements of coke smelting and puddling brought iron works to locations at coal beds. In Britain by 1850 most major coal beds were thus the site of iron smelting and fabricating industries. There can be little doubt that even before the steel inventions of the 1850s, the coal-based industrial complex was an economic unit. Because of the saving in fuel transport costs and the further advantages of agglomeration and communication in a concentrated area, its products could undersell those of producers at scattered locations. By 1850 the Industrial Revolution, as a revolution in both technology and plant and enterprise organisation, had spread from cotton-spinning to other 'light' easily mechanised industries, then to the heavier industries of transport equipment and machine production itself, the latter based also on the improvements in the iron industry that were part of the eighteenth-century development. Lodged between light and heavy industries, the machine tool companies expanded and extended the varieties and uses of their products. As these industries, particularly those using coal and iron, clustered around coal or ore mining areas, the typical industrial complex of the later nineteenth century was formed. The railroad added to the advantage of these dark and smoky districts even as it increased the demand for their products, and the Bessemer and open-hearth processes coming in after the 1850s ensured their stability for the half-century following 1870, not only in

Britain and Belgium, but around the coal beds of the valleys tributary to the lower Rhine, and the upper Ohio.

One must remember that the Industrial Revolution was based on a certain group of inventions and an accompanying organisational form, which could not spread beyond the industries where coal, iron and machinery could be introduced. In Britain in 1850 many industries and operations were not power driven, or mechanised. The largest component of the non-agricultural labour force in 1850 was, after all, domestic service, and the mechanisation of the household lay beyond anyone's imagination. Construction, including ship-building and road-building, was relatively untouched, similarly most food-processing operations and, of course, agriculture. The very growth of the larger-scale industries, and the swarming of populations to new locations gave occupation to vast numbers of small-scale producers, and furnishers of service. Office work, too – except for the development of the typewriter and the telephone and telegraph after 1850 – experienced no productivity increases, and the way was laid in all these respects not only for the perpetuation of the class of small shopkeepers and professionals, but also for the growth of the 'white collar' staff of the larger establishments. The whole society then was not industrialised, much less proletarianised. Industry, industrial capital, industrialists, and industrial wage workers assumed a place on the front bench of society and politics, constituting a special 'interest' alongside the interests of the mercantile community, the bankers, the professional and white collar class and the landed interest of ancient origin. It assumed a place beside the others, but did not crowd them off the scene.

Diffusion among regions of north-western Europe. The advances in the cotton, iron and machinery industries between 1780 and 1840 were the whole bases on which the English Midlands, with extensions in South Wales and Scotland, in 1850 rested a remarkable industrial leadership over the long-established industrial regions of the Continent. In textiles, technical obstacles which lay in the way of mechanising operations in flax, silk, or even wool did not obstruct the application of machinery to cotton. England's lead in the cotton industry then must be attributed to her superior trading position and access to markets and to raw cotton supplies. Possibly also, the long experience in wool made an easier transition to cotton than could be achieved by silk or linen producers. The development of a cotton industry on the Continent had to depend initially on the importation of English machinery and a few English workmen and plant designers – an expensive and unsatisfactory way to overtake a foreign competitor. Still machinery was eventually applied to the branches of textiles in which continental producers specialised and the slower pace of development in machinery and the iron industry on the Continent cannot be attributed to technical reasons. Clearly before 1840 the continental industrial regions – the cities with their workshops and the rather widely separated and disconnected areas of rural industry – lacked the intensity in industrial activity closely linked to machine shops, which gave British industry the critical mass necessary to a continuous and self-reinforcing economic and industrial development. The imported English spinning machinery at Ratingen in the Rhineland, in Normandy along the Seine tributaries, and later in Ghent, were sparks of modernisation which did not light a fuse to set off the fireworks.

The Belgian case is the exception on the Continent which proves the point at issue here.[13] In Flanders the dense textile industrial district lay close to a large foreign market and to the iron- and coal-based industry of Liège and the Belgian coalfield. Even under Napoleon, industrial development began, when the area was joined with the Dutch provinces in 1815, overseas markets were opened and access given to Dutch capital. The Dutch areas themselves failed to industrialise – possibly because of a history and social structure based solely on commerce, possibly also because of lack of cheap coal. Instead, the Dutch king invested in the Belgian areas which lay under his government between 1815 and 1830, and a little borrowing of workmen and machinery from England developed mechanised spinning and a domestic machinery industry. By 1850 – twenty years after the separation from Dutch rule, the Belgian Netherlands had become the world's second coal-based industrial district in which the light and heavy industries of the Industrial Revolution were joined.

The persistence of traditional and typically 'early modern' barriers to industrialisation on the Continent is shown best in France. The Revolution had swept away the remains of feudal forms – feudal land tenures and the power of the guilds in cities. It had destroyed, indeed it had confirmed, governmental centralisation and mercantilist policies of the state. The revolutionary governments and Napoleon had strengthened the state much as Louis XIV had done, though more intelligently, and much was provided for the new (and not so new) commercial and industrial bourgeoisie. Commercial and property law was regularised through the *Code Napoléon*, scientific and technical education was extended and strengthened. An educated scientific and engineering élite

[13]To the references cited in note 11, p. 45 above including an excellent chapter in Rondo Cameron's *France and the Economic Development of Europe 1800–1914* (Princeton, New Jersey, Princeton University Press, 1961) may be added a recent treatment of the Belgian 'case' by Joel Mokyr, *Industrialization in the Low Countries 1795–1850* (New Haven, Connecticut, Yale University Press, 1976).

was enlarged. The Bank of France helped stabilise the currency and brought French public finance up to the degree of modernity that England had achieved a hundred years earlier. Modernisation and regularisation of the tax system added greatly to the regularity with which business expectations could be pursued. By reducing its personal and arbitrary character, the post-Revolutionary government helped to create a climate of reduced uncertainty for mercantile and business interests.

The regimes from Napoleon I to Napoleon III offered also many direct opportunities to business enterprise. Interruption of trade with England from 1790 to 1815 reserved the Continental market to Continental – and largely to French or Belgian – producers. The inflation and the wars themselves offered the usual opportunity for short-term and individual gains. Enough venality was present, enough luxury demand, enough waste to nourish the greediest entrepreneur. Yet for all that, one cannot speak of any French government until the parliamentary democracy of the Third Republic, as an oligarchy like that in seventeenth-century Netherlands or eighteenth-century England. The peasantry, the church, the remaining aristocracy, the army, the bureaucracy – all were too strong to furnish a clear climate for modern capitalism. If the balance in England by 1780 was about right to allow an eco-technic, industrial revolution to proceed, the balance lay in France a bit too strongly on the side of what Marxists call pre-industrial economic formations. Much is made in history books of England's political gradualism, in contrast to France's recurrent revolutions. But in economic modernisation, it was in France that gradualism prevailed. Napoleonic government and its successors under the Restoration and the July Monarchy maintained a stance which combined liberalisation, protection, and paternalism until more classic liberal policies were introduced. By that time, industrialisation had developed into something different from what it had been seventy years earlier, and what France had preserved of the older forms and values – the system of technical education, the aristocratic spirit of scientific research, the balance between population growth and her own food supplies – began to pay off.[14] The nation then could lay the base for continued economic progress as a national unit, even up to the present day, through all the devastations of war and political and moral catastrophe.

The situation in the scattered German textile and metal-working re-

[14]French technical education is interestingly treated in F. B. Artz, *The Development of Technical Education in France 1500–1850* (Cambridge, Massachusetts, The M.I.T. Press, 1966). A recent dissertation by Bernard Gustin, 'The German Chemical Profession: 1824–1867' (Department of Sociology, University of Chicago, 1975) throws needed light on German chemical training and research before Liebig, and the role of the apothecaries in the development.

gions of the eighteenth century was not greatly different from that in France.[15] But they lacked two developments that English and French regions had experienced: incorporation in a national state, and within it a political revolution. Even by 1700, the physical depredations of the Thirty Years War had been repaired, but within the notoriously numerous political districts, a mixed medieval and Renaissance political economy survived and flourished. The states were not inactive in efforts to advance industry; they encouraged it by all the best principles of mercantilist economic policy. Nor was there any lack of skill or enterprise in many areas; we have seen earlier how extensive was the diffusion of the Renaissance technology in the south, central and western German states. Rhine merchants and bankers were active throughout the whole period before 1850, and in south-west Germany, the activity and ambitions of apothecaries, along with the princely sponsorship of 'pure' science in the universities, laid the foundations for Germany's later successes in chemicals. The tariff history, of which so much has been written, indicates that barriers to internal trade were overcome, but the Zollverein, too, was a mercantilist measure pursued from political motives, not only on the part of Prussia but also of the petty princes who hoped, in typically seventeenth-century fashion, to increase net revenues from a source outside the control of assemblies and nobility. Even after the industrialisation got under way in the Prussian territories and in the Empire, no one would ever have mistaken the Imperial German government for a businessman's state.

By 1850 what was lacking in both Germany and France in 1800 had been partially supplied. The social and political basis for modern capitalistic industry had inched its way toward a condition which could tolerate capitalist expansion without the continual drag of medieval or mercantilist restrictions or the unexpected and disrupting assertions of authority of divine right monarchs and their bureaucracies. Then between 1850 and 1870 two decades of classical liberal policies in both France and Prussia expanded trade, strengthened financial institutions, encouraged capital accumulation. What had been lacking in the earlier textile industrialisation was the opportunity to make the link with the iron industries, and this the railroad had only partly supplied. But the social and physical elements in modern industry were present – the in-

[15]See the articles by Herbert Kisch in note 1, p. 15 above and also Gerhard Adelmann, 'Structural Change in the Rhenish Linen and Cotton Trades at the Outset of Industrialization' in Francois Crouzet, W. H. Chaloner and W. M. Stern (eds.), *Essays in European Economic History* 1789–1914 (London, Edward Arnold Ltd., 1966). On metallurgy, see N. J. G. Pounds and W. N. Parker, *Coal and Steel in Western Europe* (Bloomington, Indiana, Indiana University Press, 1957).

The pre-history of the nineteenth century

tangible social capital of laws, skilled mechanics and engineers, educational institutions, a still disciplined labour force, and the physical capital of transport improvements. As contact between regions improved, the disadvantages of the small- and scattered-scale of the earlier textile and light machinery industries began to be overcome.

Into this atmosphere in the 1860s came as a supplement and substitute for wide geographical scale, the opportunity for heavy industry localisation. Through the accidents of politics north-western Europe's coal was distributed in bits across all of the major north-western countries. It had long been known and mined in spots – in the Saar, in the Liège region, and at a few shallow diggings in France. With the opportunity opened by market growth and the steel inventions of the 1860s, the clustering around these deposits began to take shape, and the immense industrial strength of the Franco-German–Belgian area began to make itself felt.

The development of coal, steel, chemicals and electricity on the continent belongs to the history of the latter part of the nineteenth century. It would take another chapter, or another book, to fill it in. It exhibits similarities in form and timing to the American development south of the Great Lakes, between Chicago and Pittsburgh in these same decades. But European industrial history prior to 1850 shows that it grew up on an industrial base very different from that of both England and the USA. Unlike the eighteenth- and early nineteenth-century developments in England, continental industrialisation after 1850 was not based simply on the scale economics of wide textile markets, and the accompaniment of a mechanical engineering technology. That was a combination which the continental locations, for reasons of economic and social organisation in the late eighteenth century, had not been able to achieve. But at length after half a century of sporadic, artificial, and pale imitations of British technology, continental industry hit upon a rich vein which its own tools and traditions were able to mine. In the technology of coal-based chemistry, in metallurgy scientifically developed, and in inventions leading into the lighter but even more science-based industry of the twentieth century, the Continent's long industrial traditions and its institutions of pure scientific research and of applied training could at last come into their own. With the concentration of activity around coalfields, the industrial strength was developed which, by permitting further developments away from coalfields, could lead continental industry into its upsurge after the calamities of the 1940s.

This essay develops ideas offered in lectures to graduate economics students at Yale University over a number of years, and exposed also at seminars at European and American universities since 1972. I wish to express thanks to many participants in those seminars for useful comments, and in particular to Richard Levin, Jan De Vries, Harry Miskimin and Quentin Skinner for reading and commenting on the manuscript. I am especially indebted to Peter Mathias, Eric Jones and Peter Burke for initial encouragement, even though they may, like St Peter, wish now to deny it thrice before the cock crows. The work could in any case not have been brought into this form without a generous grant from the Concilium on International and Area Studies and the Department of Economics at Yale, and the diligent and informed research assistance of Laurie Nussdorfer.

My debts to the authors in the field are only imperfectly acknowledged in the footnotes, but will be evident to anyone acquainted with some of the literature, including, I trust, the authors themselves. The footnotes indeed are intended not as source references or elaborations of the text, but as suggestions for further reading, primarily in English-language sources. A full bibliography of the main writings published in English, French, German and Italian since the publication of the major bibliographies of David Landes, *Cambridge Economic History of Europe*, vol. VI, part 2 (1965) and Maurice Lévy-Leboyer, *Les Banques Européennes et l'Industrialisation Internationale* (1964) is in preparation. I regret that I have been excluded from writings in Dutch and Swedish by 'ignorance, pure ignorance'.

Industrialization and the European Economy[1]

By SIDNEY POLLARD

I

Traditionally, the industrialization of Europe has been viewed as a series of separate "industrial revolutions". Each country has been treated as an "economy" and its progress compared separately with some kind of implied model. There can, for example, be few universities in Britain in which the annual examination ritual does not include some questions about the retardation of the industrial revolution in "France" or its acceleration in "Germany" in the nineteenth century.

There are many good reasons for making the political state, or even a notional country like Germany before 1871, the unit of investigating the process of industrialization. It follows the practice of contemporaries, who were mostly concerned with influencing economic development by the only political authority available, the state. Political historians, too, are wont to see the nineteenth-century world as made up of the individual states of Europe.[2] And in truth many issues of direct relevance to industrialization operated within political boundaries or were affected by political decisions. These included protection, the currency, access to markets and colonies, and poor laws and factory acts. Other important legal and social data, such as serfdom, the status of the businessman, primogeniture, and guild and staple rights, might in some cases coincide with the political frontiers, but as a matter of fact did not so coincide in some of the leading countries of Europe during the early stages of industrialization.

In recent years yet another reason has emerged for viewing industrialization in national terms: many of the countries attempting to industrialize today have to use the whole apparatus of state power to propel their industrial revolution forward so that it must appear as an act of political will. By analogy, this is then applied to Europe in the nineteenth century.

This article will argue that the traditional view has reached its limit as a method both of exposition and of analysis, and that further advance must come from a new starting-point. It will argue that, useful as the "national" approach has been in the past, particularly by evolving testable models, it was based on faulty observation. The industrialization of Europe did not proceed country by country. On a map of Europe in which industrialization was coloured, say, red, it would by no means be the case that an area corresponding to a country within its boundary would turn uniformly pale pink, dark pink, and so on to deepest crimson. On the contrary, industrialization would appear as red dots, surrounded by areas of lighter red diminishing to white, and with the spread of industrialization these dots would scatter across the map with little reference to political

[1] Thanks are due to the Historische Kommission of Berlin, and to the initiative of Prof. Otto Büsch, for providing me with the leisure and the facilities for some of the research on which this paper is based.

[2] A. R. Bridbury, *Historians and the Open Society* (1972).

boundaries. It would also follow that the dynamic of the infection of this rash of red dots, crossing frontiers while by-passing large areas of the home country, must clearly have a largely non-political explanation.

It is this dynamic element, the method of transmission, which should be a major object of study, but has inevitably been lost sight of in the traditional approach. To change the metaphor, we have tended to treat each country like a plant in a separate flower pot, growing independently into a recognizable industrialized society according to a genetic code wholly contained in its seed. But this is not how the industrialization of Europe occurred. Rather, it was a single process: the plants had common roots and were subject to a common climate. Further, the development and chronology of the industrial revolution in each area was vitally affected by its place in the general advance, by those ahead of it as well as those trailing behind it, and this relativist role must form part of any description or analysis.

It should be admitted at once that this article is itself open to the same objection as the traditional single-country approach: the course of the European industrial revolution in turn cannot be explained fully without its mutual interaction with other continents.[1] The story told here, it cannot be stressed too strongly, is therefore still only a partial story. But it is a beginning, and we can at least plead that the necessary and dynamic relationship between the industrialization of Europe and the rest of the world has received some attention recently[2] and is therefore in less need of emphasis. Here we shall try to make the case for a single European economy in the crucial years of industrialization, *c.* 1815–1914.

II

Before proceeding to consider Europe, an area larger than the nation state, as an appropriate framework for the study of industrialization, it is worth making a digression to a unit smaller than the nation state. It is clear that in the pioneer country itself, Great Britain, we shall miss much of the actual dynamic[3] of industrialization if we neglect the fact that the "industrial revolution" came much earlier, say, to South Lancashire and the Black Country than it did to Lincolnshire or Kent; that even concentrated manufacturing regions like Sheffield with Hallamshire or the West Country woollen areas were revolutionized generations after the cotton areas; and that some of the regions leading in the first phase, like North Wales, the Derbyshire uplands, and even Cornwall,[4] petered out as industrial concentrations in later phases.

Behind the temporal sequences lay functional relationships. Agrarian advances in some areas may have stimulated early industry in other less-favoured ones,[5] while in turn an industrial or urban expansion in areas like Lancashire

[1] "Capitalism itself is, both in the economic and sociological sense, essentially one process, with the whole world as its stage."—J. A. Schumpeter, *Business Cycles* (New York, 1939), II, 666.

[2] E.g. William Woodruff, *Impact of Western Man. A Study of Europe's Role in the World Economy, 1750–1960* (1966); Brinley Thomas, *Migration and Urban Development* (1972).

[3] Walter Isard, 'The General Theory of Location and Space Economy', *Quarterly Journal of Economics*, LXIII (1949), 479.

[4] A. H. Dodd, *The Industrial Revolution in North Wales* (Cardiff, 1933); John Rowe, *Cornwall in the Age of the Industrial Revolution* (Liverpool, 1953); S. D. Chapman, *The Early Factory Masters* (Newton Abbot, 1967).

[5] E. L. Jones, 'Agricultural Origins of Industry', *Past and Present*, XL (1968).

affected agrarian development in surrounding counties. Regional specialization associated with modernization, for example in pottery, in cutlery and toolmaking, or in woollen textiles, was tantamount to de-industrialization elsewhere. Alternatively, the introduction of advanced factories, say cotton mills, in one district, might lead to the extension of an obsolescent system, domestic labour in cotton weaving or framework knitting, in others. And in turn, the concentration of a region on the latter might help to make it miss the bus in relation to machine spinning thereafter.[1]

The point is not merely the banal one that the mechanization of one industry has consequences on others, but that around each may be a whole regional society with identifiable features appropriate to a certain stage of the industrialization process differing from the stage reached by other regions. An industrial or commercial concentration transformed its local economy and society well before it helped to transform the economic life of the country as a whole. At any given cross-section in time, let us say in 1815, some regions were more advanced than others: ultimately the whole of Britain became modernized and industrialized, but the path followed by each region was different and was determined at each point in time by its inter-relation with the rest. No history of industrialization of any one region in isolation from the others would make much sense: one of the significant features of the industrial revolution in Britain was its complex regional inter-relationships.

So it was on the Continent. A map of *c.* 1815 would have shown several distinct groups of red dots. One was shaped like a curved dagger pointing from the Channel coast to the heart of Europe, including northern France and Belgium, Rhineland–Westphalia, Alsace, and parts of Switzerland. A second group of dots would be scattered farther east, covering parts of Saxony, Bohemia, and Silesia. As if to prove the vanity of politics, both these main areas happened to straddle several major frontiers. Other secondary centres might be found in and around the large cities such as Paris, Vienna, Berlin, and St Petersburg.

Even if we agree that "the striking thing about the imitative ventures on the Continent is certainly their spottiness. One modern ironworks went a long way,"[2] the fact remains that when industrialization finally did take its leap forward, these were its jumping-off points. A map of Europe around 1865 would have shown the original areas of dots to have become solid red belts, with new dots scattered thickly all over the north-western half of Europe, and very thinly in its northern, eastern, and southern marches, and in Ireland. If we let another half-century pass and observe conditions around 1914, we note that much of the north-western half of Europe is now solid, while the scatter of dots in the periphery has become much denser. Today, after two world wars, the relative level of industrialization and of incomes across Europe still has basically the same pattern.

Is there anything to be gained by this kind of regional-local approach? Apart from its claim to greater historical accuracy,[3] its merit, if any, must lie in opening

[1] Chapman, op. cit. pp. 212–15.

[2] Goran Ohlin, 'Balanced Economic Growth in History', *American Economic Review* (1959), Supplement, p. 349.

[3] Willard Long Thorp, *Business Annals* (New York, 1926), pp. 103–4; S. B. Clough, 'The Diffusion of Industry in the Last Century and a Half', *Studi in Onore di Armando Sapori* (Milan, 1957), II; J. U. Nef, *War and Human Progress. An Essay on the Rise of Industrial Civilization* (Cambridge, Mass. 1950), p. 292;

up new and potentially fruitful questions.[1] Two types of question spring immediately to mind. The first is concerned with our concept of industrialization: how do we define, or recognize, the different stages of an industrial revolution? The national framework has left us floundering among proportions of output furnished by manufacturing industry, its share in the labour force, numbers working in factories, provision of capital or horsepower, rate of growth of industrial output, the share of capital goods in total output, and the share working for a market economy or for wages. The regional approach will at least make us wary of such definitions of industrialization as that which uses income per head as an indicator,[2] since observers from Sismondi to Villermé seem to agree that the industrial centres were noted for their poverty rather than their wealth. There is at least a chance that it may shed some light on our search for a positive identification.

The second set of questions is concerned with favourable settings and preconditions, and therefore, in at least one sense, with causes. What disposed some areas to take more quickly to the new system than others? Was it natural resources? Location and transport facilities? A developed capitalistic commerce or access to foreign technology? Regional advantages in education?[3] A background of urban handicraft, centres of consumption, or associated external economies? A flourishing domestic industry? Favourable systems of inheritance?[4] Freedom from capricious autocracy, from serfdom, or from guild obstructionism? Again, we see at once that some solutions proffered on a national scale come off badly: such are government policies like protection, subvention, or, on the contrary, free trade, when some early industrializing regions differ so widely from others subject to the same legislation, while showing great similarities to their opposite numbers existing under a very different political framework; or the doctrine of a "critical minimum effort" needed to take an economy out of its traditional balance[5] which has little persuasive power in a regional context. Perhaps here, too, we may hope to get some positive aid by operating at the micro-level in our search for causes, sequences, and the relative weight of factors.

III

However promising this regional, or for that matter the traditional national, emphasis may be, they must in the end be put into context by treating the whole of Europe as one single macro-development area. Let us begin by considering the period of early European industrialization *c.* 1815–65.

Paradoxically, whatever economic inter-relationships Europe possessed before the war, and they were many, were reinforced by blockade and counter-blockade,

W. G. Hoffman, *The Growth of Industrial Economies* (Manchester, 1958), pp. 43 ff; Carl Major Wright, *Economic Adaptation to a Changing World Market* (Copenhagen, 1939), pp. 129 ff.

[1] Compare Douglass North, 'Location Theory and Regional Economic Growth', *Journal of Political Economy*, LXIII (1955).

[2] R. M. Hartwell, *The Industrial Revolution and Economic Growth* (1971).

[3] Claude Fohlen, *Qu'est ce que la révolution industrielle?* (Paris, 1971), p. 162.

[4] H. J. Habakkuk, 'Family Structure and Economic Change in Nineteenth-Century Europe', *Journal of Economic History*, XV (1955).

[5] Harvey Leibenstein, *Economic Backwardness and Economic Growth* (New York and London, 1957), p. 185 and *passim*.

as well as, indeed, by the Napoleonic dream to turn the Continent into one single economic area. From 1815 on, the common distresses of war were followed by the common disasters of peace. First, there was the mass dumping of British manufactures on European markets. The unfortunate timing of the war, which granted the Continent artificial protection from Britain's exports and isolation from her rapid technical progress, followed by the peace, and an abrupt exposure to the full blast of her accumulated commodities and superior skills, undoubtedly constituted a traumatic setback for several continental industrial centres. And this was compounded by the artificial separation of producers from their market, as was the case with the industries of Belgium and Rhineland–Westphalia—divided from the French consumers for whose needs they had been primarily created.

Thus parts of Europe were becoming de-industrialized owing to the expansion of British industry, as East Anglia was becoming de-industrialized owing to the expansion of West Riding industry.[1] The textile and metals industries suffered most, particularly in Austria and parts of Germany. Even centres with well-established export markets, like the Silesian textile manufacture and the Rhineland mills, lost many of them to Britain.[2] In Italy French textiles were driven out by the British who, according to Talleyrand, the French ambassador in Naples, "know how to put themselves out to meet the needs, the whims and the imagination of men, making for them fabrics at all kinds of prices according to the level of poverty of the regions for which they are intended".[3]

But some firms, the best and toughest, survived. Their world lay between the more advanced British on one side and the more backward Europe on the other. Belgium was the territory which became keyed first and most successfully into such an intermediary role, and Belgian industrial history can be fully understood only within the context of this dual relationship. Though favoured by coal and iron resources, by location and transport, by relative wealth and urban and industrial traditions, Belgium's industrial revolution depended on the British. British entrepreneurs pioneered and founded her major engineering works, above all Cockerill's at Seraing, who boasted in the 1830's that they knew of all inventions within ten days of their appearance in Britain. Others helped to establish some of the leading textile works and, in the early phase of the railways, built and financed large parts of the Belgian railway network, both state-owned and private.

At the same time, Belgium became a centre of diffusion. In the words of one contemporary of 1835:[4]

> John Cockerill travels on the great highways in his coach. Here he builds furnaces and there chimney stacks. He covers fields with his tents and then when all his preparations have been made he erects the steam engines which have followed in his wake . . . and which breathe life into the great piles of bricks. And the next day the peasants hear a loud rhythmical noise coming from the factory—like the

[1] J. H. Clapham, 'The Transference of the Worsted Industry from Norfolk to the West Riding', *Economic Journal*, xx (1910).

[2] Adolf Beer, *Die österreichische Handelspolitik im 19. Jahrhundert* (Vienna, 1891), pp. 312–15; L. K. Aegidi, *Aus der Vorzeit des Zollvereins* (Hamburg, 1865), pp. 40–1.

[3] Quoted in Maurice Lévy-Leboyer, *Les banques européennes et l'industrialisation internationale dans la première moitié du XIXe siècle* (Paris, 1964), p. 151.

[4] Quoted in W. O. Henderson, *Britain and Industrial Europe, 1750–1870* (Liverpool, 1954), p. 130.

breathing of some enormous monster who, once he has begun to work, will never stop. And John Cockerill climbs back into his coach and government officials unsuspectingly sign his passport as if it referred to a consignment of wine and they do not realise that this silent man who seldom puts pen to paper is far more likely to turn their old world upside down than many a revolutionary who has his pockets stuffed with political programmes and manifestoes.

The provenance of steam engines, perhaps representing the most advanced sector of current technology, was characteristic of Belgium's position. In 1830, there were 428 steam engines in the country, of which 74 (21 per cent) were foreign built. By 1844 there were 1,606, of which only 110 (7 per cent) were foreign built. At the same time in the 1830's Belgium exported seven times as much in value of engines than she imported.[1]

The French textile industries in the northern Departments and in Alsace and the French metals and engineering industries similarly had to fit themselves into a world in which the British had established a powerful dominance. Thus, in textiles, as Lancashire–Yorkshire could not be beaten in coarse-quality mass-produced goods, the French developed higher-quality fabrics as their area of strength. Even in cottons, where the British lead was greatest, although the quantity of raw cotton consumed by Britain in 1836 was at least three times that of France, the output of French fabrics (partly, no doubt, because of higher costs and partly because of higher quality of workmanship) was worth three-fifths of that of the British. The British led in pig iron output, but the French exported large quantities of finished metal goods to the less advanced regions of Europe.[2] At home, the French might have had a protected market. Abroad, at least until the mid-century, the non-industrialized world was still too large for Britain to fill by herself, and here lay the chance for the early continental industrializers.

It was, however, in the German states that awareness of the location between the two worlds was strongest, as indeed it was also the German Confederation in which frontiers were most clearly irrelevant to economic reality. This was, perhaps, made most explicit in the debate surrounding the new Prussian Tariff Act of 1818, and its extension to a large part of Germany through the Zollverein in the course of 1828–33.

The debate of 1815–18 is usually portrayed as a simple clash of sectional interests. Thus Field Marshal H. v. Boyen in a letter to Hardenberg, the Chancellor, in 1819:

> The West, perhaps already over-supplied with manufactures, sees in England its enemy who has to be fought continuously. The East, against which each year another market for corn is being closed, longs for British manufactured products and others like them in order to get rid of its corn by these purchases.[3]

But what is significant here was the extent to which external, rather than internal, relationships were considered. The concern of the eastern landowners (as of the mercantile cities, the other large free-trade lobby) was the relationship with Britain, while western industrialists were concerned partly with Britain and partly with their export markets. In vain did the Prussian bureaucracy protest, truth-

[1] Lévy-Leboyer, op. cit. p. 364. [2] Ibid. p. 410.

[3] Quoted in W. Treue, 'Wirtschaftszustände und Wirtschaftspolitik in Preuszen 1815–1825', Supplement 31 to *Vierteljahrschrift für Sozial- und Wirtschaftsgeschichte* (1937), p. 220.

fully, that the Tariff Bill was intended to deal mainly with internal administrative and taxation problems. Its spokesmen were driven to find justification on the more general grounds of favouring Prussia's prosperity and her industrialization. The decision to go in for free trade in 1818, at a time when Prussia was the only European power to do so, was no doubt ultimately anchored in the power of the Junkers, but the arguments developed in what was a remarkably articulate debate within the hierarchy, were conducted mainly in terms of furthering industry at a stage when Prussia occupied that middle position between British industrial power and the potential markets of her backward neighbours. Here, typically, is Bülow, explaining the proposed Tariff Bill to King Frederick William III in 1817:[1]

> External competition, far from suppressing home industry, tends to perfect manufacture and induce an extension of activity. The Provinces of Lower Rhine, Westphalia and Saxony learnt by these means how to exercise their skills. While they had to stand up to foreign competition without political defences or powers of prohibition, their manufactures rose, and their industriousness (*Industriefleisz*) prospered to such an extent that they do not now need to worry about sales at home, their goods can stand up to British competition by their high quality and moderate prices.

As another example we may quote the report dated 20 June 1817 by the Taxation Commission of the Prussian *Staatsrat* which accompanied its vote for the Tariff Bill, an important stage in its progress. Raw material costs were, they noted, much the same all over Europe, though Prussia might have the edge in flax and wool; interest rates were low in Prussia (Silesian 4 per cent stocks standing at 105 being much the same as British 3 per cent stocks at $78\frac{3}{4}$); wages were lower than in England; and only British machinery was superior, but might be copied. Thus many Prussian textile districts (but not the cotton spinners of Berlin and Silesia) should be in a position to beat the British in third (export) markets, as did the unprotected mills of Saxony and Switzerland.[2] Certainly, Kunth, travelling in 1817 in the province of Saxony and to Minden and Bielefeld in the west to investigate the textile manufacturers' complaints, reported continuing exports as well as stiff British competition.[3]

Basically, they echoed Silesian provincial President Merckel and the military Governor Maj.-Gen. von Gaudi in their submission of the Silesian demands at the Peace Conference of 1814. These two spokesmen were convinced that the Silesian textiles could hold their own in South America, given fair treatment, and hoped "to ensure in the peace treaty what has never before been achieved: that Silesian linen should not be burdened in Portugal and Spain with any higher duties than French and British linen". They also noted the large markets for Silesian textiles in Russia, Poland, and Austria which should be safeguarded.[4]

[1] Memorandum dated 14 January 1817, reprinted in W. v. Eisenhart Rothe and A. Ritthaler, eds. *Vorgeschichte und Bergründung des deutschen Zollvereines 1815–1834* (Berlin, 1934), I, 40–1. Among the extensive documentation on the debate about the tariff and the customs union this is the most useful collection.

[2] Ibid. I, 64–8. Also Kunth's reply to the cotton spinners, quoted in Herrmann Freymark, *Die Reform der preussischen Handels- und Zollpolitik von 1800–1821 und ihre Bedeutung* (Jena, 1898), p. 82.

[3] Rothe and Ritthaler, op. cit. I, 43–4, 111.

[4] Wilhelm Treue, loc. cit. p. 34; Otto Linke, 'Schlesiens Wünsche bei den Friedensverhandlungen 1814', *Zeitschrift des Vereins für Geschichte und Altertumskunde Schlesiens*, 33 (1899), 193–5, 201–5.

By the time of the debates of 1828–33 which surrounded the formation of the Southern Customs Union, the Central Tax Union, and ultimately the Zollverein, the inter-relation had become much clearer. It was universally expected that within what was now a single free trade area, the more advanced industrial districts—the duchy of Berg, Saxony, Silesia, or Nuremberg, for example—would replace some British manufactures in the less industrialized regions, and these, in turn, would send exports to Switzerland and Italy, while Saxony would continue her smuggling trade into Bohemia.[1] As a matter of fact, within a few years the Zollverein had become an importer *and* exporter not only of manufactures, but also of food.[2]

Economic relations with Britain were particularly close in the subsequent twenty years or so. In the dynamic equilibrium established between the two trading areas, both natural resources and labour could be considered immobile, and German wages as a whole could stay below British wages as a whole, as long as German productivity was lower. The ratios of comparative costs (or factor endowment, if the term is preferred) favoured Britain where better machinery made British labour much more productive, but favoured Germany where comparable machinery allowed her lower wages to give her a cost advantage. Thus British yarn and German looms, British iron and German metal goods[3] were added to British ships and Baltic grain and naval stores to form a symbiosis comparable to that of American cotton and Lancashire spindles. Friedrich List[4] saw part of the truth in the danger threatening the developing German industries from the developed British ones: but he failed to see the other part, the wide open markets available to the industrializing German areas, both inside their own countries, and farther east and south, and the aid they received in their expansion by being able to use cheap British iron, yarn, or machinery. These elastic markets provided the dynamic for German growth, even though British exports were breathing down German necks, and, in principle, that dual relationship became the pattern for European industrialization. It is precisely what is not available to industrializers today.

The analysis cannot be pursued here into the next phase, *c.* 1865–1914, except for one generalization. As industrialism spread outwards, the original heartlands found the adjacent areas ever less dependent and ever more competitive, so that they had to go farther and farther afield for complementary food, raw materials, and markets, leaving Europe to the later starters. Britain was the first to begin to

[1] Bernd Bab, *Die öffentliche Meinung über den deutschen Zollverein zur Zeit seiner Entstehung* (Berlin doctoral dissertation, 1930); City Council of Nuremberg to King Ludwig I, 8 August 1829, Appendix to the Opinion of the Frankfurt/Main Treasury on the Prussian–Bavarian Customs Union, mid-August 1829, Wietersheim, director of the Department of Trade to Lindenau, Chief Minister of the Kingdom of Saxony, 14 December 1830. All in Rothe and Ritthaler, op. cit. III, 547, 549–50, 604; Heinrich von Treitschke, 'Die Anfänge des deutschen Zollvereins, I', *Preuszische Jahrbücher*, XXX, Nos. 4, 5, and 6 (1882), 659.

[2] E.g. Bülow-Cummerow, *Der Zollverein, sein System und dessen Gegner* (Berlin, 1844), pp. 74–92.

[3] Percy Ashley, *Modern Tariff History* (2nd ed. 1910), pp. 23–5; Gertrud Hermes, 'Statistische Studien zur wirtschaftlichen und gesellschaftlichen Struktur des zollvereinten Deutschlands', *Archiv für Sozielwissenschaft und Sozialpolitik*, 63 (1930), esp. 137–9; E. L. Shorter, 'Social Change and Social Policy in Bavaria, 1800–1860' (unpublished Ph.D. thesis, Harvard University, 1967), p. 26. A little later, when the spinning mills of Germany or Switzerland could produce cotton yarn of coarser counts, they still drew the finer counts from Britain, while British coarser yarns went to the peripheral areas of Europe.

[4] *Das nationale System der politischen Oekonomie* (1841, Eng. trans. 1885).

turn her back on Europe, slowly in the 1840's and more rapidly from the peak of the 1870's on,[1] but the fate of being supplanted by one's earlier industrial dependencies also began to overtake the continental heartlands in their turn in the age of "Imperialism".[2] As for the latecomers, not only is it true that the later they are, the wider the gap in technology and the greater the difference in social structure that have to be bridged; but also the larger the pack of competitors baying at their heels, and the narrower the backward world still open to them. Somewhere near the outer periphery of Europe there is the watershed between societies which on contact with the new industrialism were capable of imitating it and becoming part of it, and areas which, at least for a long period, were transformed away from it, becoming specialized as "colonial" economies and facing an ever wider gap between themselves and the advanced economies.[3] That line marks the difference between the industrialization of Europe and the industrializers of today.

It is worth repeating that we are describing not merely a sequence but also a mechanism. Looked at from the narrow limits of one country, we may grasp such partial relationships as "export-led industrialization" or "enclave" economy.[4] But the real historical relationships were multilateral and continuously changing. The British industrial revolution took the course it took, because among other reasons, elastic export markets were available for the products of the mills; German weaving could expand because the mills existed in England; and Dutch, Danish, and Swiss agriculture was built up on the nearby existence of large industrial populations. The industrial revolution in Europe was not merely the national repetition of a pattern; it was also the continuous adaptation to a continent-wide opportunity.

IV

So far we have examined industrialization in vertical cross-section, as it were: along the line of chronology. If we turn to the horizontal cross-section, observing Europe at any one point of time, two aspects stand out in importance.

The first is that Europe was not a single market merely for commodities: over much of the nineteenth century it also in effect became a single market for capital, skill, enterprise, ideas, and technical knowledge. The extent of the inter-penetration of capital, whether by the older type of private banker or by the newer joint-stock investment banker,[5] has been made clearer by recent work. Similarly

[1] Ross J. S. Hoffman, *Great Britain and the German Trade Rivalry, 1875–1914* (Philadelphia, 1933), pp. 5 ff; J. B. Condliffe, *The Commerce of Nations* (1951), p. 208.

[2] See Bernhard Harms, *Volkswirtschaft und Weltwirtschaft: Versuche der Begründung einer Weltwirtschaftslehre* (Jena, 1912), p. 200.

[3] Compare Jeffrey G. Williamson, 'Regional Inequality and the Process of National Development: A Description of the Patterns', *Economic Development and Cultural Change*, XIII (1964–5), pt II.

[4] K. Berrill, 'International Trade and the Rate of Economic Growth', *Econ. Hist. Rev.* 2nd ser. XII (1960); Otfried Fröhlich, *Die Wirkungen der Industrialisierung junger Länder auf ihre Einfuhr* (Dresden, 1936), pp. 58 ff; Amintore Fanfani, 'Considerations on the Industrial Revolution', in Marcelle Kooy, ed. *Studies in Economics and Economic History: Essays in Honour of H.M. Robertson* (1972), pp. 215 ff; J. R. T. Hughes, 'Foreign Trade and Balanced Growth: The Historical Framework', *American Economic Review*, Supplement (1959), p. 334; Ohlin, loc. cit. pp. 350–1; Ragnar Nurkse, *Patterns of Trade and Development* (Stockholm, 1959), pp. 13 ff.

[5] Harvey H. Segal and Matthew Simon, 'British Foreign Capital issues, 1865–1894', *Journal of Economic History*, XXI (1961); Leland H. Jenks, *The Migration of British Capital to 1875* (1927); Lévy-Leboyer, op. cit.; Rondo E. Cameron, *France and the Economic Development of Europe* (Princeton, 1961); Sartorius von Waltershausen, *Das volkswirtschaftliche System der Kapitalanlage im Auslande* (Berlin, 1907); Bertrand Gille,

there is now no justification for ignoring the transmission of entrepreneurship and technology from the advanced industrial regions to the developing regions across the frontiers of Europe.[1] But their significance and their role have often been misunderstood.

It is sometimes argued that the part played by foreign capital or foreign skilled labour could only have been minimal, since nowhere did it exceed a small fraction of "domestic" capital and labour available at any one time. But that line of reasoning, derived, no doubt, from the experience of present-day industrializers, misinterprets the needs of a European neighbour. European countries were not greatly dissimilar in wealth at the onset of industrialization, and they shared common traditions of settled government, urban culture, and, above all, commercial capitalism. What they required, in different degrees, was not to be pulled up from abroad, but a stimulus or trigger, and some detailed know-how in key sectors so that they could "take off" by themselves.

It is precisely that which they received from their more advanced neighbours. Thus in the post-1815 period it can be shown that every major iron and engineering works of the modern kind in Belgium and France and every early major railway, beside many textile mills and other enterprises, operated with British help. Similarly, every new German industry used British technology and many used French capital and entrepreneurship. By the mid-century, French capital and engineering skill were to be found in the main railways of Spain and Portugal, Italy, Austria, and Russia. Behind them, their very banks derived from the stimulus and the resources of the French *crédit mobilier* while, it should be noted, most of the chief London investment banks were not of British origin, nor were the Paris *Haute Banque* of French. Among the earlier industrializers at least, the initial stimulus from outside found enough capital, and a sufficiently favourable social environment (pre-existing or created by this process itself), to transfer the main burden to "native" shoulders, as had been the case in the different regions of Britain. But that does not diminish the critical role played by the fuse which in one area after another was ignited without waiting for spontaneous combustion to occur.

Thus it is the wrong counter-factual question which has often been posed in this regard. It should not be "How long would French main-line railways have been delayed without British help?" in order to prove that British help made little difference: for the British help is known to have occurred and therefore the influence of one industrialized economy upon the other does not need proof. The question should rather be: "If French capital and entrepreneurship had not flowered on the amount of British aid they received, would more have been forthcoming from Britain until they did?" In the light of the quantity of the later transfer of British capital and entrepreneurship overseas, there can surely not be much doubt about the answer.

Le crédit en France 1815–1848 (Paris, 1959); idem, *Banking and Industrialization in Europe, 1730–1914. Fontana History of Europe*, III/4 (1970); David Landes, 'The Old Bank and the New: the Financial Revolution of the Nineteenth Century', in F. Crouzet, W. H. Chaloner, and W. M. Stern, eds. *Essays in European Economic History, 1789–1914* (1969).

[1] Rondo E. Cameron, 'Some French Contributions to the Industrial Development of Germany, 1840–1870', *Journal of Economic History*, XVI (1956); Henderson, op. cit.

The second aspect is, in a way, an elaboration of the Gerschenkron concept of backwardness and the different stages of technology which each country has to tackle when it arrives at its "take-off" point.[1] Europe was a sufficiently close community for major events to have affected it simultaneously, even while its component industrial regions were at very different stages of development and relationship to each other. This applied, for example, to the impact of war and blockade in 1800–15, to railways in the 1840's and 1850's, socialism in the 1860's. cheap agricultural imports in the 1870's, and the scramble for colonies in the last years of the century.

Thus railways came to England at the end of her industrial revolution, to France as part of hers, to Russia in advance of hers. They used existing engineering talent in Britain, developed such talent in France and Germany, brought in foreign talent in Austria and Russia. And they were the basis of capital exports from Britain and France, internal capital development in Germany, and massive capital imports into Russia. The cheap grain of the 1870's, on the other hand, divided Europe into those areas which could cash in on the boom by exporting grain, like Russia and Hungary, and those which used grain as feedstuff to convert their agriculture to animal husbandry and dairying, like parts of Britain, the Netherlands, Denmark, or Switzerland. Germany was caught out by being still dominated by an anachronistic ruling class, the Junkers, who set out to stem the tide of history. The Junkers, indeed, did not believe their own propaganda about the importance of the German frontier which enclosed the Reich: they knew that the most important German frontier was the one which ran along the Elbe, separating the agrarian Germany from industrialized Germany. And they kept German economic life divided along this line by their victory in the "war of the canals" around 1900, which prevented the obvious link-up of German waterways and preserved a gap, the only such major gap in Europe, running through the middle of Germany, between the linked Danube–Rhine–Rhône–Seine–Ems–Weser system, and the Elbe–Oder system linking up with the east.[2] Moreover, they also achieved an unparalleled exploitation of the rest of Germany through the ingenious "export certificate",[3] by which the urban consumer was obliged not only to pay more for the bread he consumed, but also to subsidize the eastern rye he did not want to consume.

Thus the structural similarities between the English Corn Law of 1815 and the German tariff of 1879 were accompanied also by vital differences. The timing effect discussed here is another reason why European industrialization should not be seen as the repetition of a model, but as a single, if complex, process.

V

From the 1860's on, the needs of the advanced economies called forth a rapidly growing phalanx of international bodies and forms of collaboration, from the

[1] Alexander Gerschenkron, *Economic Backwardness in Historical Perspective: A Book of Essays* (Cambridge, Mass. 1962). Also see David S. Landes, *The Unbound Prometheus* (Cambridge, 1969), pp. 335–6.

[2] Louis Marlo, *La politique allemande et la navigation intérieure* (2nd ed. Paris, 1908); Richard Hennig, *Die Hauptwege des Weltverkehrs* (Jena, 1913), pp. 82–3.

[3] Alexander Gerschenkron, *Bread and Democracy in Germany* (Berkeley, 1943); Wilhelm Röpke, *German Commercial Policy* (1934), pp. 54–6; Cobden Club, *The Influence of Protection on Agriculture in Germany* (1910).

Universal Postal Union and the Berne copyright convention to international railway agreements and international monetary harmonization based on the gold standard. The logic of events fostered an internationalization of economic life, primarily within Europe but ultimately also across the world, which raised the hopes of the Peace Movement, expressed most clearly by Norman Angell's *The Great Illusion* of 1910, that economic interdependence was making European war impossible.

But in the last quarter of the century there began to be found, together with these developments, also those of an opposite tendency. The relationship of the nineteenth-century state, particularly the state based on national revolution, with the industrial revolution with its changing class structure, are complex and contradictory.[1] At an earlier stage the state had been a progressive force, reacting favourably to the elemental needs and pressures of industry by creating large and free internal markets, by abolishing serfdom and feudal tax privileges, by limiting the civil rights of labour or manipulating taxes in favour of capital. But now the tendency of its actions was increasingly to inhibit progress by disrupting the international economy. Barriers between naturally complementary regions were raised ever higher, labour migration impeded, and foreign ownership of capital restricted as capital exports became a political weapon. Real costs rose as subsidies encouraged the building of sailing ships in France in the 1890's and the growing of sugar all over Europe. After the brief liberal interlude of *c.* 1860–79 with its great expansion of international trade, countries turned to Neo-Mercantilism, meaning to become strong by becoming autarkic.

While the positive influence of protection on industrialization is much in doubt, its negative influence on foreign trade and the international division of labour is not in dispute. Given the nature of the political state in its competitive setting in Europe, its obstructive role in the evolution of an integrated European economy is as "natural" as that evolution itself. But it is surely not entirely fanciful, though it would be going much beyond the scope of this article, to see in the opposition of the political authority to the underlying economic logic of European industrial development one of the roots of tensions and wars of the period *c.* 1865–1945. It may even be that there is a similar link between the astonishing progress of the European economy since that date and the decision of the political authorities, both in the east and the west, to work within rather than against, the integrating tendencies of European industrialization.

VI

It is the argument of this article that the industrialization of nineteenth-century Europe was a single process and had an economic logic of its own. Like an epidemic, it took little note of frontiers, crossing them with ease while leaving neighbouring home territories untouched. The political and legal base in each region was not without influence on it, but it is clear that the factors which made one area more susceptible to infection than another included locational advantages, resources and, above all, preceding economic development bringing in its train a favourable social structure.

[1] They are explored brilliantly by E. J. Hobsbawm, *The Age of Revolution: Europe, 1789–1848* (1962).

It follows that the study of industrialization in any given European country will remain incomplete unless it incorporates a European dimension: any model of a closed economy would lack some of its basic and essential characteristics. These include relationships with economies which were more advanced, with those at similar stages, and with those still awaiting development. They also include differentiation of phasing as various common European experiences touch all economies simultaneously. Finally, they must include an absolute time scale: as the boundary line between industrialized heartland and underdeveloped periphery shifted across Europe, so the basic operative technology and the necessary costs and the nature of social dislocation required for industrialization underwent continuous change. Moreover, the later the start, the larger the existing industrialized world against which the industrial revolution is played out and the farther away the periphery. In the case of today's industrializers, the industrialized heartland is so large as to be in effect uninfluenced by their actions, while the technological gap is so great that government has to play a role that would have been quite inappropriate in Europe.[1] But this should not blind us to the mutual inter-actions of the nineteenth-century industrial revolutions in Europe.

This article has discussed only some aspects of these problems. In this sense it is not intended to present a conclusion, but a programme of work. Its object is not to draw on existing research, but to stimulate new study. While it may help towards a more realistic evaluation of some activities which have had the limelight in the past, as for example the policies of government, its main intention is to urge the re-examination of the impact of the total European situation, in each phase, on any given segment of it.

University of Sheffield

[1] A. G. Kenwood and A. L. Lougheed, *The Growth of the International Economy 1820–1960* (1971), p. 134; Talcott Parsons, 'Some Reflections on the Institutional Framework of Economic Development', in *Structure and Process in Modern Societies* (Glencoe, Ill. 1960), p. 116; P. Streeten, 'The Frontiers of Development Studies: Some Issues of Development Policy', *Journal of Development Studies*, IV (1967).

The Take-Off into Self-Sustained Growth

by

WALT W. ROSTOW

[*Editorial note:* The following study is taken from Chapters III and IV of Professor Rostow's *The Stages of Economic Growth.* In order to emphasize those parts which seemed most relevant to the rest of this book, I reluctantly decided not to include direct mention of the full sweep of Professor Rostow's framework. Nevertheless, it should be remembered that this framework is much broader than the following excerpt suggests.

In summary, Professor Rostow identifies five stages of growth: (1) the traditional society, (2) the preconditions for take-off, (3) the take-off, (4) the drive to maturity, and (5) the age of high mass consumption. The traditional society is one in which techniques of production are of a low order of efficiency, in which scientific knowledge is limited and unsystematic, in which the bulk of society's resources

Reprinted from *The Stages of Economic Growth,* by Walt W. Rostow (pp. 17–58, with omissions), copyright © 1960 by Cambridge University Press, by permission of the publishers and author.

and efforts are devoted to agriculture, in which the social structure is hierarchical and rigid, and in which there is a low ceiling on productivity. The second and third stages are dealt with in detail in this selection. The fourth stage, the drive to maturity, is one in which sustained expansion over a long period extends modern technology over the entire range of a country's economic activity. Professor Rostow suggests that maturity is normally attained about sixty years after the take-off begins. The fifth stage, the age of high mass consumption, is one in which there is an increasing emphasis upon durable consumers' goods and services. It has also frequently been associated with the devotion of large amounts of resources to social welfare and security.

For a critique of Professor Rostow's analysis of the take-off, see Selection 22.]

I. THE PRECONDITIONS FOR TAKE-OFF

We consider in this [section] the preconditions for take-off: the transitional era when a society prepares itself—or is prepared by external forces—for sustained growth.

It is necessary to begin by distinguishing two kinds of cases history has to offer.

There is first what might be called the general case. This case fits not merely the evolution of most of Europe but also the greater part of Asia, the Middle East, and Africa. In this general case the creation of the preconditions for take-off required fundamental changes in a well-established traditional society: changes which touched and substantially altered the social structure and political system as well as techniques of production.

Then there is the second case. This case covers the small group of nations that were, in a sense, "born free":[1] the United States, Australia, New Zealand, Canada, and, perhaps, a few others. These nations were created mainly out of a Britain already far along in the transitional process. Moreover, they were founded by social groups—usually one type of non-conformist or another—who were at the margin of the dynamic transitional process slowly going forward within Britain. Finally their physical settings—of wild but abundant land and other natural resources—discouraged the maintenance of such elements in the traditional structure as were transplanted, and they accelerated the transitional process by offering extremely attractive incentives to get on with economic growth. Thus the nations within the second case never became

[1] A phrase used by Louis Hartz in *The Liberal Tradition in America* (New York, 1955).

so deeply caught up in the structures, politics and values of the traditional society; and, therefore, the process of their transition to modern growth was mainly economic and technical. The creation of the preconditions for take-off was largely a matter of building social overhead capital—railways, ports and roads—and of finding an economic setting in which a shift from agriculture and trade to manufacture was profitable; for, in the first instance, comparative advantage lay in agriculture and the production of food-stuffs and raw materials for export.

The distinction between the two cases is real enough; but looked at closely the lines of demarcation turn out to be not all that sharp. The United States, for example, created for itself a kind of traditional society in the South, as an appendage to Lancashire, and then New England's cotton mills; and the long, slow disengagement of the South from its peculiar version of a traditional society belongs clearly in the general rather than the special case. Canada, moreover, has had its regional problem of a sort of traditional society in Quebec. The take-off of the American South is a phenomenon of the last two decades; while the take-off in Quebec may only now be getting whole-heartedly under way.

There are other types of fuzziness as well. Are the Latin American states to be regarded as in the general case, or among the lucky offspring of already transitional Europe? On the whole, we would judge, they belong in the general case; that is, they began with a version of a traditional society—often a merging of traditional Latin Europe and native traditional cultures—which required fundamental change before the mixed blessings of compound interest could be attained; but the Latin American cases vary among themselves. Similarly, Scandinavia, somewhat like Britain itself, faced less searching problems than many other parts of Europe in shaking off the limiting parameters of the traditional society. Sweden is almost in the second rather than the first category. . . .

This [section] is concentrated on the general case; that is, on the process, within a traditional society, by which the preconditions for take-off are created.

The Nature of the Transition

The transition we are examining has, evidently, many dimensions. A society predominantly agricultural—with, in fact, usually 75% or more of its working force in agriculture—must shift to a predominance for industry, communications, trade and services.

A society whose economic, social and political arrangements are built around the life of relatively small—mainly self-sufficient—regions must orient its commerce and its thought to the nation and to a still larger international setting.

. . . . .

The income above minimum levels of consumption, largely concentrated in the hands of those who own land, must be shifted into the hands of those who will spend it on roads and railroads, schools and factories rather than on country houses and servants, personal ornaments and temples.

Men must come to be valued in the society not for their connexion with clan or class, or, even, their guild; but for their individual ability to perform specific, increasingly specialized functions.

And, above all, the concept must be spread that man need not regard his physical environment as virtually a factor given by nature and providence, but as an ordered world which, if rationally understood, can be manipulated in ways which yield productive change and, in one dimension at least, progress.

. . . . .

The Analysis of the Transition

The modern economist—or perhaps one should say, given the recent shift of interest to growth, the modern economist of a decade ago—might have been inclined to say to the historian something of this sort: "This complexity about whole societies is all very well; and it is no doubt of some interest to you and your kind; but don't make such heavy weather of it. What you are talking about is a rise in the rate of investment and in the *per capita* stock of capital. Get the investment-rate up to the point where the increase in output outstrips the rate of population increase—to, say, a rate of investment over 10% of national income—and the job is done. The difference between a traditional and a modern society is merely a question of whether its investment-rate is low relative to population increase—let us say under 5% of national income; or whether it has risen up to 10% or over. With a capital/output ratio of about 3, a 10% investment-rate will outstrip any likely population growth; and there you are, with a regular increase in output per head."

And what the old-fashioned modern economist might have said was, of course, quite true.

But to get the rate of investment up some men in the society must be able to manipulate and apply—and in a closed system they must be able to create—modern science and useful cost-reducing inventions.

Some other men in the society must be prepared to undergo the strain and risks of leadership in bringing the flow of available inventions productively into the capital stock.

Some other men in the society must be prepared to lend their money on long term, at high risk, to back the innovating entrepreneurs—not in money-lending, playing the exchanges, foreign trade or real estate—but in modern industry.

And the population at large must be prepared to accept training for—and then to operate—an economic system whose methods are subject to regular change, and one which also increasingly confines the individual in large, disciplined organizations allocating to him specialized narrow, recurrent tasks.

In short, the rise in the rate of investment—which the economist conjures up to summarize the transition—requires a radical shift in the society's effective attitude toward fundamental and applied science; toward the initiation of change in productive technique; toward the taking of risk; and toward the conditions and methods of work.

. . . . .

Having peered briefly inside the process of investment in a world of changing production functions, we can conclude by agreeing that, in the end, the essence of the transition can be described legitimately as a rise in the rate of investment to a level which regularly, substantially and perceptibly outstrips population growth; although, when this is said, it carries no implication that the rise in the investment-rate is an ultimate cause.

Two Sectoral Problems

The rise of the investment-rate, as well as reflecting these more profound societal changes, is also the consequence of developments in particular sectors of the economy, where the transformation of the economy actually takes place. The analysis of economic growth can, then, proceed only a short and highly abstracted way without disaggregation.

To illustrate the need to pierce the veil of aggregative analysis in the transitional period we shall look briefly now at two particular problems shared, in one way or another, by all societies which have learned how to grow: the problem of increased productivity in agriculture and the extractive industries; and the problem of social overhead capital.

Agriculture and the Extractive Industries

Although a good deal of the early growth process hinges on the food-supply, the first of these two sectoral problems is properly to be defined as that of agriculture and the extractive industries. The general requirement of the transition is to apply quick-yielding changes in productivity to the most accessible and naturally productive resources. Generally, this means higher productivity in food-production. But it may also mean wool, cotton, or silk—as in nineteenth-century New Zealand, the American South, and Japan. And in Sweden it meant timber; in Malaya, rubber; in the Middle East, oil; and in certain American regions, Australia, and Alaska, gold helped to do the trick.

The point is that it takes more than industry to industrialize. Industry itself takes time to develop momentum and competitive competence; in the meanwhile there is certain to be a big social overhead capital bill to meet; and there is almost certain to be a radically increased population to feed. In a generalized sense modernization takes a lot of working capital; and a good part of this working capital must come from rapid increases in output achieved by higher productivity in agriculture and the extractive industries.

More specifically the attempt simultaneously to expand fixed capital —of long gestation period—and to feed an expanding population requires both increased food output at home and/or increased imports from abroad. Capital imports can help, of course, but in the end loans must be serviced; and the servicing of loans requires enlarged exports.

It is, therefore, an essential condition for a successful transition that investment be increased and—even more important—that the hitherto unexploited back-log of innovations be brought to bear on a society's land and other natural resources, where quick increases in output are possible.

Having made the general case in terms of requirements for working capital, look for a moment more closely at the question of agriculture and the food-supply. There are, in fact, three distinct major roles agriculture must play in the transitional process between a traditional society and a successful take-off.

First, agriculture must supply more food. Food is needed to meet the likely rise in population, without yielding either starvation or a depletion of foreign exchange available for purposes essential to growth. But increased supplies and increased transfers of food out of rural areas are needed for another reason: to feed the urban populations which are certain to grow at a disproportionately high rate during the transition. And, in most cases, increased agricultural supplies are needed as well to help meet the foreign exchange bill for capital development: either positively by earning foreign exchange, as in the United States, Russia, Canada, and several other nations which generated and maintained agricultural surpluses while their populations were growing (and their urban populations growing faster than the population as a whole); or negatively, to minimize the foreign exchange bill for food—like a whole series of nations from Britain in the 1790's to Israel in the 1950's.

The central fact is that, in the transitional period, industry is not likely to have established a sufficiently large and productive base to earn enough foreign exchange to meet the increments in the nation's food bill via increased imports. Population increases, urbanization, and increased foreign exchange requirements for fixed and working capital are all thus likely to conspire to exert a peculiar pressure on the agricultural sector in the transitional process. Put another way, the rate of increase in output in agriculture may set the limit within which the transition to modernization proceeds.

But this is not all. Agriculture may enter the picture in a related but quite distinctive way, from the side of demand as well as supply. Let us assume that the governmental sector in this transitional economy is not so large that its expanded demand can support the rapid growth of industry. Let us assume that some of the potential leading sectors are in consumers' goods—as, indeed, has often been the case: not only cotton textiles—as in England and New England—but a wide range of import substitutes, as in a number of Latin American cases. In addition, the modern sector can—and often should—be built in part on items of capital for agriculture: farm machinery, chemical fertilizers, diesel pumps, etc. In short, an environment of rising real incomes in agriculture, rooted in increased productivity, may be an important stimulus to new modern industrial sectors essential to the take-off.

The income side of the productivity revolution in agriculture may be important even in those cases where the transition to industrialization is not based on consumers' goods industries; for it is from rising rural incomes that increased taxes of one sort or another can be drawn —necessary to finance the government's functions in the transition—without imposing either starvation on the peasants or inflation on the urban population.

And there is a third distinctive role for agriculture in the transitional period which goes beyond its functions in supplying resources, effective demand or tax revenues: agriculture must yield up a substantial part of its surplus income to the modern sector. At the core of the *Wealth of Nations*—lost among propositions about pins and free trade—is Adam

Smith's perception that surplus income derived from ownership of land must, somehow, be transferred out of the hands of those who would sterilize it in prodigal living into the hands of the productive men who will invest it in the modern sector and then regularly plough back their profits as output and productivity rise.

In their nineteenth-century land-reform schemes this is precisely what Japan, Russia, and many other nations have done during the transition in an effort to increase the supply of capital available for social overhead and other essential modernizing processes.

It is thus the multiple, distinctive, but converging consequences of the revolution in agriculture which give to it a peculiar importance in the period of preconditions. Agriculture must supply expanded food, expanded markets, and an expanded supply of loanable funds to the modern sector.

Social Overhead Capital

Where data exist on the level and pattern of capital formation in pre-take-off societies—and for the take-off as well—it is clear that a very high proportion of total investment must go into transport and other social overhead outlays.

Aside from their quantitative importance, social overhead outlays have three characteristics which distinguish them from investment in general, as usually presented in aggregative models. First, their periods of gestation and of pay-off are usually long. Unlike double-cropping or the application of chemical fertilizers, a railway system is unlikely to yield its results in a year or two from the time its construction is undertaken, although it will yield large benefits over a very long time. Second, social overhead capital is generally lumpy. You either build the line from, say, Chicago to San Francisco or you do not: an incomplete railway line is of limited use, although many other forms of investment —in industry and agriculture—can proceed usefully by small increments. Third, of its nature, the profits from social overhead capital often return to the community as a whole—through indirect chains of causation—rather than directly to the initiating enterpreneurs.

Taken together, these three characteristics of social overhead capital —the long periods of gestation and pay-off, the lumpiness, and the indirect routes of pay-off—decree that governments must generally play an extremely important role in the process of building social overhead capital; which means governments must generally play an extremely important role in the preconditions period. Put another way, social overhead capital cannot be formed—in some of its most essential forms —by an enlarging flow of ploughed-back profits from an initially small base. You cannot get well started unless you can mobilize quite large initial capital sums.

Thus, even in so highly capitalist a transitional society as the United States between 1815 and 1840, state and local governments played a major role in initiating the build-up of social overhead capital. The Erie Canal was built by the New York State legislature; and the great American continental railway networks were built with enormous federal subsidies in the form of land grants.

The argument about agriculture and social overhead capital in transitional societies underlies a point of method and a point of substance. The point of method is that orderly disaggregation is necessary for an analysis of economic growth that comes to grips with the key strategic factors. Aggregates which may be useful for purposes of short-run income analysis conceal more than they illuminate when carried over into the analysis of growth. The point of substance is that the preparation of a viable base for a modern industrial structure requires that quite revolutionary changes be brought about in two non-industrial sectors: agriculture and social overhead capital, most notably in transport.

Non-Economic Change

We turn, now, to the non-economic side of the preconditions for take-off.

The broad lines of societal change necessary to prepare a traditional society for regular growth are becoming familiar enough. It would be widely agreed that a new élite—a new leadership—must emerge and be given scope to begin the building of a modern industrial society; and, while the Protestant ethic by no means represents a set of values uniquely suitable for modernization, it is essential that the members of this new élite regard modernization as a possible task, serving some end they judge to be ethically good or otherwise advantageous.

Sociologically this new élite must—to a degree—supersede in social and political authority the old land-based élite, whose grasp on income above minimum levels of consumption must be broken where it proves impossible simply to divert that income smoothly into the modern sector.

And more generally—in rural as in urban areas—the horizon of expectations must lift; and men must become prepared for a life of change and specialized function.

Something like this group of sociological and psychological changes would now be agreed to be at the heart of the creation of the preconditions for take-off. But this is an insufficient view. While in no way denying the significance of some such changes in attitude, value, social structure and expectations, we would emphasize, in addition, the role of the political process and of political motive in the transition.

As a matter of historical fact a reactive nationalism—reacting against intrusion from more advanced nations—has been a most important and powerful motive force in the transition from traditional to modern societies, at least as important as the profit motive. Men holding effective authority or influence have been willing to uproot traditional societies not, primarily, to make more money but because the traditional society failed—or threatened to fail—to protect them from humiliation by foreigners. Leave Britain aside for a moment and consider the circumstances and motives that set traditional societies in other regions on the road to modernization.

In Germany it was certainly a nationalism based on past humiliation and future hope that did the job: the memory of Napoleon, and the Prussian perception of the potentialities for power of German unity and German nationalism. It was German nationalism which stole the revolution of 1848 at Frankfurt and made the framework within which the German take-off occurred—the Junkers and the men of the East, more than the men of trade and the liberals of the West. In Russia it was a series of military intrusions and defeats, stretching out over a century, which was the great engine of change: Napoleon's invasion, the Crimean War, the Russo-Japanese War, and then, finally, the First World War. In Japan it was the demonstration effect not of high profits or manufactured consumers' goods, but of the Opium War in China in the early 1840's and Commodore Perry's seven black ships a decade later that cast the die for modernization. And in China, the deeply entrenched traditional society yielded only slowly and painfully; but it did, in the end, yield to a century of humiliations from abroad that it could not prevent.

And so also, of course, with the colonial areas of the southern half of the world. But there, in the colonies, a dual demonstration effect operated.

Although imperial powers pursued policies which did not always optimize the development of the preconditions for take-off, they could not avoid bringing about transformation in thought, knowledge, institutions and the supply of social overhead capital which moved the colonial society along the transitional path; and they often included modernization of a sort as one explicit object of colonial policy.

In any case, the reality of the effective power that went with an ability to wield modern technology was demonstrated and the more thoughtful local people drew appropriate conclusions. Ports, docks, roads, and later, railways were built; a centralized tax system was imposed; some colonials were drawn into those minimum modern economic activities necessary to conduct trade to produce what the colonial power wished to export and what could profitably be produced locally for the expanding urban and commercialized agricultural markets; some modern goods and services were diffused sufficiently to alter the conception of an attainable level of consumption; the opportunity for a Western education was opened to a few, at least; and a concept of nationalism, transcending the old ties to clan or region, inevitably crystallized around an accumulating resentment of colonial rule.

In the end, out of these semi-modernized settings, local coalitions emerged which generated political and, in some cases, military pressure capable of forcing withdrawal; coalitions created by both the positive and negative types of demonstration.

Xenophobic nationalism or that peculiar form of it which developed in colonial areas has not, of course, been a unique motive in bringing about the modernization of traditional societies. The merchant has been always present, seeing in modernization not only the removal of obstacles to enlarged markets and profits but also the high status denied him—despite his wealth—in the traditional society. And there have almost always been intellectuals who saw in modernization ways of increasing the dignity or value of human life, for individuals and for the nation as a whole. And the soldier—an absolutely crucial figure of the transition—often brought much more to the job than resentment of foreign domination and dreams of future national glory on foreign fields of battle.

. . . . .

The Alternative Directions of Nationalism

Now we come to the crux of the matter. Nationalism can be turned in any one of several directions. It can be turned outward to right real or believed past humiliations suffered on the world scene or to exploit real or believed opportunities for national aggrandizement which appear for the first time as realistic possibilities, once the new modern state is established and the economy develops some momentum; nationalism can be held inward and focused on the political consolidation of the victory won by the national over the regionally based power; or nationalism can be turned to the tasks of economic, social, and political modernization which have been obstructed by the old regionally based, usually aristocratic societal structure, by the former colonial power, or by both in coalition.

. . . . it is clear that the length of time and the vicissitudes of transition from traditional to modern status depend substantially on the degree to which local talent, energy, and resources are channelled on to the domestic tasks of modernization as opposed to alternative possible objectives of nationalism; and this channelling must, in the general case, be in substantial part a function of political leadership.

This is so because the central government has essential, major technical tasks to perform in the period of preconditions. There is no need for the government to own the means of production; on the contrary. But the government must be capable of organizing the nation so that unified commercial markets develop; it must create and maintain a tax and fiscal system which diverts resources into modern uses, if necessary at the expense of the old rent-collectors; and it must lead the way through the whole spectrum of national policy—from tariffs to education and public health—toward the modernization of the economy and the society of which it is a part. For, as emphasized earlier, it is the inescapable responsibility of the state to make sure the stock of social overhead capital required for take-off is built; and it is likely as well that only vigorous leadership from the central government can bring about those radical changes in the productivity of agriculture and the use of other natural resources whose quick achievement may also constitute a precondition for take-off.

II. THE TAKE-OFF

We turn now to analyze narrowly that decisive interval in the history of a society when growth becomes its normal condition. We consider how it comes about that the slow-moving changes of the preconditions period, when forces of modernization contend against the habits and institutions, the values and vested interests of the traditional society, make a decisive break-through; and compound interest gets built into the society's structure.

As suggested in [the previous section], take-offs have occurred in two quite different types of societies; and, therefore, the process of establishing preconditions for take-off has varied. In the first and most general case the achievement of preconditions for take-off required major changes in political and social structure and even in effective social values. In the second case take-off was delayed not by political, social and cultural obstacles but by the high (and even expanding) levels of welfare that could be achieved by exploiting land and natural resources. In this second case take-off was initiated by a more narrowly economic process as, for example, in the northern United States, Australia and, perhaps, Sweden. And, you will recall, as one would expect in the essentially biological field of economic growth, history offers mixed as well as pure cases.

The beginning of take-off can usually be traced to a particular sharp stimulus. The stimulus may take the form of a political revolution which affects directly the balance of social power and effective values, the character of economic institutions, the distribution of income, the pattern of investment outlays and the proportion of potential innovations actually applied. Such was the case, for example, with the German revolution of 1848, the Meiji restoration in Japan of 1868, and the more recent achievement of Indian independence and the Communist victory in China. It may come about through a technological (including transport) innovation, which sets in motion a chain of secondary expansion in modern sectors and has powerful potential external economy effects which the society exploits. It may take the form of a newly favourable international environment, such as the opening of British and French markets to Swedish timber in the 1860's or a sharp relative rise in export prices and/or large new capital imports, as in the case of the United States from the late 1840's, Canada and Russia from the mid-1890's; but it may also come as a challenge posed by an unfavourable shift in the international environment, such as a sharp fall in the terms of trade (or a war-time blockage of foreign trade) requiring the rapid development of manufactured import substitutes, as with the Argentine and Australia from 1930 to 1945.

What is essential here is not the form of stimulus but the fact that the prior development of the society and its economy result in a positive, sustained, and self-reinforcing response to it: the result is not a once-over change in production functions or in the volume of investment, but a higher proportion of potential innovations accepted in a more or less regular flow, and a higher rate of investment.

The use of aggregative national-income terms evidently reveals little of the process which is occurring. It is nevertheless useful to regard as a necessary but not sufficient condition for the take-off the fact that the proportion of net investment to national income (or net national product) rises from, say, 5% to over 10%, definitely outstripping the likely population pressure (since under the assumed take-off circumstances the capital/output ratio is low),[2] and yielding a distinct rise in

[2] Capital/output ratio is the amount by which a given increase in investment increases the volume of output: a rough—very rough—measure of the productivity of capital investment; but since the arithmetic of economic growth requires some such concept, implicitly or explicitly, we had better refine the tool rather than abandon it. In the early stages of economic development two contrary forces operate on the capital/output ratio. On the one hand there is a vast requirement of basic overhead capital in transport, power, education etc. Here, due mainly to the long period over which investment yields its return, the apparent (short-run) capital/output ratio is high. On the other hand, there are generally large unexploited back-logs of

real output *per capita*. Whether real consumption *per capita* rises depends on the pattern of income distribution and population pressure, as well as on the magnitude, character and productivity of investment itself.

As indicated in the accompanying table, we believe it possible to identify at least tentatively such take-off periods for a number of countries which have passed into the stage of growth.

TABLE 1

SOME TENTATIVE, APPROXIMATE TAKE-OFF DATES

Country	Take-Off	Country	Take-Off
Great Britain	1783–1802	Russia	1890–1914
France	1830–60	Canada	1896–1914
Belgium	1833–60	Argentina	1935–
United States[a]	1843–60	Turkey	1937–
Germany	1850–73	India	1952–
Sweden	1868–90	China	1952–
Japan	1878–1900		

[a] The American take-off is here viewed as the upshot of two different periods of expansion: the first, that of the 1840's, marked by railway and manufacturing development, mainly confined to the East—this occurred while the West and South digested the extensive agricultural expansion of the previous decade; the second the great railway push into the Middle West during the 1850's marked by a heavy inflow of foreign capital. By the opening of the Civil War the American economy of North and West, with real momentum in its heavy-industry sector, is judged to have taken off.

The Take-Off Defined and Isolated

The take-off is such a decisive transition in a society's history that it is important to examine the nature of our definition and the inner mechanism of take-off somewhat more closely.

There are several problems of choice involved in defining the take-off with precision. We might begin with one arbitrary definition and consider briefly the two major alternatives.

For the present purposes the take-off is defined as requiring all three of the following related conditions:

(1) a rise in the rate of productive investment from, say 5% or less to over 10% of national income (or net national product (NNP));

known techniques and available natural resources to be put to work; and these back-logs make for a low capital/output ratio. We can assume formally a low capital/output ratio for the take-off period because we are assuming that the preconditions have been created, including a good deal of social overhead capital. In fact, the aggregate marginal capital/output ratio is likely to be kept up during the take-off by the requirement of continuing large outlays for overhead items which yield their returns only over long periods. Nevertheless, a ratio of 3:1 or 3.5:1 for the incremental capital/output ratio seems realistic as a rough bench-mark until we have learned more about capital/output ratios on a sectoral basis.

(2) the development of one or more substantial manufacturing[3] sectors, with a high rate of growth;

(3) the existence or quick emergence of a political, social and institutional framework which exploits the impulses to expansion in the modern sector and the potential external economy effects of the take-off and gives to growth an on-going character.

The third condition implies a considerable capability to mobilize capital from domestic sources. Some take-offs have occurred with virtually no capital imports, for example, Britain and Japan. Some take-offs have had a high component of foreign capital, for example, the United States, Russia and Canada. But some countries have imported large quantities of foreign capital for long periods, which undoubtedly contributed to creating the preconditions for take-off without actually initiating take-off, for example the Argentine before 1914, Venezuela down to recent years, the Belgian Congo currently.

In short, whatever the role of capital imports, the preconditions for take-off include an initial ability to mobilize domestic savings productively, as well as a structure which subsequently permits a high marginal rate of savings.

This definition is designed to isolate the early stage when industrialization takes hold rather than the later stage when industrialization becomes a more massive and statistically more impressive phenomenon. In Britain, for example, there is no doubt that it was between 1815 and 1850 that industrialization fully took hold. If the criterion chosen for take-off was the period of most rapid overall industrial growth, or the period when large-scale industry matured, all our take-off dates would have to be set later; Britain, for example, to 1819–48; the United States, to 1868–93; Sweden, to 1890–1920; Japan, to 1900–20; Russia, to 1928–40. The earlier dating is chosen here because it is believed that the decisive transformations (including a decisive shift in the investment-rate) occur in the first industrial phases; and later industrial maturity can be directly traced back to foundations laid in these first phases.

This definition is also designed to rule out from the take-off the quite substantial economic progress which can occur in an economy before a truly self-reinforcing growth process gets under way. Consider, for example, British economic expansion between, say, 1750 and 1783; Russian economic expansion between, say, 1861 and 1890, Canadian economic expansion between 1867 and the mid-1890's. Such periods

[3] In this context "manufacturing" is taken to include the processing of agricultural products or raw materials by modern methods: for example, timber in Sweden, meat in Australia, dairy products in Denmark. The dual requirement of a "manufacturing" sector is that its processes set in motion a chain of further modern sector requirements and that its expansion provides the potentiality of external economy effects, industrial in character.

—for which there is an equivalent in the economic history of almost every growing economy—were marked by extremely important, even decisive, developments. The transport network expanded, and with it both internal and external commerce; a revolution in agricultural productivity was, at least, begun; new institutions for mobilizing savings were developed; a class of commercial and even industrial entrepreneurs began to emerge; industrial enterprise on a limited scale (or in limited sectors) grew. And yet, however essential these pre-take-off periods were for later development, their scale and momentum were insufficient to transform the economy radically or, in some cases, to outstrip population growth and to yield an increase in *per capita* output.

With a sense of the considerable violence done to economic history, we are here seeking to isolate a period when the scale of productive economic activity reaches a critical level and produces changes which lead to a massive and progressive structural transformation in economies and the societies of which they are a part, better viewed as changes in kind than merely in degree.

Evidence on Investment-Rates in the Take-Off

The case for the concept of take-off hinges, in part, on quantitative evidence on the scale and productivity of investment in relation to population growth. Here we face a difficult problem; for investment data are not now generally available for early stages in economic history. Below is set out such a case as there is for regarding the shift from a productive investment-rate of about 5% of NNP to 10% or more as central to the process.

1. ***A Prima Facie Case*** If we take the marginal capital/output ratio for an economy in its early stages of economic development at 3.5:1 and if we assume, as is not abnormal, a population rise of 1–15% per annum it is clear that something between 3.5 and 5.25% of NNP must be regularly invested if NNP *per capita* is to be sustained. An increase of 2% per annum in NNP *per capita* requires, under these assumptions, that something between 10.5 and 12.5% of NNP be regularly invested. By definition and assumption, then, a transition from relatively stagnant to substantial, regular rise in NNP *per capita*, under typical population conditions, requires that the proportion of national product productively invested should move from somewhere in the vicinity of 5% to something in the vicinity of 10%.

2. ***The Swedish Case*** In the appendix to his paper on international differences in capital formation,[4] Kuznets gives gross and net capital

[4] [In Universities-National Bureau Committee for Economic Research, *Capital Formation and Economic Growth* (Princeton, N.J., 1955).—*Ed.*]

formation figures in relation to gross and net national product for a substantial group of countries where reasonably good statistical data exist. Excepting Sweden, these data do not go back clearly to pre-take-off stages. The Swedish data begin in the decade 1861–70; and the Swedish take-off is to be dated from the latter years of the decade, as shown in Table 2. (GCF: Gross Capital Formation; GNP: Gross National Product; NCF: Net Capital Formation; DGCF: Domestic GCF.)

TABLE 2

KUZNETS' TABLE OF CALCULATIONS FOR SWEDEN

Decade	Domestic GCF/GNP (%)	Domestic NCF/NNP (%)	Depreciation to DGCF (%)
1. 1861–70	5.8	3.5–	(42)
2. 1871–80	8.8	5.3	(42)
3. 1881–90	10.8	6.6	(42)
4. 1891–1900	13.7	8.1	43.9
5. 1901–10	18.0	11.6	40.0
6. 1911–20	20.2	13.5	38.3
7. 1921–30	19.0	11.4	45.2

. . . . .

3. *The Canadian Case* The data developed by O. J. Firestone[5] for Canada indicate a similar transition for net capital formation in its take-off (say 1896–1914); but the gross investment proportion in the period from Confederation to the mid-1890's was higher than appears to have marked other periods when the preconditions were established, due to investment in the railway network (abnormally large for a nation of Canada's population), and to relatively heavy foreign investment, even before the great capital import boom of the pre-1914 decade (see Table 3).

4. *The Pattern of Contemporary Evidence in General*[6] In the years after 1945 the number of countries for which reasonably respectable national income (or product) data exist has grown; and with such data

[5] O. J. Firestone, *Canada's Economic Development, 1867–1952, with Special Reference to Changes in the Country's National Product and National Wealth*, paper prepared for the International Association for Research in Income and Wealth (1953), to which Mr. Firestone has kindly furnished me certain revisions, shortly to be published. By 1900 Canada already had about 18,000 miles of railway line; but the territory served had been developed to a limited degree only. By 1900 Canada already had a net balance of foreign indebtedness of over $1 billion. Although this figure was almost quadrupled in the next two decades, capital imports represented an important increment to domestic capital sources from the period of Confederation down to the pre-1914 Canadian boom, which begins in the mid-1890's.

[6] I am indebted to Mr. Everett Hagen for mobilizing the statistical data in this section, except where otherwise indicated.

TABLE 3

CANADA: GROSS AND NET INVESTMENT IN DURABLE PHYSICAL ASSETS AS PERCENTAGE OF GROSS AND NET NATIONAL EXPENDITURE (FOR SELECTED YEARS)

	GCF/GNP	NCF/NNP	Capital Consumption as Percentage of Gross Investment
1870	15.0	7.1	56.2
1900	13.1	4.0	72.5
1920	16.6	10.6	41.3
1929	23.0	12.1	53.3
1952	16.8	9.3	49.7

there have developed some tolerable savings and investment estimates for countries at different stages of the growth process. Within the category of nations usually grouped as "underdeveloped" one can distinguish four types.[7]

(*a*) Pre-take-off economies, where the apparent savings and investment-rates, including limited net capital imports, probably come to under 5% of net national product. In general, data for such countries are not satisfactory, and one's judgment that capital formation is low must rest on fragmentary data and partially subjective judgment. Examples are Ethiopia, Kenya, Thailand, Cambodia, Afghanistan and perhaps Indonesia.

(*b*) Economies attempting take-off, where the apparent savings and investment-rates, including limited net capital imports, have risen over 5% of net national product.[8] For example, Mexico (1950), Net Capital

[7] The percentages given are of net capital formation to net domestic product. The latter is the product net of depreciation of the geographic area. It includes the value of output produced in the area, regardless of whether the income flows abroad. Since indirect business taxes are not deducted, it tends to be larger than national income; hence the percentages are lower than if national income was used as the denominator in computing them.

[8] The Department of State estimates [Office of Intelligence Research, Report No. 6672, August 25, 1954.—*Ed.*] for economies which are either attempting take-off or which have, perhaps, passed into a stage of regular growth include:

	%		%
Argentina	13	Colombia	14
Brazil	14	Philippines	8
Chile	11	Venezuela	23

Venezuela has been for some time an "enclave economy," with a high investment-rate concentrated in a modern export sector whose growth did not generate general economic momentum in the Venezuelan economy; but in the past few years Venezuela may have moved over into the category of economies experiencing an authentic take-off.

Formation/Net Domestic Product 7.2%; Chile (1950), NCF/NDP 9.5%; Panama (1950), NCF/NDP 7.5%; Philippines (1952), NCF/NDP 6.4%; Puerto Rico (1952), NCF (private)/NDP 7.6%; India (1953), NCF/NDP perhaps about 7%. Whether the take-off period will, in fact, be successful remains in most of these cases still to be seen; although Mexico, at least, would appear to have passed beyond this historical watershed.

(*c*) Growing economies, where the apparent savings and investment-rates, including limited net capital imports, have reached 10% or over; for example, Colombia (1950), NCF/NDP 16.3%.

(*d*) Enclave economies: (i) cases where the apparent savings and investment-rates, including substantial net capital imports, have reached 10% or over, but the domestic preconditions for sustained growth have not been achieved. These economies, associated with major export industries, lack the third condition for take-off suggested above (pp. 89–90). They include the Belgian Congo (1951), NCF/NDP 21.7%; Southern Rhodesia (1950), GCF/GDP 45.5%, (1952) GCF/GDP 45.4%.

(ii) Cases where net capital exports are large. For example, Burma (1938), NCF/NDP, 7.1%; net capital exports/NDP 11.5%; Nigeria (1950–1), NCF/NDP 5.1%; net capital exports/NDP 5.6%.

5. ***The Cases of India and Communist China*** The two outstanding contemporary cases of economies attempting purposefully to take off are India and Communist China, both operating under national plans. . . . The Indian Planning Commission estimated investment as 5% of NNP in the initial year of the plan, 1950–1. Using a 3.1 marginal capital/output ratio, they envisaged a marginal savings rate of 20% for the First Five Year Plan, a 50% rate thereafter, down to 1968–9, when the average proportion of income invested would level off at 20% of NNP. As one would expect, the sectoral composition of this process is not fully worked out in the initial plan; but the Indian effort may well be remembered in economic history as the first take-off defined *ex ante* in national product terms.

So far as the aggregates are concerned, what we can say is that the Indian planned figures fall well within the range of prima facie hypothesis and historical experience, if India in fact fulfils the full requirements for take-off. The Chinese Communist figures are somewhat more ambitious in both agriculture and industry.

As of 1959, the momentum achieved over the past six years in China appears somewhat greater than that in India; but it will be some time before the accounts of progress in the two countries can be cast up with confidence—notably, with respect to agricultural development, which must play so large a role in each. What can be said is that the plans of both countries, in their overall investment goals and sectoral composition, are consistent with the take-off requirements; and, perhaps more important, the commitment of both societies to modernization appears too deep to permit more than temporary set-backs.

The Inner Structure of the Take-Off

Whatever the importance and virtue of viewing the take-off in aggregative terms—embracing national output, the proportion of output invested, and an aggregate marginal capital/output ratio—that approach tells us relatively little of what actually happens and of the causal processes at work in a take-off; nor is the investment-rate criterion conclusive.

Following the definition of take-off, we must consider not merely how a rise in the investment-rate is brought about, from both supply and demand perspectives, but how rapidly growing manufacturing sectors emerged and imparted their primary and secondary growth impulses to the economy.

Perhaps the most important thing to be said about the behavior of these variables in historical cases of take-off is that they have assumed many different forms. There is no single pattern. The rate and productivity of investment can rise, and the consequences of this rise can be diffused into a self-reinforcing general growth process by many different technical and economic routes, under the aegis of many different political, social and cultural settings, driven along by a wide variety of human motivations.

The purpose of the following paragraphs is to suggest briefly, and by way of illustration only, certain elements of both uniformity and variety in the variables whose movement has determined the inner structure of the take-off.

The Supply of Loanable Funds

By and large, the loanable funds required to finance the take-off have come from two types of source: from shifts in the control of income flows, including income-distribution changes and capital imports; and from the plough-back of profits in rapidly expanding particular sectors.

The notion of economic development occurring as a result of income shifts from those who will spend (hoard[9] or lend) less produc-

[9] Hoarding can, of course, be helpful in the growth process by depressing consumption and freeing resources for investment, if, in fact, non-hoarding persons or institutions acquire the resources and possess the will to expand productive investment. A direct transfer of income is evidently not required.

tively to those who will spend (or lend) more productively is one of the oldest and most fundamental notions in economics. It is basic, for example, to the *Wealth of Nations*.[10]

Historically, income shifts conducive to economic development have assumed many forms. In Meiji Japan and also in Czarist Russia the substitution of government bonds for the great landholders' claims on the flow of rent payments led to a highly Smithian redistribution of income into the hands of those in the modern sector. In both cases the real value of the government bonds exchanged for land depreciated; and, in general, the feudal landlords emerged with a less attractive arrangement than had first appeared to be offered. Aside from the confiscation effect, two positive impulses arose from land reform: the State itself used the flow of payments from peasants, now diverted from landlords' hands, for activity which encouraged economic development; and a certain number of the more enterprising former landlords directly invested in commerce and industry. In contemporary India and China we can observe quite different degrees of income transfer by this route. India is relying to only a very limited extent on the elimination of large incomes unproductively spent by large landlords; although this element figures in a small way in its programme. Communist China has systematically transferred all non-governmental pools of capital into the hands of the State, in a series of undisguised or barely disguised capital levies; and it is drawing heavily for capital resources on the mass of middle and poor peasants who remain.[11]

In addition to confiscatory and taxation devices, which can operate effectively when the State is spending more productively than the taxed individuals, inflation has been important to several take-offs. In Britain of the late 1790's, the United States of the 1850's, Japan of the 1870's there is no doubt that capital formation was aided by price inflation, which shifted resources away from consumption to profits.

The shift of income flows into more productive hands has, of course, been aided historically not only by government fiscal measures but also by banks and capital markets. Virtually without exception, the take-off periods have been marked by the extension of banking institutions which expanded the supply of working capital; and in most cases also by an expansion in the range of long-range financing done by a central, formally organized, capital market.

[10] See, especially, Smith's observations on the "perversion" of wealth by "prodigality"—that is, unproductive consumption expenditures—and on the virtues of "parsimony" which transfers income to those who will increase "the fund which is destined for the maintenance of productive hands." Routledge edition (London, 1890), pp. 259-60.

[11] W. W. Rostow *et al.*, *Prospects for Communist China* (New York and London, 1954), Part 4.

Although these familiar capital-supply functions of the State and private institutions have been important to the take-off, it is likely to prove the case, on close examination, that a necessary condition for take-off was the existence of one or more rapidly growing sectors whose entrepreneurs (private or public) ploughed back into new capacity a very high proportion of profits. Put another way, the demand side of the investment process, rather than the supply of loanable funds, may be the decisive element in the take-off, as opposed to the period of creating the preconditions, or of sustaining growth once it is under way. The distinction is, historically, sometimes difficult to make, notably when the State simultaneously acts both to mobilize supplies of finance and to undertake major entrepreneurial acts. There are, nevertheless, periods in economic history when quite substantial improvements in the machinery of capital supply do not, in themselves, initiate a take-off, but fall within the period when the preconditions are created: for example, British banking developments in the century before 1783 and Russian banking developments before 1890.

One extremely important version of the plough-back process has taken place through foreign trade. Developing economies have created from their natural resources major export industries; and the rapid expansion in exports has been used to finance the import of capital equipment and to service the foreign debt during the take-off. United States, Russian and Canadian grain fulfilled this function, Swedish timber and pulp, Japanese silk, etc. Currently Chinese exports to the Communist bloc, wrung at great administrative and human cost from the agricultural sector, play this decisive role. It should be noted that the development of such export sectors has not in itself guaranteed accelerated capital formation. Enlarged foreign-exchange proceeds have been used in many familiar cases to finance hoards (as in the famous case of Indian bullion imports) or unproductive consumption outlays.

One possible mechanism for inducing a high rate of plough-back into productive investment is a rapid expansion in the effective demand for domestically manufactured consumers' goods, which would direct into the hands of vigorous entrepreneurs an increasing proportion of income flows under circumstances which would lead them to expand their own capacity and to increase their requirements for industrial raw materials, semi-manufactured products and manufactured components.

A final element in the supply of loanable funds is, of course, capital imports. Foreign capital has played a major role in the take-off stage of many economies: for example the United States, Russia, Sweden, Canada. The cases of Britain and Japan indicate, however, that it cannot be regarded as an essential condition. Foreign capital was notably useful when the construction of railways or other large overhead capital items with a long period of gestation played an important role in the take-off or the late preconditions period. Whatever its strategic role, the proportion of investment required for growth which goes into industry is relatively small compared to that required for utilities, transport and the housing of enlarged urban populations. And foreign capital can be mightily useful in helping carry the burden of these overhead items either directly or indirectly.

What can we say, in general, then, about the supply of finance during the take-off period? First, as a precondition, it appears necessary that the community's surplus above the mass-consumption level does not flow into the hands of those who will sterilize it by hoarding, luxury consumption or low-productivity investment outlays. Second, as a precondition, it appears necessary that institutions be developed which provide cheap and adequate working capital. Third, as a necessary condition, it appears that one or more sectors of the community must grow rapidly, inducing a more general industrialization process; and that the entrepreneurs in such sectors plough back a substantial proportion of their profits in further productive investment, one possible and recurrent version of the plough-back process being the investment of proceeds from a rapidly growing export sector.

. . . . .

The Sources of Entrepreneurship

It is evident that the take-off requires the existence and the successful activity of some group in the society which is prepared to accept innovations. As noted above, the problem of entrepreneurship in the take-off has not been profound in a limited group of wealthy agricultural nations whose populations derived by emigration mainly from north-western Europe. There the problem of take-off was primarily economic; and when economic incentives for industrialization emerged commercial and banking groups moved over easily into industrial entrepreneurship. In many other countries, however, the development of adequate entrepreneurship was a more searching social process.

Under some human motivation or other, a group must come to perceive it to be both possible and good to undertake acts of capital investment; and, for their efforts to be tolerably successful, they must act with approximate rationality in selecting the directions toward which their enterprise is directed. They must not only produce growth but tolerably balanced growth. We cannot quite say that it is necessary for them to act as if they were trying to maximize profit; for the criteria for private-profit maximization do not necessarily converge with the criteria for an optimum rate and pattern of growth in various sectors. But in a growing economy, over periods longer than the business cycle, economic history is reasonably tolerant of deviations from rationality, in the sense that excess capacity is finally put to productive use. Leaving aside here the question of ultimate human motivation, and assuming that the major overhead items are generated, if necessary, by some form of State initiative (including subsidy), we can say as a first approximation that some group must successfully emerge which behaves as if it were moved by the profit motive, in a dynamic economy with changing production functions.

In this connexion it is increasingly conventional for economists to pay their respects to the Protestant ethic.[12] The historian should not be ungrateful for this light on the grey horizon of formal growth models. But the known cases of economic growth which theory must seek to explain take us beyond the orbit of Protestantism. In a world where Samurai, Parsees, Jews, North Italians, Turkish, Russian, and Chinese civil servants (as well as Huguenots, Scotsmen and British north-countrymen) have played the role of a leading élite in economic growth, John Calvin should not be made to bear quite this weight. More fundamentally, allusion to a positive scale of religious or other values conducive to profit-maximizing activities is an insufficient sociological basis for this important phenomenon. What appears to be required for the emergence of such élites is not merely an appropriate value system but two further conditions: first, the new élite must feel itself denied the conventional routes to prestige and power by the traditional less acquisitive society of which it is a part; second, the traditional society must be sufficiently flexible (or weak) to permit its members to seek material advance (or political power) as a route upwards alternative to conformity.

Although an élite entrepreneurial class appears to be required for take-off, with significant power over aggregate income flows and industrial investment decisions, most take-offs have been preceded or accompanied by radical change in agricultural techniques and market organization. By and large the agricultural entrepreneur has been the individual landowning farmer. A requirement for take-off is, therefore, a class of farmers willing and able to respond to the possibilities opened up for them by new techniques, landholding arrangements, transport facilities, and forms of market and credit organization. A small purposeful élite can go a long way in initiating economic growth; but, especially in agriculture (and to some extent in the industrial working force), a wider based revolution in outlook must come about.

[12] See, for example, N. Kaldor, "Economic Growth and Cyclical Fluctuations," *Economic Journal* (March 1954), p. 67.

Whatever further empirical research may reveal about the motives which have led men to undertake the constructive entrepreneurial acts of the take-off period, this much appears sure: these motives have varied greatly, from one society to another; and they have rarely, if ever, been motives of an unmixed material character.

Leading Sectors in the Take-Off

. . . The overall rate of growth of an economy must be regarded in the first instance as the consequence of differing growth rates in particular sectors of the economy, such sectoral growth-rates being in part derived from certain overall demand factors (for example population, consumers' income, tastes etc.); in part, from the primary and secondary effects of changing supply factors, when these are effectively exploited.

On this view the sectors of an economy may be grouped in three categories:

(1) Primary growth sectors, where possibilities for innovation or for the exploitation of newly profitable or hitherto unexplored resources yield a high growth-rate and set in motion expansionary forces elsewhere in the economy.

(2) Supplementary growth sectors, where rapid advance occurs in direct response to—or as a requirement of—advance in the primary growth sectors; for example coal, iron and engineering in relation to railroads. These sectors may have to be tracked many stages back into the economy.

(3) Derived-growth sectors, where advance occurs in some fairly steady relation to the growth of total real income, population, industrial production or some other overall, modestly increasing variable. Food output in relation to population and housing in relation to family formation are classic derived relations of this order.

In the earlier stages of growth, primary and supplementary growth sectors derive their momentum essentially from the introduction and diffusion of changes in the cost-supply environment (in turn, of course, partially influenced by demand changes); while the derived-growth sectors are linked essentially to changes in demand (while subject also to continuing changes in production functions of a less dramatic character). In the age of high mass-consumption leading sectors become more dependent on demand factors than in the earlier stages . . .

At any period of time it appears to be true even in a mature and growing economy that forward momentum is maintained as the result of rapid expansion in a limited number of primary sectors, whose expansion has significant external economy and other secondary effects. From this perspective the behavior of sectors during the take-off is merely a special version of the growth process in general; or, put another way, growth proceeds by repeating endlessly, in different patterns, with different leading sectors, the experience of the take-off. Like the take-off, long-term growth requires that the society not only generate vast quantities of capital for depreciation and maintenance, for housing and for a balanced complement of utilities and other overheads, but also a sequence of highly productive primary sectors, growing rapidly, based on new production functions. Only thus has the aggregate marginal capital/output ratio been kept low.

Once again history is full of variety: a considerable array of sectors appears to have played this key role in the take-off process.

The development of a cotton-textile industry sufficient to meet domestic requirements has not generally imparted a sufficient impulse in itself to launch a self-sustaining growth process. The development of modern cotton-textile industries in substitution for imports has, more typically, marked the pre-take-off period, as for example in India, China and Mexico.

There is, however, the famous exception of Britain's industrial revolution. Baines's table[13] on raw-cotton imports and his comment on it are worth quoting, covering as they do the original leading sector in the first take-off (see Table 4).

TABLE 4

RATE OF INCREASE IN THE IMPORT OF COTTON-WOOL, IN PERIODS OF TEN YEARS FROM 1741 TO 1831

	%		%
1741–51	81	1791–1801	67.5
1751–61	21.5	1801–11	39.5
1761–71	25.5	1811–21	93
1771–81	75.75	1821–31	85
1781–91	319.5		

From 1697 to 1741 the increase was trifling; between 1741 and 1751 the manufacture, though still insignificant in extent, made a considerable spring; during the next twenty years, the increase was moderate; from 1771 to 1781, owing to the invention of the jenny and the water-frame, a rapid increase took place: in the ten years from 1781 to 1791, being those which immediately followed the invention of the mule and the expiration of Arkwright's patent, the rate of advancement was prodigiously accelerated, being nearly 320%: and from that time to the present, and especially since the close of the war, the increase, though considerably moderated, has been rapid and steady far beyond all precedent in any other manufacture.

Why did the development of a modern factory system in cotton textiles lead on in Britain to a self-sustaining growth process, whereas it failed to do so in other cases? Part of the answer lies in the fact that by the late eighteenth century the preconditions for take-off in Britain were

[13] E. Baines, *History of the Cotton Manufacture* (London, 1835), p. 348.

very fully developed. Progress in textiles, coal, iron and even steam power had been considerable throughout the eighteenth century; and the social and institutional environment was propitious. But two further technical elements helped determine the upshot. First, the British cotton-textile industry was large in relation to the total size of the economy. From its modern beginnings, but notably from the 1780's forward, a very high proportion of total cotton-textile output was directed abroad, reaching 60% by the 1820's.[14] The evolution of this industry was a more massive fact, with wider secondary repercussions, than if it were simply supplying the domestic market. Industrial enterprise on this scale had secondary reactions on the development of urban areas, the demand for coal, iron and machinery, the demand for working capital and ultimately the demand for cheap transport, which powerfully stimulated industrial development in other directions.[15]

Second, a source of effective demand for rapid expansion in British cotton textiles was supplied, in the first instance, by the sharp reduction in real costs and prices which accompanied the technological developments in manufacture and the cheapening real cost of raw cotton induced by the cotton-gin. In this Britain had an advantage not enjoyed by those who came later; for they merely substituted domestic for foreign-manufactured cotton textiles. The substitution undoubtedly had important secondary effects by introducing a modern industrial sector and releasing, on balance, a pool of foreign exchange for other purposes; but there was no sharp fall in the real cost of acquiring cotton textiles and no equivalent rise in real income.

The introduction of the railroad has been historically the most powerful single initiator of take-offs.[16] It was decisive in the United States, France, Germany, Canada, and Russia; it has played an extremely important part in the Swedish, Japanese and other cases.

The railroad has had three major kinds of impact on economic growth during the take-off period. First, it has lowered internal transport costs, brought new areas and products into commercial markets and, in general,

---

[14] The volume (official value) of British cotton-goods exports rose from £355,060 in 1780 to £7,624,505 in 1802 (Baines, *op. cit.* p. 350). See also the calculation of R. C. O. Matthews, *A Study in Trade Cycle History* (Cambridge, 1954), pp. 127–9.

[15] If we are prepared to treat New England of the first half of the nineteenth century as a separable economy, its take-off into sustained growth can be allocated to the period, roughly, 1820–50; and, again, a disproportionately large cotton-textile industry based substantially on exports (that is, from New England to the rest of the United States) is the regional foundation for sustained growth.

[16] For a detailed analysis of the routes of impact of the railroad on economic development see Paul H. Cootner, *Transport Innovation and Economic Development: The Case of the U.S. Steam Railroads* (1953), unpublished doctoral thesis, M.I.T. (Cambridge, Mass.).

performed the Smithian function of widening the market. Second, it has been a prerequisite in many cases to the development of a major new and rapidly enlarging export sector which, in turn, has served to generate capital for internal development, as, for example, the American railroads before 1914. Third, and perhaps most important for the take-off itself, the development of railways has led on to the development of modern coal, iron and engineering industries. In many countries the growth of modern basic industrial sectors can be traced in the most direct way to the requirements for building and, especially, for maintaining substantial railway systems. When a society has developed deeper institutional, social and political prerequisites for take-off, the rapid growth of a railway system, with these powerful triple effects, has often served to lift it into self-sustained growth. Where the prerequisites have not existed, however, very substantial railway building has failed to initiate a take-off, as for example in India, China, pre-1895 Canada, pre-1914 Argentina, etc.

It is clear that an enlargement and modernization of armed forces could play the role of a leading sector in take-off. It was a factor in the Russian, Japanese and German take-offs; and it figures heavily in current Chinese Communist plans. But historically the role of modern armaments has been ancillary rather than central to the take-off.

Quite aside from their role in supplying foreign exchange for general capital-formation purposes, raw materials and food-stuffs can play the role of leading sectors in the take-off if they involve the application of modern processing techniques. The timber industry, built on the steam-saw, fulfilled this function in the first phase of Sweden's take-off, to be followed shortly by the pulp industry. Similarly, the shift of Denmark to meat and dairy products, after 1873, appears to have reinforced the development of a manufacturing sector in the economy, as well as providing a major source of foreign exchange. And as Lockwood notes, even the export of Japanese silk thread had important secondary effects which developed modern production techniques.[17]

> To satisfy the demands of American weaving and hosiery mills for uniform, high-grade yarn, however, it was necessary to improve the quality of the product, from the silkworm egg on through to the bale of silk. In sericulture this meant the introduction of scientific methods of breeding and disease control; in reeling it stimulated the shift to large filatures equipped with machinery; in marketing it led to large-scale organization in the collection and sale of cocoons and raw silk . . . it exerted steady pressure in favor

[17] W. W. Lockwood, *The Economic Development of Japan* (Princeton, 1954), pp. 338–9.

of the application of science, machinery, and modern business enterprise.

The role of leading sector has been assumed, finally, by the accelerated development of domestic manufacture of consumption goods over a wide range in substitution for imports, as, for example, in Australia, the Argentine and, perhaps, in contemporary Turkey.

What can we say, then, in general about these leading sectors? Historically, they have ranged from cotton textiles, through heavy-industry complexes based on railroads and military end-products, to timber, pulp, dairy products and finally a wide variety of consumers' goods. There is, clearly, no one sectoral sequence for take-off, no single sector which constitutes the magic key. There is no need for a growing society to recapitulate, for example, the structural sequence and pattern of Britain, the United States or Russia. Four basic factors must be present:

(1) There must be enlarged effective demand for the product or products of sectors which yield a foundation for a rapid rate of growth in output. Historically this has been brought about initially by the transfer of income from consumption or hoarding to productive investment; by capital imports; by a sharp increase in the productivity of current investment inputs, yielding an increase in consumers' real income expended on domestic manufactures; or by a combination of these routes.

(2) There must be an introduction into these sectors of new production functions as well as an expansion of capacity.

(3) The society must be capable of generating capital initially required to detonate the take-off in these key sectors; and especially there must be a high rate of plough-back by the (private or state) entrepreneurs controlling capacity and technique in these sectors and in the supplementary growth sectors they stimulated to expand.

(4) Finally, the leading sector or sectors must be such that their expansion and technical transformation induce a chain of requirements for increased capacity and the potentiality for new production functions in other sectors, to which the society, in fact, progressively responds.

The Take-Off in Perspective

This view of the take-off is, then, a return to a rather old-fashioned way of looking at economic development. The take-off is defined as an industrial revolution, tied directly to radical changes in methods of production, having their decisive consequence over a relatively short period of time.

This view would not deny the role of longer, slower changes in the whole process of economic growth. On the contrary, take-off requires the massive set of preconditions, going to the heart of a society's economic organization, its politics, and its effective scale of values, considered in [the previous section].

What this argument does assert is that the rapid growth of one or more new manufacturing sectors is a powerful and essential engine of economic transformation. Its power derives from the multiplicity of its forms of impact, when a society is prepared to respond positively to this impact. Growth in such sectors, with new production functions of high productivity, in itself tends to raise output per head; it places incomes in the hands of men who will not merely save a high proportion of an expanding income but who will plough it into highly productive investment; it sets up a chain of effective demand for other manufactured products; it sets up a requirement for enlarged urban areas, whose capital costs may be high, but whose population and market organization help to make industrialization an on-going process; and, finally, it opens up a range of external economy effects which, in the end, help to produce new leading sectors when the initial impulse of the take-off's leading sectors begins to wane.

In non-economic terms, the take-off usually witnesses a definitive social, political, and cultural victory of those who would modernize the economy over those who would either cling to the traditional society or seek other goals; but—because nationalism can be a social solvent as well as a diversionary force—the victory can assume forms of mutual accommodation, rather than the destruction of the traditional groups by the more modern; see, for example, the role of the Junkers in nascent industrial Germany, and the persistence of much in traditional Japan beyond 1880. By and large, the maintenance of momentum for a generation persuades the society to persist, and to concentrate its efforts on extending the tricks of modern technology beyond the sectors modernized during take-off.

E. Anthony Wrigley

Urban Growth and Agricultural Change: England and the Continent in the Early Modern Period

The complexity and contingency of any relationship between economic growth and urban growth should need no stressing. It is clearly hazardous to undervalue, still more to ignore, the difficulties attending any examination of this topic. In the early part of this article I prefer to sketch out an initial thesis rather starkly at the risk of oversimplifying "reality." Later the limitations of the initial formulation become clear.

A rising level of real income per head and a rising proportion of urban dwellers, other things being equal, are likely to be linked phenomena in a preindustrial economy. If income elasticity of demand for food is less than unity, then, with rising real incomes, demand for secondary and tertiary products will grow more rapidly than that for primary products, and will therefore cause employment in secondary and tertiary industries to rise more rapidly than in agriculture. Such employment is likely to be higher in towns, especially in the case of tertiary employment, and will result in an increase in the proportion of the total population living in towns. There may be an important feedback element in this relationship, since the growth of towns may help to further agricultural investment and specialization and so carry forward the rising trend in real incomes. Declining real incomes will have an opposite effect.[1]

In an economy which meets its own food needs, urban growth may not only be a symptom of rising real incomes: it

E. Anthony Wrigley is Professor of Population Studies at the London School of Economics and Associate Director of the Economic and Social Research Council Cambridge Group for the History of Population and Social Structure.

The author is grateful to Penelope Corfield, Jack Goldstone, and Tony Sutcliffe for comments on an earlier version of this article.

0022-1953/85/040683-46 $02.50/0

1 Adam Smith stressed the mutual stimulus which urban growth and agricultural improvement might afford each other in *An Enquiry into the Nature and Causes of the Wealth of Nations* (Edinburgh, 1863), 181–187. He included a chapter entitled, "How the commerce of towns contributed to the improvement of the country."

may also be a rough measure of the level of productivity per worker in the agricultural sector of the economy. If productivity per head in agriculture is sufficiently low, the surplus of food available after meeting the needs of the agricultural population may be enough to sustain only a tiny urban sector. At the other extreme, if agricultural productivity is high, the economy may be able to support a third or a half of the population in towns without prejudicing nutritional levels elsewhere. In a closed economy, therefore, a substantial rise in the proportion of the population living in towns is strong presumptive evidence of a significant improvement in production per head in agriculture, and may provide an indication of the scale of the change.

Sufficient information is now available to justify an initial application of this line of thought to early modern England.

THE PACE OF URBAN GROWTH IN ENGLAND Table 1 sets out some estimates of the size of the populations of leading English towns in about 1520, 1600, 1700, 1750, and 1800. Table 2 provides estimates of the total population of England at each of these dates and of the population of London and other urban centers with 5,000 or more inhabitants.

As the notes to the tables make clear, all of the data given are subject to a substantial measure of uncertainty. Their sources are various and, apart from those taken from the 1851 census, most have been obtained by inflating and adjusting the raw data because the latter cover only a proportion of the total population. Some figures were originally suspiciously rounded and many incorporate alterations made in the light of subjective assessments of their deficiencies. Moreover, it is entirely arbitrary to draw a dividing line between urban and non-urban at 5,000. It was done on the supposition that only a tiny fraction of the inhabitants of towns larger than 5,000 would have been principally engaged in agriculture, but a plausible case might be made for a significantly lower dividing line.

To add still further to the crudity of the exercise, it is questionable whether the same dividing line should be used over a long period of time, during which the population increased greatly. For example, a moderate-sized market town with a population of, say, 3,000 at the start of the period, serving a hinterland in which the population doubled, would itself grow in size and at some point exceed 5,000 in population even though the functions it discharged did not alter. If this pattern were widely repeated, it would result in an upward drift in the overall urban percentage, but such movement would have occurred only because the total population had risen. It would not imply any structural change in the economy. Nevertheless, certain features of change over time are so prominent that they would remain clear cut, or might even be more pronounced, if less crude information were available.

During the sixteenth century, urban growth, relative to national population trends, was largely confined to London, the population of which quadrupled. London's share of the national total rose from 2.25 to 5 percent. The percentage of the population living in other towns, however, rose only modestly from 3 to 3.25 percent. Even this rise was largely because ten towns crept over the 5,000 mark (Plymouth, King's Lynn, Gloucester, Chester, Hull, Great Yarmouth, Ipswich, Cambridge, Worcester, and Oxford). Most large provincial centers were growing *less* quickly than the national population as a whole, as appears to have been true of Norwich, Bristol, Exeter, Canterbury, Coventry, Colchester, and Salisbury. Of the initial list of towns above 5,000 in population, excluding London, only Newcastle increased its share of the national total (Table 1). Its surge in growth was no doubt partly due to London's extraordinary rise, since the coal trade down the east coast flourished as London grew and brought with it prosperity to Newcastle.[2]

The effect of the artificial boost given to urban growth outside London by the inclusion of several new towns which had reached 5,000 in population between 1520 and 1600 can be estimated either by ignoring the new entrants on the 1600 list, or by basing the calculation on the full 1600 list. The group of nine provincial towns on the 1520 list displayed a collective growth of

2 The modesty of urban growth outside London may be overstated by concentrating on places with 5,000 or more inhabitants. There is evidence to suggest that the smaller towns were growing more rapidly than the provincial centers. Using a total of 2,000 rather than 5,000 inhabitants to divide urban populations from the remainder would probably have resulted in an impression of greater buoyancy for the uban population in the sixteenth century. Charles Phythian-Adams, "Urban Decay in Late Medieval England," in Philip Abrams and Wrigley (eds.), *Towns in Societies. Essays in Economic History and Historical Sociology* (Cambridge, 1978), 171–172.

Table 1 Urban Populations ('000s)

c. 1520	c. 1600	c. 1670	c. 1700		c. 1750	1801
London 55	London 200	London 475	London 575		London 675	London 959
Norwich 12	Norwich 15	Norwich 20	Norwich 30		Bristol 50	Manchester 89[a]
Bristol 10	York 12	Bristol 20	Bristol 21		Norwich 36	Liverpool 83
York 8	Bristol 12	York 12	Newcastle 16		Newcastle 29	Birmingham 74
Salisbury 8	Newcastle 10	Newcastle 12	Exeter 14		Birmingham 24	Bristol 60
Exeter 8	Exeter 9	Colchester 9	York 12		Liverpool 22	Leeds 53
Colchester 7	Plymouth 8	Exeter 9	Gt Yarmouth 10		Manchester 18	Sheffield 46
Coventry 7	Salisbury 6	Chester 8	Birmingham		Leeds 16	Plymouth 43[b]
Newcastle 5	King's Lynn 6	Ipswich 8	Chester		Exeter 16	Newcastle 42[c]
Canterbury 5	Gloucester 6	Gt Yarmouth 8	Colchester		Plymouth 15	Norwich 36
	Chester 6	Plymouth 8	Ipswich	8–9	Chester 13	Portsmouth 33
	Coventry 6	Worcester 8	Manchester		Coventry 13	Bath 33
	Hull 6	Coventry 7	Plymouth		Nottingham 12	Hull 30
	Gt Yarmouth 5	King's Lynn 7	Worcester		Sheffield 12	Nottingham 29
	Ipswich 5	Manchester 6	Bury St Edmunds		York 11	Sunderland 26
	Cambridge 5	Canterbury 6	Cambridge		Chatham 10	Stoke 23[d]
	Worcester 5	Leeds 6	Canterbury		Gt Yarmouth 10	Chatham 23[e]
	Canterbury 5	Birmingham 6	Chatham		Portsmouth 10	Wolverhampton 21[f]
	Oxford 5	Cambridge 6	Coventry		Sunderland 10	Bolton 17
	Colchester 5	Hull 6	Gloucester		Worcester 10	Exeter 17
		Salisbury 6	Hull			Leicester 17
		Bury St Edmunds 5	King's Lynn			Great Yarmouth 17
		Leicester 5	Leeds	5–7		Stockport 17
		Oxford 5	Leicester			York 16
		Shrewsbury 5	Liverpool			Coventry 16
		Gloucester 5	Nottingham			Chester 16
			Oxford			Shrewsbury 15
			Portsmouth			
			Salisbury			
			Shrewsbury			
			Sunderland			
			Tiverton			

NOTES:
a Including Salford.
b Including Devonport
c Including Gateshead
d Stoke and Burslem
e The Medway towns: Chatham, Rochester, and Gillingham
f Wolverhampton, Willenhall, Bilston, and Wednesfield

SOURCES: The following sources were used in compiling the table. (1) Penelope J. Corfield, "Urban Development in England and Wales in the Sixteenth and Seventeenth Centuries," in Donald C. Coleman and Arthur H. John (eds.), *Trade, Government and Economy in Pre-industrial England* (London, 1976), 214–247; (2) Corfield, *The Impact of English Towns, 1700–1800* (Oxford, 1982); (3) Frank V. Emery, "England circa 1600," in H. Clifford Darby (ed.), *A New Historical Geography of England before 1600* (Cambridge, 1976), 248–301; (4) Charles Phythian-Adams, *Desolation of a City. Coventry and the Urban Crisis of the Late Middle Ages* (Cambridge, 1979); (5) John Patten, *English Towns 1500–1700* (Folkestone, 1978); (6) *idem*, "Population Distribution in Norfolk and Suffolk during the Sixteenth and Seventeenth Centuries," *Transactions of the Institute of British Geographers*, LXV (1975), 45–65; (7) 1851 Census, Population tables I, Numbers of inhabitants 1801–51, I, II, *Parliamentary Papers* (1852–3), LXXXV, LXXXVI; (8) C. M. Law, "Some Notes on the Urban Population of England and Wales in the Eighteenth Century," *The Local Historian*, X (1972), 142–147; (9) Roger Finlay, *Population and Metropolis. The Demography of London, 1580–1650* (Cambridge, 1981); (10) 1801 Census, Enumeration, *Parliamentary Papers* (1802), VII.

The data were abstracted as follows with the numbers given in brackets relating to the sources listed above. London 1520, 1600: (1), 217. London 1660: (9), 60. London 1700, 1750: (2), 8. Other towns 1520: (4), 12 and (1), 222. Other towns 1600: (3), 294–298 and (5), 115. Other towns 1670: (5), 106; 109–110; 114, 116, 120 and (1), 239, 241. Other towns 1700: (1), 223. Other towns 1750: (8), 22–26. All towns 1801: (7).

Some of the estimates may refer to dates a dozen or more years away from the date at the head of the column, except in 1750 and 1801. Often different scholars suggest different totals, or the same scholar may present more than one estimate for the same place and period. An element of judgment and selection is therefore unavoidable in compiling a table of this type. In one instance a quoted estimate seemed improbable and has been changed: (5), Table 9, 109 gives a figure of 9,000 for Cambridge in the 1670s, which seems implausibly high in view of the estimates for 1600 and 1700. In one or two cases (for example, Plymouth, 1670), I could find no figure for a year close to the target year and the total was therefore estimated.

Table 2 National, London, and Other Urban Population Estimates (thousands)

	C. 1520	C. 1600	C. 1670	C. 1700	C. 1750	1801
Population of England	2400	4110	4980	5060	5770	8660
London	55	200	475	575	675	960
Other urban population in towns with 5,000 or more inhabitants	70	135	205	275	540	1420
Total urban	125	335	680	850	1215	2380
Urban populations as a percentage of the national total						
London	2.25	5.0	9.5	11.5	11.5	11.0
Other urban	3.0	3.25	4.0	5.5	9.5	16.5
Total urban	5.25	8.25	13.5	17.0	21.0	27.5

NOTES: The population totals and percentages have been rounded to emphasize the approximate nature of the calculations. National population totals refer to England, excluding Monmouth, and those for c. 1600, c. 1670, c. 1700, and c. 1750 relate to the years 1601, 1671, 1701, and 1751.

All the estimates of urban populations given in this table are subject to substantial margins of error. This is true even for those for 1801 derived from the 1801 census. For example, Law's careful examination of the 1801 material in the general context of nineteenth-century censuses led him to suggest a total urban population in England and Wales of 3,009,260 (in towns of 2,500 inhabitants or more) compared with Corfield's estimate of 2,725,171, using the same definition of urban. The former figure is 10% larger than the latter. All earlier totals are subject to far wider margins of error. Law, "The Growth of Urban Population in England and Wales, 1801–1911," *Transactions of the Institute of British Geographers*, XLI (1976), 141; Corfield, *Impact of English Towns*, 8.

SOURCES: English population totals 1600 to 1801: Wrigley and Roger S. Schofield, *The Population History of England, 1541–1871: A Reconstruction* (Cambridge, Mass., 1981), 208–209. The figure for 1520 is an estimate based on the discussions in *ibid.*, 565–568,

London totals: Table 1.

Other urban totals: 1520, 1600, 1670, Table 1; 1700, 1750, and 1801, Corfield, *Impact of English Towns*, 8. The totals have been rounded and, in the case of the 1801 total, slightly reduced to reflect the fact that Corfield's estimates refer to England and Wales (by this date 3 Welsh towns exceeded 5,000 in population).

only about 15 percent, sluggish growth during a period when the national total population rose by about 70 percent. If, alternatively, the list of nineteen large provincial towns in 1600 is made the basis of measurement, a rather higher percentage growth figure results. In 1520 these towns housed a total population of about 107,000 (see Table 1, sources). By 1600 the total had risen to about 137,000, a rise of 28 percent. Both calculations underline the point that the doubling of the population living in provincial towns shown in Table 2 is misleading, since it was preponderantly due to the recruitment of new towns into the category, and not to growth within the individual towns.

If London's growth is ignored, the urban growth pattern elsewhere conforms well to expectation. In the course of the sixteenth century, real incomes in England fell substantially. If the Phelps Brown and Hopkins index of real wages is used as a guide to the extent of the fall in living standards, it suggests a decline between 1520 and 1600 of over 40 percent. Even if the index overstates the change, a significant deterioration is nonetheless probable and ought, in conformity with the model sketched earlier, to have acted as a brake on urban growth and to have reduced urban population in percentage terms. London remains an exception so important as to outweigh in aggregate faltering urban growth elsewhere, but its overall dominance should not be allowed to obscure the significance of trends in the provinces.[3]

In the seventeenth century circumstances changed greatly. England's population grew by less than a quarter over the century as a whole and was falling gently during its third quarter. The Phelps Brown and Hopkins index suggests that real wages bottomed out early in the century and had risen substantially by its end.[4] Urban growth went on apace whether judged in absolute or percentage terms. London continued to dominate the picture. By 1700 the capital housed about 11 percent of the total national population, more than double the percentage of a century earlier. It had become the largest city in Western Europe and continued to dwarf all local rivals. But other towns also began to grow vigorously. They grew rather slowly in the first half of the cen-

3 E. Henry Phelps Brown and Sheila V. Hopkins, "Seven Centuries of the Price of Consumables, Compared with Builders' Wages," *Economica*, XXIII (1956), 296–314. The construction of a slightly modified version of the Phelps Brown and Hopkins index is described in Wrigley and Schofield, *Population History*, Appendix 9. The 25-year centered moving average of the annual figures given in Table A9.2 fell by 44% between 1520 and 1600. David M. Palliser, "Tawney's Century: Brave New World or Malthusian Trap?" *Economic History Review*, XXXV (1982), 349–351. Palliser suggests reasons to suspect that the Phelps Brown and Hopkins index overstates the extent of the fall in real incomes.

4 The centered 25-year moving average suggests a rise of 27% between 1600 and 1700. Both the extent of the rise and the timing of the end of the long fall in Tudor and early Stuart times are debatable. Bowden's index of the purchasing power of agricultural wages in southern England suggests that the beginning of a recovery may have been as late as the 1640s. From a stable plateau in the 1520s and 1530s (the index figure is 80 in both decades where 1460–9 = 100), his index falls by 35–40% by the 1590s and shows no subsequent decisive trend until the series ends in 1640–9. Peter Bowden, "Statistical Appendix," in Joan Thirsk (ed.), *The Agrarian History of England and Wales: IV, 1500–1640* (Cambridge, 1967), 865.

tury, but after 1670 their relative growth was at least as rapid as that of London. The smaller urban centers were now increasing far faster than the country as a whole. Their population more than doubled during the century, a rate of growth more than four times that of the national aggregate.

Measurement of urban growth is less bedevilled in the seventeenth century than it was in the previous century by the problem of 'drift' across the arbitrary 5,000 dividing line between urban and rural. In the sixteenth century, the increase of population in towns of 5,000 or more was about 95 percent, judged crudely, but the increase in the towns on the 1600 list was only 28 percent and only 15 percent on the 1520 list. The comparable figures for the seventeenth century were more closely bunched at about 105, 60, and 46 percent respectively.[5] Several major provincial centers, notably Norwich, Bristol, Newcastle, and Exeter, increased by between 50 and 100 percent. A striking portent for the future was the appearance on the list for the first time of towns never previously of much note but later to herald a new age. In 1670 Birmingham, Manchester, and Leeds appear for the first time, and in 1700 Liverpool.

If the seventeenth century saw a notable acceleration of growth within an urban system still consisting largely of towns with long-familiar names, the eighteenth brought radical reordering of the urban hierarchy and further rapid urban growth. London, moreover, although still vastly larger than any other city, no longer stood out because of its rate of growth. In 1800 it comprised much the same proportion of the national population as it had 100 years before. Meanwhile, the share of other towns larger than 5,000 in population increased dramatically, rising from 6 to 17 percent of the national total, and for the first time surpassing London's share.

Growth was widely but very unevenly spread. London's old rivals fared less well than London in the main. Bristol grew rapidly, riding on the back of buoyant Atlantic trade, but several cities which had once figured prominently in the English urban hierarchy, including Norwich, Exeter, and York, grew less quickly than the population overall and ended the century with

5 In order to calculate the 60% figure some town populations had to be estimated (i.e. for some of the towns appearing on the 1700 list but not the 1600 list in Table 1).

smaller fractions of the national total than at the beginning. For many centuries such towns had exchanged places in the premier urban league, but the league's membership had not greatly altered. By 1800, however, only Bristol, Newcastle, and Norwich of the old major regional centers remained among the country's ten largest towns. Manchester, Liverpool, and Birmingham stood second, third, and fourth after London. They ranged between 70,000 and 90,000 in population, having grown fiftyfold or more since the early sixteenth century.

Lower down the list a host of new names appeared. Several were the seats of new industry—Leeds, Sheffield, Stoke, Wolverhampton, Bolton, and Stockport; but others reflected changing social customs and new forms of expenditure. Bath, for example, with 33,000 inhabitants, was the twelfth largest town in England, a gracious monument to the new ways in which the wealthy and well born found it convenient to make or maintain contacts with each other or to pass their hours of leisure. Ports and dockyard towns also enjoyed vigorous growth. Plymouth and Portsmouth were among the country's largest towns, and Hull, Sunderland, Chatham, and Great Yarmouth all exceeded 15,000 in population at the time of the 1801 census (although not all of the towns in this category were new names).

The simplest model connecting real income and urban growth will no longer "save the phenomena" for the eighteenth century. The sustained momentum of urban growth, accelerating toward the century's end, would suggest a parallel rise in real incomes, but in some parts of England the long-sustained rise in real wages had ceased before the middle of the eighteenth century. It had probably halted nationally by 1780, to be succeeded by a substantial fall lasting about thirty years.[6]

By the eighteenth century the assumption of a closed economy is even more unrealistic than for the sixteenth. External

6 The behavior of wages and prices and, *a fortiori*, of real wages, both regionally and nationally between 1700 and 1850 has been the subject of much controversy. There have been several valuable surveys of the issues recently, and also some new empirical work. Michael W. Flinn, "Trends in Real Wages, 1750–1850," *Economic History Review*, XXVII (1974), 395–411. G. Nicholas von Tunzelmann, "Trends in Real Wages, 1750–1850, Revisited," *ibid.*, XXXII (1979), 33–49. Peter H. Lindert and Jeffrey G. Williamson, "English Workers' Living Standards during the Industrial Revolution: A New Look," *ibid.*, XXXVI (1983), 1–25.

demand represented a substantial fraction of total demand in many industries, although it is easy to exaggerate the importance of overseas markets. Any increase in the relative importance of overseas trade, however, would stimulate urban growth, conspicuously in the case of ports like Bristol and Liverpool, but in a lesser degree also elsewhere. Equally, transport improvements within England increased the scale of internal trade. The average distance travelled by goods between producer and consumer probably also increased. Both trends must have stimulated employment in the urban centers through which goods passed.[7] The pattern of urban growth suggests that the deceleration in urban growth which might have been expected on the simplest possible view of the link between real incomes and urban growth affected the older centers in the expected fashion, but that the new features of the English economy imparted impetus to those towns most caught up in the new developments. London, affected by all of the various and conflicting influences on urban growth, occupied an intermediate position.

The extent of the contrast among the fortunes of different types of towns in the course of the early modern period is illustrated in Table 3. The choice of towns in each group is inevitably arbitrary, both in the sense that other sets might have been made up to represent the type in question, and in the sense that 'pure' types are rare in large towns since size and complexity of function are closely linked. The balance of market functions, craft industry, administrative services, and professional employment varied considerably among the set of ten historic regional centers, and two of them, Chester and Exeter, also had important port functions. Nevertheless, the contrast among the groups is sufficiently marked to make it unlikely that other choices made with the same distinction in mind would have produced a very different result.

The historic regional centers did not keep pace with national population growth over the early modern period as a whole. In the two middle periods, when real wages were rising, they experienced a faster population growth than the national average;

7 On overseas trade see, e.g., R. P. Thomas and Donald N. McCloskey, "Overseas Trade and Empire, 1700–1860," in Roderick Floud and McCloskey (eds.), *The Economic History of Britain since 1700* (Cambridge, 1981), I, 87–102. An excellent summary of knowledge about inland transport may be found in John A. Chartres, *Internal Trade in England, 1500–1700* (London, 1977), 13–46.

Table 3 Patterns of Growth. Urban Growth in Early Modern England (thousands)

	POPULATION TOTALS					PERCENTAGE GROWTH			
	C. 1520	C. 1600	C. 1700	C. 1750	1801	1600/1520	1700/1600	1750/1700	1801/1750
England	2400	4110	5060	5770	8660	71	23	14	50
London	55	200	575	675	960	264	188	17	42
10 historic regional centers[a]	62	73	107	126	153	18	47	18	21
8 established ports[b]	38	53	81	128	190	39	53	58	48
4 'new' manufacturing towns[c]	6	11	27	70	262	83	145	159	274

NOTES:
a Norwich, York, Salisbury, Chester, Worcester, Exeter, Cambridge, Coventry, Shrewsbury, Gloucester.
b Bristol, Hull, Colchester, Newcastle, Ipswich, Great Yarmouth, King's Lynn, Southampton.
c Birmingham, Manchester, Leeds, Sheffield.
SOURES: See source note to Table 1.

in the sixteenth and later eighteenth centuries, however, when real wages were probably declining, their rate of growth fell well below the national average. Thus they exhibited what might be termed the classic pattern of relative growth in terms of the model of the relationship between income and urban growth described earlier.

At the other extreme, the four towns which in 1801 were the largest manufacturing towns in England were always growing faster than any other groups in the table, apart from London until 1700. Their rate of growth accelerated steadily throughout the three centuries, becoming so hectic in the last half-century that their population almost quadrupled in fifty years.

The established ports also grew with increasing speed, except in the last period, outstripping the national growth rate, except in the first and last periods. The slight fall in growth rate in this group in the later eighteenth century was due in part to the extraordinarily rapid growth of Liverpool. If Liverpool were included in the group, the percentage growth of population in the group would rise to 72 between 1700 and 1750 and to 82 between 1750 and 1801.

London contrasts sharply with each of the other three groups, growing far more quickly until the end of the seventeenth century. It became so large that in 1700 it housed more than two-and-a-half times as many people as the other three groups combined. Thereafter, however, London was outpaced by both the ports and the new manufacturing towns, and did not even match the national average growth rate.

The foregoing is both compressed and simplistic. Uncertainties of definition, estimation, and periodization have been dealt with summarily or ignored. Nor has the nature as opposed to the quantity of urban growth been explored. It is possible, for example, that much of the growth of sixteenth-century London was a "push" phenomenon linked to the scale and depth of rural poverty, and due to what Clark termed "subsistence" migration. Later, in contrast, movement to the capital may have had a greater "pull" element as living standards rose and "betterment" migration came to predominate.[8]

8 Peter Clark, "The Migrant in Kentish Towns, 1580–1640," in *idem* and Paul Slack (eds.), *Crisis and Order in English Towns, 1500–1700* (London, 1972), 117–163.

It remains reasonable to argue, however, that there were important links between some types of urban growth and real income trends in early modern England, although urban growth is not to be explained solely in this way. Neither the headlong growth of London in the sixteenth and early seventeenth centuries, nor the acceleration of urban growth in the new manufacturing towns in the later eighteenth century, is easily explicable in terms of the behavior of domestic real income per head, yet both were developments of massive importance.

URBAN GROWTH AND AGRICULTURE Even if the causes of urban growth are elusive, the fact of growth remains, and some of its implications can be examined. In any preindustrial community, agriculture is the dominant form of economic activity and the levels of productivity per head achieved in agriculture necessarily govern the growth opportunities of other industries. This point was so well known to those living in preindustrial economies as scarcely to warrant remark, and, when political economy reached its first great statement in the *Wealth of Nations,* Smith made the examination of this issue one of his chief concerns.[9]

But agricultural productivity has proved difficult to measure directly. One way of measuring it indirectly is to consider the extent of the rise in agricultural productivity suggested by the course of urban growth in England, while also taking into account changes in the occupational structure of the rural component of the total population.

To simplify calculations I have assumed that consumption of food per head did not vary between 1520 and 1800 and that England was neither a net importer nor a net exporter of food. The first assumption is doubtful and the second is demonstrably false, especially during the eighteenth century. In its early decades England was a substantial net exporter of grain, and toward the century's end large quantities of meat and grain reached the En-

9 Smith's chapter, "Of the Different Employment of Capitals," includes a strong plea for agricultural investment as the ultimate basis of national productive capacity. "The capital employed in agriculture, therefore, not only puts into motion a greater quantity of productive labour than any capital employed in manufactures, but in proportion, too, to the quantity of productive labour which it employs it adds a much greater value to the annual produce of the land and labour of the country, to the real wealth and revenue of its inhabitants. Of all the ways in which a capital can be employed, it is by far the most advantageous to the society." Smith, *Wealth of Nations*, 161–162.

glish market from Ireland. Yet it is convenient to begin with such simple assumptions.[10]

In 1520 the urban percentage was 5.25: in 1801, 27.5 (Table 2). This movement in itself suggests a useful gain in agricultural productivity. In 1520, 100 rural families fed 106 families in all; in 1801, 138 (100 × 100/[100 − 5.25] = 106: 100 × 100/[100 − 27.5] = 138). The level of productivity in 1800 was 30 percent higher than in 1520, far from a negligible increase, even if scarcely sensational. But any such exercise must understate the extent of the increase in agricultural productivity if there also is a decline in the proportion of the rural labor force engaged in agriculture.

There can be no reasonable doubt that such a decline occurred. In certain rural areas in the eighteenth century the growth in non-agricultural employment was so great as to dwarf the remaining agricultural population. Framework knitting became the dominant source of employment in many Leicestershire villages. In parts of Warwickshire and Staffordshire there was very rapid growth in the manufacture of small metal wares—nails, chains, buckets, etc. In much of south Lancashire and the West Riding of Yorkshire, the textile industry, whether cotton or wool, provided income for many more men and women than did agriculture. The steady growth of coal production in Northumberland and Durham produced the same result in substantial tracts of these counties. Even in more strongly agricultural counties in the south, lace making, straw plaiting, and the like provided much employment for women.

Moreover, in areas which attracted little industry there was often a growth in employment in service industries. In the rare

10 Brinley Thomas has recently estimated the relationship between the value of imports of grain, meat, and butter and the income of British agriculture. From 1814 to 1816, total imports of the three commodities represented 6.4% of British agricultural income and of these imports 70% came from Ireland. See his article in this issue, "Escaping from Constraints: The Industrial Revolution in a Malthusian Context," Table 2. Jones recently estimated that 90% of the population of Great Britain was fed from domestic agricultural production in 1800. Eric L. Jones, "Agriculture, 1700–80," in Floud and McCloskey (eds.), *Economic History of Britain*, I, 68. In contrast, in the first half of the eighteenth century, English net grain exports were a substantial fraction of total production, reaching perhaps 6% of gross domestic grain output about the mid-century. Phyllis Deane and W. A. Cole, *British Economic Growth, 1688–1959* (Cambridge, 1962), 65; David Ormrod, "Dutch Commercial and Industrial Decline and British Growth in the Late Seventeenth and Early Eighteenth Centuries," in Frederick Krantz and Paul M. Hohenberg (eds.), *Failed Transitions to Modern Industrial Society: Renaissance Italy and Seventeenth-Century Holland* (Montreal, 1975), 36–43.

cases where parish registers provide data on occupation over long periods of time, it is a commonplace to note a growth in specialist employments not previously encountered, especially during the eighteenth century. A small town like Colyton, for instance, even provided a living for a peruke maker in the 1760s.

Almost everywhere the proportion of men described as laborers, husbandmen, yeomen, or farmers declined as a proportion of all the occupations mentioned. Many craftsmen also owned scraps of land and their produce was of crucial significance in their domestic economy. Even those who worked no land might nevertheless have been drawn into the labor of harvest. However, the reverse was often also true. Those to whom an agricultural occupation was attributed might turn their hands to craftwork during the seasonal slack periods on the farms.

Ideally, it would be preferable to measure hours worked in different forms of employment rather than to treat each member of the work force as uniquely engaged in a single occupation, but for the present purpose it is enough to show that there was a major fall in the proportion of the rural labor force in agricultural occupations.

By 1800 a tolerably accurate picture of rural employment structure can be drawn. Deane and Cole estimated that 35.9 percent of the labor force in 1801 was engaged in agriculture, forestry, and fishing. If we assume for simplicity's sake that none of these workers was living in towns, then it follows that only some 50 percent of the rural population were engaged in agriculture, given that the rural population comprised 72.5 percent of the whole ($35.9/72.5 \times 100 = 49.5$). The comparable figure at earlier periods is difficult to establish, although useful clues may be found in the work of King and Massie. In order to make revised estimates of the changes in agricultural productivity, I have assumed that 80 percent of the rural labor force was in agriculture in 1520; that this figure declined very slowly to 70 percent by 1670, with the bulk of the fall occurring after 1600; that it fell more quickly to 66 percent in 1700; and then moved linearly in the eighteenth century to 50 percent in 1801.[11]

11 Deane and Cole's estimate of the percentage of the labor force, quoted above, may be found in their *British Economic Growth*, 142. The first serious attempt to quantify the occupational and social structure of England before 1801 was that carried out by Gregory King in 1688. King, *Natural and Political Observations and Conclusions upon the State and Condition of England, 1696*, reprinted in *The Earliest Classics*, with an introduction by Peter

These assumptions allow the population to be subdivided into three groups rather than two: the urban population, the rural population engaged in agriculture, and the rural population dependent on employment other than in agriculture. At the same time, changes in output per head can be calculated. The results are set out in Table 4.

The table suggests that the rural agricultural population scarcely changed in size for a century and a half between 1600 and 1750 and that even in 1800 it was only a tenth larger than 200 years earlier. It is not surprising, therefore, that the table should show a striking rise in agricultural productivity, and indeed, to the degree that the assumptions which have been made in constructing the table are justified, the conclusions are inescapable. The crucial assumptions are that England was not a signficant net importer or exporter of food; that the growth of the percentage of the population living in towns followed the pattern set out in Table 2; and that the proportion of the rural population engaged in non-agricultural production rose from 20 to 50 percent between 1520 and 1800. For the second and third assumptions the central issue is the extent of change over the period as a whole; the question of its timing, although fascinating, is of lesser importance.[12]

Laslett (London, 1973), 48–49. When Deane and Cole considered King's estimates they suggested that "between 75 and 80 percent of the occupied population was primarily engaged in agriculture." Deane and Cole, *British Economic Growth*, 3. This figure is implausibly high. Having regard to the total national populations in 1688 and 1801 (4.90 and 8.66 millions) and the proportions in agriculture (say 75% and 35.9%), such a high proportion in 1688 would imply a fall in the absolute scale of agricultural employment of more than 15% over the intervening century.

Lindert has recently reexamined changes in English occupational structure in the eighteenth century, but the results, at least for agriculture, do not appear convincing. The percentage of the male labor force employed in agriculture is estimated at 22.3 in 1700, 26.1 in 1740, and 13.7 in 1811. Lindert, *English Occupations, 1670–1811* (Davis, 1980), 46–47. See also note 13 below.

12 In connection with the argument of this paragraph it may be noted that Jones estimated that the agricultural population of England and Wales increased by 8.5% in the eighteenth century. Jones, "Agriculture, 1700–80," 71. Crafts' recent calculations produce results broadly similar to those implied by Table 4 in relation to agricultural productivity. He estimated that agricultural output rose at the following percentage rates per annum over the periods 1710–1740, 1740–1780, and 1780–1800: 0.9, 0.5, and 0.6. This change implies a total increase in output of almost exactly 80% over the 90-year period as a whole. Assuming that the agricultural labor force grew by 13% over the century, as Table 4 suggests, the increase in output per man, at 59%, is closely similar to that which may be calculated from column 7 of Table 4 (52%). Nicholas F. R. Crafts, "The Eighteenth

Precision about any of the three basic assumptions is beyond reach. Regarding the first, it is worth noting that because England was a substantial food exporter in the early eighteenth century, agricultural productivity was probably understated at that period. Equally, by the end of the century, England had become a net importer, especially from Ireland, which implies the opposite. The second set of assumptions concerning urban growth is probably sufficiently accurate to avoid significant error. The third assumption relates to the agricultural proportion of the rural population. The figure for 1800 is fairly firmly grounded in the evidence of the early censuses, but earlier estimates become increasingly fallible. Any figure for the early sixteenth century must be largely guesswork. A lower figure would reduce the apparent gain in agricultural productivity, but only a radically lower figure would greatly change the general picture.[13]

Century: A Survey," in Floud and McCloskey (eds.), *Economic History of Britain*, I, Tables 1.1, 1.2, 2–3.

13 It is possible to use the work of King and of Joseph Massie to test how far the estimates made by contemporaries agree with those in Table 4. In 1760, out of a total of 1,472,000 families, Massie supposed that 210,000 were freeholders, 155,000 farmers, and 200,000 husbandmen, making a total of 565,000 clearly engaged in agriculture. In addition he estimated that there were 200,000 families of laborers outside London. Not all of these would be agricultural laborers. Massie reckoned that there were 20,000 laborers in London, only slightly short of the number to be expected on the assumption that laborers were as numerous per 1,000 population inside cities as in the countryside. We may assume, therefore, that a further 15,000–20,000 were to be found in other urban centers, and it is probable that a further group, although living in the countryside, were employed outside agriculture, especially in the building trades (which were not separately itemized by Massie). Assume that these, too, were about 20,000 in number; then the overall total of those families engaged in agriculture would be c.730,000, or fractionally less than one half of the national total. The comparable figure in Table 4, that for 1750, is 46%, a broadly similar figure.

A comparison with King (1688) is more difficult to make because of the form in which he drew up his famous table. King's totals for families of freeholders and farmers are not greatly different from Massie's (180,000 and 150,000 respectively), but he gave no separate figure for husbandmen. Instead, all other families which might be engaged in agriculture appear in two categories: laboring people and outservants (364,000); and cottagers and paupers (400,000). Most of those engaged in industrial crafts—weavers, glovers, knitters, tanners, carpenters, coopers, sawyers, thatchers, smiths, wrights, cordwainers, and so on—must have been included in one of these two categories (King lists only 60,000 families of artisans or handicraft workers in the entire country), together with all laboring families living in London or other towns. Assuming that half of the cottars were what Massie would have termed husbandmen and half were engaged in industrial crafts, a split similar to that found in Massie's separate tabulation of the two categories; that 15% of all laborers were in towns (where at this date about 15% of the population lived); and that of the remaining 309,000 laborers, 25,000 were employed outside agriculture—the total

Table 4 Urban, Rural Agricultural, and Rural Non-Agricultural Populations (millions)

	(1) TOTAL POPULATION	(2) URBAN POPULATION	(3) RURAL POPULATION	(4) PROPORTION OF RURAL POPULATION IN AGRICULTURE	(5) RURAL AGRICULTURAL POPULATION (3) × (4)	(6) RURAL NON-AGRICULTURAL POPULATION (3) − (5)	(7) TOTAL POPULATION PER 100 RURAL AGRICULTURAL (1)/(5)	(8) COLUMN (7) 1520 = 100
1520	2.40	0.13	2.27	0.80	1.82	0.45	132	100
1600	4.11	0.34	3.77	0.76	2.87	0.90	143	108
1670	4.98	0.68	4.30	0.70	3.01	1.29	165	125
1700	5.06	0.85	4.21	0.66	2.78	1.43	182	138
1750	5.77	1.22	4.55	0.58	2.64	1.91	219	166
1801	8.66	2.38	6.28	0.50	3.14	3.14	276	209

Percentages of total population in major categories

	(1) URBAN	(2) RURAL AGRICULTURAL	(3) RURAL NON-AGRICULTURAL	(4) TOTAL
1520	5.5	76.0	18.5	100
1600	8.0	70.0	22.0	100
1670	13.5	60.5	26.0	100
1700	17.0	55.0	28.0	100
1750	21.0	46.0	33.0	100
1801	27.5	36.25	36.25	100

Populations relative to 1800 total (1800 = 100)				
1520	5.5	58	14	28
1600	14	91	29	47
1670	29	96	41	58
1700	36	89	46	58
1750	51	84	61	67
1801	100	100	100	100
Relative population growth by period (100 × total at later date/total at earlier date)				
1600/1520	262	158	200	171
1670/1600	197	105	143	121
1700/1670	127	92	111	102
1750/1700	144	95	134	114
1801/1750	195	119	164	150

SOURCE: For population totals in top panel see Table 3 and discussion in text.

Other assumptions could be used to construct Table 4. For example, if one were to assume that 15 percent of English food requirements were met by imports in 1800, and that agriculture employed 55 percent of the rural labor force rather than 50 percent, while other assumptions were unchanged, the estimated overall rise in agricultural productivity would be reduced from 109 to 61 percent. If one were to go still further, reducing the 1520 figure in column 4 from 0.80 to 0.70 and retaining the modified assumptions for 1801, the figure would drop even more to 41 percent, but this figure is improbably low. On present evidence, therefore, although the particular figure given in Table 4 is arbitrary, there is a strong likelihood that the true figure lies between 60 and 100 percent.[14]

The phasing of changes in the proportion of the rural labor force engaged in agriculture, no less than their scale, is also largely a matter of judgment, rather than a demonstrable pattern. The same pressures which kept urban growth outside London at such modest proportions in the sixteenth century are likely to have restricted employment opportunities outside agriculture. It therefore seems appropriate to make only a small reduction in the proportion of the rural population in agriculture between 1520 and 1600. Thereafter the pace of change probably accelerated.

---

of families employed in agriculture is 814,000 or 60% of the national total of 1,361,000 families. The comparable figure in Table 4 is about 58% (1670, 60.5%; 1700, 55.0%). There is a convenient reproduction of both King's and Massie's tables and a discussion of the difficulties in comparing them in Peter Mathias, "The Social Structure in the Eighteenth Century: A Calculation by Joseph Massie," in *idem, The Transformation of England. Essays in the Economic and Social History of England in the Eighteenth Century* (London, 1979), 171–189. See note 11 for the provenance of King's table. Massie's table is derived from a broadsheet dated Jan. 10, 1760, entitled *A computation of the money that hath been exorbitantly raised upon the people of Great Britain by the sugar planters in one year from January 1759 to January 1760; shewing how much money a family of each rank, degree or class hath lost by that rapacious monopoly.*

14 Sixteenth-century estimates of occupational structure are inevitably insecurely based, and the dominance of agriculture is probably overstated. The muster roll taken in 1608 for Gloucestershire, excluding Bristol, for example, suggests that only 46.2% of the adult male population between the ages of 20 and 60 were engaged in agriculture. The comparable figure when the 1811 census was taken was virtually the same at 45.8%. The economic history of Gloucestershire is far from typical of the country as a whole, but Gloucestershire affords an example which should caution us against assuming too readily that rural non-agricultural employment was always very limited in Tudor or early Stuart England. A. J. and R. H. Tawney, "An Occupational Census of the Seventeenth Century," *Economic History Review*, V (1934), 25–46. 1811 Census, Enumeration, *Parliamentary Papers* (1812), XI, 121.

One further implicit assumption deserves discussion. Individual intake of food measured in calories varies within fairly narrow limits. There is little evidence of widespread malnutrition so extreme as to cause death in early modern England. With rare and usually local exceptions even severe harvest failure did not provoke heavy mortalities. It is improbable (although also undemonstrable) that mean personal daily calorie intake varied in a manner which would significantly undermine the line of argument followed above. Nevertheless, periods of rising real income must have been periods in which food consumption per head rose, with the opposite happening in times of declining living standards. In addition there were changes in the composition of individual diet as incomes rose or fell. Meat was a luxury to the pauper but a commonplace to more prosperous members of society, and there were secular shifts in the relative prices of grain and meat which reflected the long-term trends of the average real income.[15]

Since such foods as meat and dairy produce needed larger inputs of labor, as well as land and capital, to produce the same number of calories of food as grain, it might seem that an allowance should be made for the impact of secular real income trends in attempting an individual measure of agricultural productivity. This consideration in turn would imply that agricultural productivity per head was rising faster during the seventeenth and early eighteenth century than suggested by Table 4, but less quickly during the later eighteenth century. It also suggests that it may

15 On harvest fluctuations and mortality, see Wrigley and Schofield, *Population History*, 320–340, 370–382, appendix 10. The principal exception to the rule that harvest failure did not provoke big mortalities was northwestern England in the sixteenth and early seventeenth centuries. Andrew B. Appleby, *Famine in Tudor and Stuart England* (Stanford, 1978). Estimating income elasticity of demand for food in eighteenth-century England presents great difficulties because of lack of relevant data. In a recent review of the limited evidence, Crafts concluded that the most probable figure for the late eighteenth and early nineteenth centuries was 0.7. Crafts, "Income Elasticities of Demand and the Release of Labour by Agriculture during the Industrial Revolution," *Journal of European Economic History*, IX (1980), 154–159. Kussmaul provides a convenient graph of the relative prices of grain and meat. It bears a strong resemblance to an inverted graph of the Phelps Brown and Hopkins real wage series, although the match is by no means perfect, especially in the early seventeenth and late eighteenth centuries. Ann Kussmaul, *Servants in Husbandry in Early Modern England* (Cambridge, 1981), Fig. 6.3, 104. Real wage data based on Phelps Brown and Hopkins are set out in Wrigley and Schofield, *Population History*, appendix 9, and shown graphically in *ibid.*, Fig. 10.5, 414.

have been falling in the sixteenth century, a finding in keeping with common sense, since the rural agricultural population is estimated to have risen by almost 60 percent between 1520 and 1600, a scale of increase likely to have involved a falling marginal productivity of labor and much concealed or overt underemployment.

Making an explicit allowance for real income changes, however, presents problems. At present, real income data are based on slender foundations and involve wide marigns of uncertainty. Little is known about any changes which may have occurred over time in the income elasticity of demand for food. The only existing real wage series covering the whole period, that of Phelps Brown and Hopkins, stood higher in 1800 than it had two centuries earlier. The apparent gains in agricultural productivity over the seventeenth and eighteen centuries as a whole are therefore unlikely to be overstated because of a failure to take real income explicitly into account.[16]

Labor released from agriculture is available to increase other forms of production. The gross changes were striking. The rural agricultural population in 1520 was 76 percent of the total population in 1520 but only 36 percent in 1800 (Table 4). Non-agricultural employment therefore grew from 24 to 64 percent of the whole. This difference may overstate the extent of the change in that some of the growth of non-agricultural employment represented jobs created by increased specialization of function. A carter, for example, making his living by moving to market goods previously taken there by local farmers, may be placed outside agriculture in an occupational breakdown, but undertakes a task previously performed by the farmer.

Yet the change was great. Smith argued that a surplus of agricultural production over the food needs of the farming population might either be consumed unproductively by, say, retinues of servants, or productively, if the surplus maintained an army of "manufacturers" whose output added to the wealth of the community as a whole. In early modern England the growth of employment in industry and commerce is a testimony to the

16 The Phelps Brown and Hopkins index stood at 409 on average for the 11 harvest years 1595–96 to 1605–06 compared with 507 in the period 1795–96 to 1805–06. *Ibid.*, 642–644.

predominantly "productive" use to which the growing relative surpluses in the agricultural sector were put. Smith considered that the scale of such growth was largely conditioned by the extent of the rise in agricultural productivity. He did not envisage the much more radical type of change which has come to be called an industrial revolution, nor is there any compelling reason to suppose that even increases in agricultural productivity as striking as those achieved in England between 1600 and 1800 must necessarily engender an industrial revolution. Yet the scale of change in early modern England bears stressing. It stands out more clearly if comparison is made with other countries.[17]

URBAN GROWTH ON THE CONTINENT De Vries has recently published an informative analysis of urban growth patterns between 1500 and 1800 for Europe as a whole and for some major regional subdivisions. In it he makes use of an empirical relationship between the sizes of the towns and their position in a rank ordering whereby the difference in their populations is proportional to the difference in their rank orders. Thus $p_i = p_1/i$ where p_1 is the population of the largest town and i refers to the rank order of a town after all towns have been arranged in descending order of size. The population of the fifth largest town may therefore be expected to be a fifth of that of the largest, and so on. If logarithmic scales are used to plot the population and rank coordinates of each town, the resulting distribution in the archetypal case will fall on a straight line with a slope of 45 degrees. In practice the slope of the line may vary somewhat from the 45-degree slope suggested by the strictly proportional relationship, and the first few points plotted are sometimes aberrant, as when the largest settlement is much bigger than the second largest. But regularities in town size distributions are often impressive, and any anomalies within a data set or changes in the angle of the plotted slope over time may prove illuminating.[18]

17 On Smith's views on productive and unproductive labor, see the chapter in the *Wealth of Nations*, "Of the accumulation of capital, or of productive and unproductive labor." An attempt to clarify the contingent nature of the circumstances which preceded and accompanied the industrial revolution may be found in Wrigley, "The Process of Modernization and the Industrial Revolution in England," *Journal of Interdisciplinary History*, III (1972), 225–259.

18 Jan de Vries, "Patterns of Urbanization in Preindustrial Europe, 1500–1800," in H. Schmal (ed.), *Patterns of European Urbanization since 1500* (London, 1981), 77–109.

By compiling rank-size data for European towns at intervals over a three-century period starting in 1500, de Vries was able to establish changes in the slope and shape of the urban hierarchy sufficiently pronounced and consistent to distinguish three major periods, as shown in Figure 1. At the beginning of the sixteenth century the slope was gentle and there was a flat 'top' to the distribution. He attributed the latter to the still strongly regional character of the European economy, which was insufficiently articulated to produce leading urban centers of the size implied by the slope of the lower part of the distribution. During the sixteenth century the rank-size plot gradually straightened. The continent-wide economy was becoming more integrated, and cities such as London, Paris, and Amsterdam grew rapidly as they assumed urban functions over wide hinterlands.

A second period then supervened, running from about the early seventeenth to the mid-eighteenth century. In general, in this period, the larger the town, the more rapidly it grew, so that

Fig. 1 Urbanization in Europe

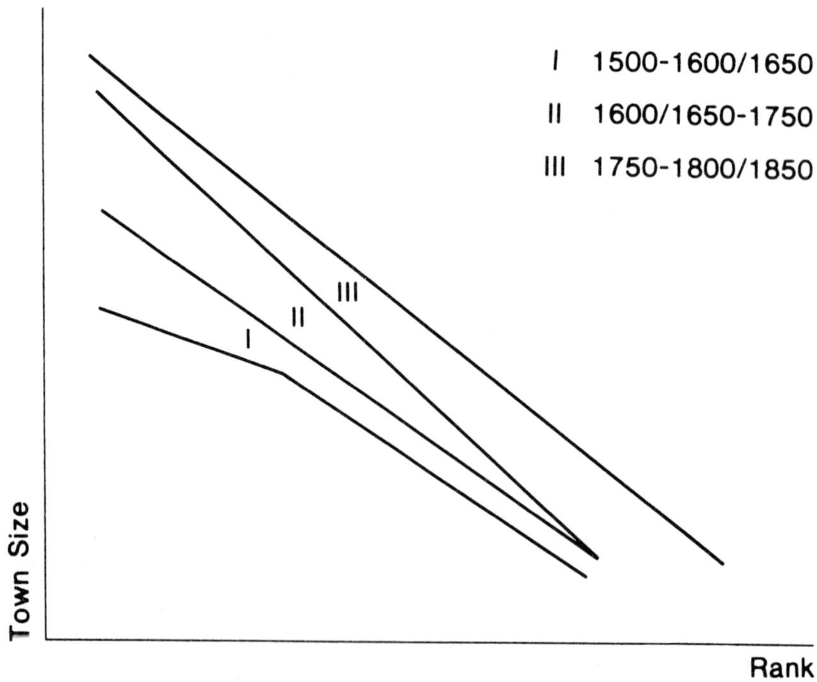

SOURCE: de Vries, "Patterns of Urbanization."

a disproportionate part of the overall rise in urban population took place in the larger towns. The rank-size line pivoted slowly round a point close to its lowest reading and thus grew steadily steeper. After about 1750, however, urban growth changed in character again and a third period started. The rank-size line moved outward from the origin of the graph, implying a rise in the number of towns in all size categories, but there was more rapid growth in the number of smaller towns than in the number of larger ones. The slope became less steep once more. The sequence of changes was shared by the major subdivisions of Europe used by de Vries, although the timing of the changes varied somewhat. The pattern of change in England, however, was unlike that on the continent.

As Table 5 shows, England began the period with an unusually slight proportion of her population living in large towns, but passed the European average in the mid-seventeenth century and the northwest European average by 1700. By the beginning of the nineteenth century England was relatively heavily urbanized. Events there, however, were so different from those elsewhere as to distort patterns of change when England is included in some larger grouping.

It is instructive to remove England from both the urban and the overall population totals used to generate urban percentages for continental areas. The result is shown in Table 6. Urbanization

Table 5 Urbanization in England and the Continent

Percentage of total populations in towns with 10,000 or more people

	1500	1600	1650	1700	1750	1800
England	3.2[a]	6.1	10.8[b]	13.4	17.5	24.0
North and West Europe[c]	6.0	8.1	10.7	13.0	13.8	14.7
Europe[d]	6.1	8.0	9.3	9.5	9.9	10.6

NOTES:
a c. 1520
b c. 1670
c Scandinavia, Netherlands, Belgium, England, Scotland, Wales, and Ireland.
d The countries included in Europe are Germany, France, Switzerland, Italy, Spain, and Portugal together with those listed in note c.

SOURCES: For England see source notes to Tables 1 and 2. Other data from de Vries, "Patterns of Urbanization," 88.

Table 6 Urbanization in England and the Continent (revised)

Percentage of total population in towns with 10,000 or more people

	1600	1700	1750	1800
England	6.1	13.4	17.5	24.0
North and West Europe minus England	9.2	12.8	12.1	10.0
Europe minus England	8.1	9.2	9.4	9.5

NOTES: De Vries provides estimates of population totals for north and west Europe and for Europe as a whole for 1600, 1700, 1750, and 1800 in the work listed below (although not for the other dates in Table 5). This information and the urban percentages given in Table 5 make it possible to calculate the size of the urban populations. The English totals can then be removed from both the urban and total populations and the percentages recalculated.

SOURCES: For England see source notes to Tables 1 and 2. For other data see source notes to Table 5; de Vries, *The Economy of Europe in an Age of Crisis* (Cambridge, 1976), 5.

in northern and western Europe recedes rather than advancing substantially in the eighteenth century, although the scale of advance in the seventeenth century is little altered. In Europe as a whole the exclusion of England slows the increase in urbanization in the seventeenth century and brings it almost to a halt in the eighteenth.[19] Over the full 200-year period the urban percentage quadrupled in England, scarcely changed in the rest of north-western Europe, and advanced rather modestly on the continent as a whole. The English experience appears to be unique.

The extent of the contrast also comes home forcefully if de Vries' estimates are reworked to permit another comparison to be made, as in Table 7. The urban population of Europe more than doubled between 1600 and 1800, but the total population rose by almost 60 percent, so that much of the rise in urban population was caused by the rise in overall numbers rather than by an increase in the proportion of the population living in towns. Column 4 shows how the urban population would have grown if the urban percentage had stayed at the level prevailing in 1600. The totals in column 5 show the net gain in urban population at

19 This finding might be thought spurious if de Vries' estimates of the population of English towns and those shown in Table 1 were widely divergent. But a detailed comparison of his data with those of Table 1 suggests that this is a groundless fear.

Table 7 The English Share of European Urban Growth, 1600–1800 (populations in millions)

EUROPE[a]	(1) TOTAL POPULATION	(2) URBAN PROPORTION	(3) URBAN TOTAL (1) × (2)	(4) URBAN TOTAL AT 1600 PROPORTION	(5) NET GAIN ON 1600 (3) − (4)	(6) NET GAIN ON LAST DATE (DIFFERENCE BETWEEN SUCCESSIVE TOTALS IN (5))
1600	70.6	0.080	5.65	5.65	0.00	
1700	75.0	0.095	7.13	6.00	1.13	1.13
1750	86.6	0.099	8.57	6.93	1.64	0.51
1800	111.8	0.106	11.85	8.94	2.91	1.27
ENGLAND						
1600	4.11	0.061[b]	0.249	0.249	0.000	
1700	5.06	0.134[b]	0.680	0.309	0.371	0.371
1750	5.77	0.175[b]	1.012	0.352	0.661	0.290
1800	8.66	0.240[b]	2.079	0.528	1.551	0.890

English Percentage of Total European Net Urban Gain

1600/1700	100(0.371/1.13) = 33
1700/1750	100(0.290/0.51) = 57
1750/1800	100(0.890/1.27) = 70
1600/1800	100(1.551/2.91) = 53

NOTES: a For a list of the countries comprising "Europe," see notes to Table 5.
b These proportions are derived from the column (3) totals rather than vice-versa as in the top panel.

SOURCES: For England see source notes to Tables 1 and 2.
For Europe see source notes to Table 5.

each later date compared with 1600: that is, the number by which the urban population exceeded what would have obtained had the urban percentage not changed in the interim. The second panel of the table repeats the calculation for England.

By combining information from the two upper panels, the proportion of total European urban growth that occurred in England can be calculated for the periods 1600 to 1700, 1700 to 1750, and 1750 to 1800. The proportion rises steadily from a third in the seventeenth century to 70 percent in the second half of the eighteenth. Over the two-century period as a whole the proportion exceeded one half. Since England contained only 5.8 percent of the population of Europe in 1600 and only 7.7 percent even in 1800, these proportions are an extraordinary testimony to the extent of the difference in urban growth between the island and the continent.

De Vries placed particular emphasis on the absence of growth in the smaller towns between 1600 and 1750, a period which he defined as the age of the rural proletariat. He argued that urban growth was almost entirely confined to very large cities, with 80 percent of all growth taking place in towns with 40,000 inhabitants or more. These towns were almost all capital cities or large ports, the development of which was stimulated by the growth in administrative, military, and legal employment in both absolutist and constitutional states, or by the development of long-distance, often extra-European trade. Smaller towns stagnated or lost population, afflicted by loss of political autonomy and by the "abandonment of cities as locations for many of the most labour-absorbing industries."[20] The balance of advantage favored protoindustrial development in the countryside.

There was unquestionably much growth in industrial employment in rural England during this period. At the same time London, combining administrative, commercial, and trading dominance, enjoyed an astonishing expansion. But neither the development of rural industry nor the growth of London precluded an equally remarkable surge of urban growth elsewhere in England. Other towns grew by much the same absolute amount as London between 1600 and 1750 and proportionally were grow-

20 De Vries, "Patterns of Urbanization," 101.

ing even faster than London (Table 2). Towns with between 5,000 and 10,000 inhabitants doubled in number from fifteen to thirty-one, while over the same period the number in Europe minus England fell from 372 to 331. Figure 2 shows rank-size plots for England. Like Figure 1 it is schematically drawn: empirical data plots would show points falling on either side of the straight lines for each date. Although the number of towns is small and the presence of London severely distorts the top of the distribution,

Fig. 2 Urbanization in England

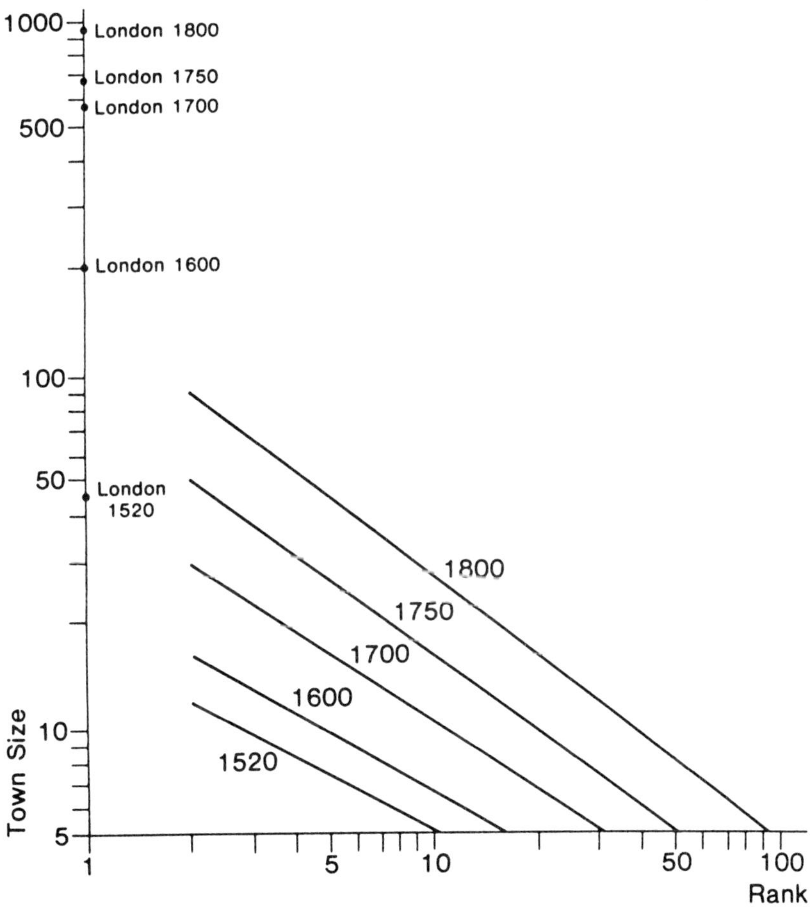

the lines emphasize how different England was from the continent (compare Figure 1).[21]

De Vries' study has the admirable virtue of providing a framework of comparison for smaller units by epitomizing the characteristics of Europe as a whole. This framework helps to bring out the distinctiveness of English history, and it may be further examined by considering comparative data about England's two greatest rivals, Holland and France.

HOLLAND AND FRANCE Even in 1800 England was a less urbanized country than Holland. Already in the early sixteenth century many Dutch people were town dwellers and their numbers increased rapidly as the century wore on. In this period the course of real wages in Holland was almost the reverse of the comparable pattern in England: real wages rose to a peak around 1610, the date when the English series approached its nadir, and urbanization continued. The percentage of the population living in towns grew from about 21 to about 29 percent, and since the total population of Holland was rising moderately quickly at the time, the absolute number of town dwellers rose even faster, from about 260,000 to about 435,000. There followed a period of half a century during which Dutch real wages fell back somewhat, just as a recovery was starting in England, maintaining the inverse movement in the two countries for a full century. After the 1650s, however, Dutch real wages rose rapidly once more to a new and substantially higher peak in the 1690s, after which they fell, uncertainly at first, but more quickly and without interruption from the 1740s until the end of the century. The trend of urban percentages moved broadly in sympathy with real wage movements, reaching a high point around 1700, when about 39 percent of the Dutch population was urban. It then declined slowly until, by 1800, it was not much higher than the English figure. Some Dutch towns experienced severe falls in population, and, since the population as a whole grew only modestly in the eighteenth century, the total number of town dwellers fell slightly.[22]

21 On the growth of London, see Wrigley, "A Simple Model of London's Importance in Changing English Society and Economy," *Past & Present*, 37 (1967), 44–70. The totals of towns with between 5,000 and 10,000 inhabitants are taken from de Vries, "Patterns of Urbanization," 93. For sources of data about English towns see source notes to Table 1.

22 For Dutch real wages, see de Vries, "The Population and Economy of the Preindustrial

Dutch rural population trends also warrant notice. During the long period of urban growth from 1550 to 1700 the rural population grew only modestly. It was 17 percent larger at the latter than the former date. In the eighteenth century, however, when there was urban decline and real wages languished, rural population numbers rose by 20 percent. In England in the eighteenth century the rural population grew by 49 percent overall, but the rural agricultural population by only 13 percent, even though the overall population growth in England during the century reached 71 percent compared with a Dutch figure of only 11 percent. If, therefore, it is safe to assume that the Dutch rural agricultural population was rising as fast as the rural population as a whole—a plausible supposition at a time of falling real wages—rural agricultural population totals in the two countries must have moved almost exactly in step in the eighteenth century, even though their overall rates of population growth diverged so markedly. Conversely, in the sixteenth century, while impoverishment increased in England and its rural agricultural population grew fast (by 58 percent between 1520 and 1600), in Holland there can have been little comparable increase, although the population of Holland was rising fairly rapidly. In columns 5 to 7 of Table 8 the rural population is subdivided between agricultural and other employment in conformity with arbitrary assumptions, the basis of which is explained in the notes to the table. To the degree that the figures mirror reality, they underline the points just made. The golden age of the Dutch rural economy was clearly one free from increasing pressure on the land, but it was succeeded by more trying times.[23]

A comparison of the urban sectors in England and Holland should not be pressed too far. It would make no sense, for example, to use estimates of urban percentages in Holland as an indirect measure of agricultural productivity, since Holland was a large importer of agricultural products, especially Baltic grain,

---

Netherlands," in this issue, Fig. 2. See Table 8 notes for sources for the estimates of Dutch urban populations. A. van der Woude provides much detail of urban fortunes in Holland betwen 1525 and 1795 in "Demografische ontwikkeling van de Noordelijke Nederlanden 1500–1800," in *Algemene Geschiedenis der Nederlanden*, V (1980), 134–139.

23 De Vries, "Population and Economy of Preindustrial Netherlands," Fig. 2, charts the fall in real wages from a late-seventeenth-century peak which, in a period of static or falling population, was eloquent testimony to Dutch difficulties.

Table 8 Dutch Urban and Rural Population Estimates (thousands)

	(1) TOTAL POPULATION	(2) PROPORTION URBAN	(3) POPULATION (1) × (2)	(4) RURAL POPULATION (1) − (3)	(5) RURAL AGRICULTURAL PROPORTION	(6) RURAL AGRICULTURAL POPULATION (4) × (5)	(7) RURAL NON-AGRICULTURAL POPULATION (4) − (6)
1550	1250	0.21	260	990	0.75	745	245
1600	1500	0.29	435	1065	0.705	750	315
1650	1875	0.37	700	1175	0.66	775	400
1700	1900	0.39	740	1160	0.66	768	395
1750	1925	0.35	675	1250	0.66	825	425
1800	2100	0.33	700	1400	0.66	925	475

NOTES: Urban is defined as relating to towns of 5,000 inhabitants or more.

The population totals in columns (3) and (6) have been rounded to emphasize the approximate and arbitrary nature of the calculation.

The proportion urban in column (2) was estimated from data given by de Vries. His data refer to the percentages of the population living in towns containing 2,500 inhabitants or more, and in towns containing 10,000 inhabitants or more. I assumed that half of those living in towns with 2,500 to 9,999 inhabitants were living in towns of 5,000 to 9,999 inhabitants. For some dates de Vries provides data only for the urban percentage in towns of 10,000 or more inhabitants, but sufficient paired estimates are given for other dates to make it feasible to provide plausible estimates at half-century intervals.

The rural agricultural proportions (column 5) stem from the following observations and assumptions. Mitchell provides data which show that 44.1% of the total Dutch labor force in 1849 was in agriculture. At that date the urban proportion was the same as in 1800 (de Vries, Table 1), and therefore the proportion of the *rural* labor force engaged in agriculture was 66% (0.441/0.67 = 0.66). I have assumed arbitrarily that in Holland in the mid-sixteenth century the comparable figure was 75%, a slightly lower figure than that used in the parallel calculation for England, because Holland was far more urbanized than England in the sixteenth century and non-agricultural rural employment was probably more common. I have assumed that it fell to 0.66 linearly over the next century but did not change thereafter as the great Dutch growth spurt ended.

SOURCES:

Column 1: J. A. Faber, H. K. Roessingh, B. H. Slicher Van Bath, Van der Woude, and H. J. van Xanten, "Population Changes and Economic Developments in the Netherlands: A Historical Survey," *A. A. G. Bijdragen*, XII (1965), 110.

Column 2: de Vries, "Population and the Labor Market in the Netherlands, 1500–1850," paper given at the Conference on British Demographic History (1982), Table 1.

Columnn 5: Brian R. Mitchell, *European Historical Statistics, 1750–1975* (Cambridge, 1981; 2nd rev. ed.), 167 (see also notes above).

and was not broadly self-sufficient as was England in the early modern period. Yet the beneficial effects of urban growth on Dutch agriculture in promoting specialization and making it easier to achieve higher production per man and per farm reflect the same processes at work in Holland as London's growth produced in England. Again, the Dutch passenger canal network was the transport wonder of its age, a response to urban growth and the closely associated rise in living standards, just as similar developments in eighteenth-century England promoted the investment of capital in turnpike roads and the construction of a new canal network.[24]

Both Holland and England, therefore, vividly illustrate the beneficial interactions among urban growth, rising living standards, and a surge in agricultural productivity which was possible within the context of an early modern economy. The food needs of towns were met through the operation of a commercial market in foodstuffs. Farming units benefited from specialization and avoided the subdivision and fragmentation of holdings which were commonly the bane of peasant societies when population increased. By the late eighteenth century yields per acre were substantially higher in England and the Netherlands than elsewhere in Europe, having roughly doubled over the previous two centuries. Yet in neither country was the agriculturally employed population much larger in 1800 than in 1600.[25]

The example of Holland also shows, however, that such progress was not necessarily a passport to further success. Smith, aware like so many of his contemporaries of the exceptional economic achievements of Holland, used it as an example of the limits to growth which must beset any country at some stage, making it difficult to sustain the gains of the past, much less secure further advances. The tide began to ebb in eighteenth-century Holland. Urban population fell slightly, even though numbers rose modestly overall. Real wages sagged. Far from forming part of the vanguard of the industrial revolution, Holland moved into the industrial era later than most of Western Europe,

24 *Idem*, "Barges and Capitalism. Passenger Transportation in the Dutch Economy, 1632–1839," *A. A. G. Bijdragen*, XXI (1978), 33–398.

25 For yield data, see Slicher van Bath, *The Agrarian History of Western Europe, A.D. 500–1850* (London, 1963), 280–282.

and even as late as 1850 was no more urbanized than it had been in the later seventeenth century.[26]

France provides a very different contrast to the case of England. Some relevant data are given in Table 9. The steps leading to the estimates given in the table are explained in the table notes. As with the English and Dutch estimates, they are subject to a margin of error which in some cases may be substantial because the empirical data base is insecure or inconsistent, because some of the assumptions are questionable, or because the chains of reasoning used to produce some of the estimates were long and there may have been a compounding of errors. Despite such uncertainties, the contrast between France and England stands out so strongly that no reasonable change in the assumptions used would significantly change the picture.

In 1500 France was not only a far more populous country than England, but it was also more urbanized. Within the present borders of France there were already fourteen towns with a population of more than 20,000 and twenty-one with populations between 10,000 and 20,000, while in England only London exceeded 20,000, and probably only two other towns, Norwich and Bristol, had 10,000 or more inhabitants. London was not only smaller than Paris but was also smaller than Lyon. Overall France was almost twice as urban as England, although much less urban than Holland, but, in the next three centuries, French towns grew hardly more quickly than the French population as a whole. France closely resembled de Vries' 'Europe' both in level of urbanization and in its change over time (Table 5). By 1800 the degree of urbanization reached in England was approaching three times the French level. Moreover, the proportion of the English rural population employed outside agriculture grew much faster than in France, and agricultural productivity, using the rough measure employed *faute de mieux* in this article, progressed far more rapidly in England than in France.[27]

26 Smith makes the general observation that, "In a country which had acquired that full complement of riches which the nature of its soil and climate, and its situation with respect to other countries allowed it to acquire, which could, therefore, advance no further, and which was not going backwards, both the wages of labour and the profits of stock would, probably, be very low." He subsequently suggested that Holland was approaching this state. Smith, *Wealth of Nations*, 43–44.

27 The French town population estimates are taken from de Vries, "Patterns of Urbanization," 82–85. In some instances the effects of revolution and war caused a sharp drop in French city populations in the last decade of the eighteenth century. Jacques Dupâquier, *La population française aux XVIIe et XVIIIe siècles* (Paris, 1979), 91–92.

Population always grew faster in England than in France, at times dramatically so (Tables 2 and 9), yet, because of the far faster urban growth in England and the swifter rise in the proportion of the rural population employed outside agriculture, the growth of rural agricultural employment was actually slower in England than in France, except in the sixteenth century. Between 1600 and 1800 the number of those dependent upon agriculture for a living in France grew by 30 percent, whereas in England the comparable figure was 9 percent. (Over the same period the national population growth rates were 53 and 111 percent respectively.) These apparently exact percentages should be viewed with reserve, but it is probably safe to assert that the rural agricultural population in England grew no faster than its counterpart in France. If the data used are taken on trust, inferences about comparative trends in agricultural productivity in the two countries are possible.

The upper panel of Table 10 sets out information for England and France in an indexed form to make it easier to appreciate the divergent course of development which characterized them. The value for 1600 has in each case been made equal to 100. In the sixteenth century, the differences between the two countries were chiefly related to the much higher rate of growth of the English population rather than to differences in the relative proportions of the populations falling into the urban, rural agricultural, and rural non-agricultural categories. After 1600, the contrasts between the two countries grew more pronounced. England continued to grow more quickly overall, but the rural agricultural element in its population grew little over the next two centuries. Indeed, it appears to have been falling for much of the time, whereas urban growth was meteoric; the rise in the rural non-agricultural component was also pronounced. Growth in France was much more balanced and, as a result, although the total population grew less than in England, the rural agricultural population grew appreciably faster.

If it were safe to assume that the level of food consumption per head was much the same in the two countries, that the estimates of rural agricultural population are accurate, and that each country may be regarded as having supplied its own food needs (except that in this table imports were assumed to meet 10 percent of English food requirements in 1800), then it is a simple matter to compare both levels of productivity in agriculture and their

Table 9 French Urban and Rural Population Estimates (thousands)

	(1) TOTAL POPULATION	(2) URBAN POPULATION	(3) PERCENTAGE URBAN	(4) RURAL POPULATION	(5) RURAL AGRICULTURAL PROPORTION	(6) RURAL AGRICULTURAL POPULATION (4) × (5)	(7) RURAL NON-AGRICULTURAL POPULATION (4) − (6)
1500	15,500	1,410	9.1	14,090	0.80	11,270	2,820
1600	19,000	1,660	8.7	17,340	0.755	13,100	4,240
1700	21,500	2,350	10.9	19,150	0.71	13,600	5,550
1750	24,500	2,530	10.3	21,970	0.685	15,050	6,920
1800	29,100	3,220	11.1	25,880	0.66	17,080	8,800

NOTES: Urban is defined as relating to towns of 5,000 inhabitants or more. France was treated as if it occupied its present territory; therefore towns such as Lille were included in urban population totals. The urban population in column (2) was estimated in several stages. Chandler and Fox provide estimates of urban populations of towns of 20,000 inhabitants or more for the dates shown. De Vries gives totals of towns in several size categories ('000s) 20–29, 30–39, 40–49, 50–99, and 100 and over. Where the total for a town given by Chandler and Fox falls into the size category given by de Vries, this total was used. Otherwise de Vries' calculations were preferred and a town population was assumed to be at the centerpoint of the size range (none fell into the open-ended top category). The discrepancies between the two sources, as might be expected, were much more marked at the earlier than the later dates. The number of towns of 20,000 or more inhabitants in the two sources were as follows at the five successive dates, Chandler and Fox in parentheses: 14 (20), 20 (15), 24 (26), 31 (32), 32 (36). De Vries' totals of towns between 10,000 and 20,000 were multiplied by 13,500 to give a population total for this size range of towns. These two operations provide a figure for all towns with populations of 10,000 or more, leaving the problem of estimating a figure for towns of between 5,000 and 10,000 people. De Vries provides estimates of the number of towns in this size range in Europe as a whole by extrapolating from the rank-size distributions of towns of 10,000 inhabitants and above. It can be shown that the rank-size distribution of French towns was similar to that for Europe as measured by the ratio of the number of all towns above 10,000 in population to the number in the size range between 10,000 and 20,000. The ratio of the number of towns in the size range between 5,000 and 10,000 to the number of towns larger than 10,000 derived from de Vries' data was used to estimate the total of smaller towns in France. (The ratios in question at the five successive dates are 2.15, 1.76, 1.60, 1.43, and 1.56; of these the third is interpolated from the second and fourth since de Vries provides no data for 1700.) The total of towns in the size range between 5,000 and 10,000 estimated in this fashion was multiplied by 6,500 to give a population total for each date. The breakdown of the individual totals was as follows ('000s):

Population totals by town size class

	20 and over	10–19	5–9	Total
1500	637	284	488	1,409
1600	866	311	481	1,658
1700	1,389	405	559	2,353
1750	1,691	324	514	2,529
1800	1,810	621	793	3,224

The proportion of the rural population engaged in agriculture is based on the following data and assumptions. In 1851 the proportion of the population living in towns of 5,000 or more inhabitants was 17.9% (Pouthas, 22, 76). In 1856 the proportion of the total French labor force engaged in agriculture was 51.7%. In the mid-nineteenth century, therefore, about 63% of the rural labor force worked on the land (100 × (0.517/0.821) = 63). I have assumed that the comparable figure in 1800 was 66%. As in the case of England, I assumed that in France the comparable figure in the early sixteenth century was 80%. The proportions at intermediate dates were obtained by rough linear interpolation.

SOURCES:

Column (1): Jacques Dupâquier, *La population française aux XVII*[e] *et XVIII*[e] *siècles* (Paris, 1979), 11, 34, 81; M. Reinhard, A. Armengaud, and Dupâquier, *Histoire générale de la population mondiale* (Paris, 1968; 3rd ed.), 108, 119–120; C. McEvedy and Richard Jones, *Atlas of World Population History* (London, 1978), 55–60.

Column (2): T. Chandler and G. Fox, *3000 Years of Urban Growth* (New York, 1974), 15–19, 21; de Vries, "Patterns of Urbanization," Tables 3.2, 3.4, 3.7.

Column (5): C. H. Pouthas, *La population française pendant la première moitié du XIX*[e] *siècle* (Paris, 1965); Mitchell, *European Historical Statistics*, 163.

Table 1e English and French Growth Patterns Compared

	ENGLAND					FRANCE			
	POPULATIONS RELATIVE TO 1600 TOTAL (1600 = 100)								
	URBAN	RURAL AGRICULTURAL	RURAL NON-AGRICULTURAL	TOTAL		URBAN	RURAL AGRICULTURAL	RURAL NON-AGRICULTURAL	TOTAL
1520	38	63	50	58	1500	85	86	67	82
1600	100	100	100	100	1600	100	100	100	100
1700	250	97	159	123	1700	142	104	131	113
1750	359	92	212	140	1750	152	115	163	129
1750	359	92	212	140	1800	194	130	208	153

TOTAL POPULATION PER 100 AGRICULTURAL

	ENGLAND (1)		FRANCE (2)		ENGLAND COLUMN (1) 1520 = 100 (3)		FRANCE COLUMN (2) 1500 = 100 (4)
1520	132	1500	138	1520	100	1500	100
1600	143	1600	145	1600	108	1600	105
1700	182	1700	158	1700	138	1700	114
1750	219	1750	163	1750	166	1750	118
1801	248	1800	170	1801	188	1800	123

NOTE: See text for modification made to 1801 figure in bottom panel, columns (1) and (3)
SOURCES: Tables 4 and 9

rates of change over time in the two countries. They are given in the lower panel of the table. The figures of columns 1 and 2 represent an absolute measure of productivity per head of the rural agricultural population whereas those in columns 3 and 4 index changes over time.

The extent of the contrast between England and France may well be understated by the table, both because it is reasonable to suppose that consumption of food per head was higher in England than in France since English real incomes appear to have been the higher of the two, and also because a higher proportion of the output of English agriculture may have been used as industrial raw material rather than food. Land and labor used for the raising of sheep for wool, for example, cannot also be used to grow corn. The effect, however, is difficult to quantify since output in such cases often consisted both of food and of industrial raw materials. Producing wool for cloth or hides for the leather industry also meant producing meat for human consumption. Pure industrial land usage, as with flax for linen, was a rarer phenomenon. The use of the land to provide "fuel" for horses or oxen also suggests that the contrast between the two countries may be greater than appears in the table, since in England, by the later eighteenth century, a substantial proportion of horses were employed outside agriculture on turnpike roads, on canal towpaths, and particularly for urban transportation of men and goods.[28]

In England, as in Holland at an earlier date, urban growth, rising productivity in agriculture, and improving real incomes were interwoven with one another, and were mutually reinforcing for long periods. The French economy did not follow the same

28 Thirsk presents good reasons to suppose that the value of industrial crops rose rapidly in the later sixteenth and seventeenth centuries, and also provides abundant evidence of the enormous increase in rural employment outside agriculture. Thirsk, *Economic Policy and Projects. The Development of a Consumer Society in Early Modern England* (Oxford, 1978). There are some guarded attempts to quantify the scale of industrial crops and new food crops in *ibid.*,177–178.

Thompson estimates that there were almost half a million (487,000) horses in use outside agriculture in 1811 (riding horses, carriage horses, post horses, trade horses, and those in the stage coach trade). At the level of consumption of oats and hay per animal which Thompson thinks appropriate, and on reasonable asumptions about yields of oats and hay per acre, sustaining such a large population must have meant devoting about 2 million acres to their fodder. Agriculture at that date is estimated to have required about 800,000 horses. F. M. L. Thompson, "Nineteenth-Century Horse Sense," *Economic History Review*, XXIX (1976), 78, 80.

pattern. On the assumptions used in constructing Table 9, there is evidence of increased pressure on the land in the eighteenth century. In a peasant economy this pressure may provoke serious economic difficulties and increase the proportion of excessively subdivided holdings, conforming to de Vries' peasant model of a rural economy. Both England and Holland accord better with his specialization model.[29]

Arguments couched in this manner are always too simplistic. It is misleading, for example, to treat preindustrial countries as if they were homogeneous units. There were major differences among different areas, even within the smallest of the three countries considered. Sometimes even intraprovincial differences in Holland were marked: interprovincial contrasts could rival those among countries. The regional variety of a country as large as France was still more notable. It could scarcely be otherwise when it is recalled that France contained more than ten times as many people as Holland and had an area fifteen times greater. Furthermore, better knowledge would no doubt cause significant alterations to be made to many of the tables on which the arguments advanced in this article are based. Yet the contrasts and similarities between the three countries would probably remain even if perfect knowledge were available.[30]

It is not difficult to find convincing reasons why a major advance in agricultural productivity is a prerequisite of industrial growth. Only if resources can be spared from the task of ensuring an adequate supply of foodstuffs can a larger scale of industrial production be attempted. It may seem quaint that Smith should have associated the prosperity of, say, Sheffield's industry with the efficiency of agriculture in the vicinity of the town, but his remark is a parable of the circumstances needed for industrial success in a preindustrial age.[31]

England was singularly fortunate in this regard. Output per head in agriculture appears to have risen by three quarters between

29 De Vries, *The Dutch Rural Economy in the Golden Age, 1500–1700* (New Haven, 1974), 1–21.

30 For national population totals see Tables 8 and 9. The area of France today is 212,209 square miles; that of Holland 13,967. Their areas varied somewhat over the early modern period.

31 See Smith, *Wealth of Nations*, 180–181.

the beginning of the seventeenth and the end of the eighteenth century. The annual gain in agricultural productivity per head was modest—about 0.3 percent—yet it permitted a more rapid overall rate of population growth than that found in either France or Holland, while simultaneously releasing into secondary and tertiary employment a far higher proportion of the active population than in any other European country. By 1800 little more than a third of the English labor force was engaged in agriculture at a time when it is improbable that the comparable figure elsewhere in Europe, apart from the Netherlands, was less than 55 to 60 percent. In many countries it was substantially higher.[32]

The pace of English population growth in the early modern period was exceptional. Between the mid-sixteenth and the early nineteenth century England's population grew by about 280 percent. The population of other major European countries rose much less rapidly. Germany, France, the Netherlands, Spain, and Italy all grew by between 50 and 80 percent over the same period. The demographic mechanisms by which the growth of England came about are now fairly well understood, but at one remove the problem of explaining the contrast remains. In this connection it is important to pay heed to the probability that there was nothing unusual in the rate of growth of English *rural agricultural* population, unless it was that it grew so slowly, especially after 1600. The great bulk of the overall increase took place in that part of the population which made its living *outside* agriculture. This was what made England so distinctive.[33]

In one sense an explanation of English agricultural improvement may be sought in the relatively rapid and uninterrupted growth in the urban sector of the English economy, well able to

32 In Finland in 1754, 79% of all economically active heads of families were engaged in agriculture, forestry, or fishing; in 1805, 82%. There are data for many European countries from about the middle of the nineteenth century. In the following list the percentages refer to the proportion of the total labor force (male and female) engaged in agriculture, forestry, or fishing, unless otherwise indicated: Belgium 1846, 51%; Denmark 1850, 49% (family heads); France 1856, 52%; Ireland 1851, 48%; Netherlands 1849, 44%; Spain 1860, 70% (males only); Sweden 1860, 64%; Great Britain 1851, 22%. Mitchell, *European Historical Statistics 1750–1975*, Table C1. Half a century earlier it is virtually certain that each of the comparable percentages would have been higher, sometimes substantially so.

33 Wrigley, "The Growth of Population in Eighteenth-Century England: a Conundrum Resolved," *Past & Present*, 98 (1983), 122–125. The secular changes in fertility and morality which largely governed population growth rates in England are described in Wrigley and Schofield, *Population History*, 192–284.

afford food but producing none itself. The extraordinary stimulus to improvement afforded by the growth of London, which in 1500 was not even one of the dozen largest cities in Europe, but by 1700 was larger than any other, was probably the most important single factor in engendering agricultural improvement. The momentum first given by the growth of London was carried forward later by the more general urban growth in the later seventeenth and eighteenth centuries. It encouraged both change within agriculture itself and in a host of associated institutions and activities: transport improvement, credit facilities, capital markets, and commercial exchange. The joint effect of rising productivity and increasing urbanization also fostered a great expansion in rural employment outside agriculture. The combination of a steadily rising demand for goods and services other than food with a far more sophisticated market mechanism for exciting and satisfying such a demand was the basis of prosperity for the industries in the countryside.

Yet the reasons for the exceptional growth of London from the early sixteenth to the late seventeenth century remain imperfectly understood. Its rise both deserves and demands much more attention than it has received. Understanding of the phenomenon is likely to benefit especially from study of the connection between the development of the nation state under the Tudors and Stuarts and the growth of the capital city, and from considering London's fortunes within a wide framework of analysis of European commercial exchange.

Urban growth toward the end of the early modern period presents problems of interpretation no less difficult than those at its beginning. During the eighteenth century the urban hierarchy of England was turned upside down. The new industrial towns and port cities of the north and the midlands thrust their way past all rivals other than London. Only Bristol and Newcastle, among the traditional regional centers, matched the new challenge, and then only because they could benefit from the stimuli which had forced the pace of the growth in places like Liverpool and Sunderland. Many once-great centers were on the way to the pleasant obscurity of county rather than national fame—York, Exeter, Chester, Worcester, and Salisbury.

The upsetting of the old urban hierarchy in England was, at

the time, an event without recent precedent in European history. Elsewhere the exact ranking of major cities in each country varied from time to time, but it was rare for tiny settlements to develop into major centers. The same lists of large urban centers may be found century after century only slightly rearranged. Occasionally one of the smaller centers made progress through the ranks. Atlantic and colonial trade brought Bordeaux and Nantes rapid advancement between 1500 and 1800, for example, but revolutionary changes were rare. The progress of the new centers in England was such, however, that not merely had Liverpool and Manchester outpaced all of their English rivals other than London in 1800, but by 1850 they were the seventh and ninth largest cities in Europe, and the largest anywhere in Europe other than those which were capital cities, an extraordinary tribute to their economic vitality unassisted by the employment in government, the professions, and the arts associated with capital cities.[34]

The history of urban growth in England was distinctive. The contrast with the course of events elsewhere in Europe was especially notable between 1600 and 1750, since the paralysis which affected all but the largest towns on the continent was absent in England. The expansion of secondary employment in the countryside was not a bar to urban growth in England in the manner hypothesized by de Vries for Europe as a whole. This fact may afford an important clue to the differences between England and her continental neighbors in real income levels or trends, in relative wage levels in town and country, in the terms of trade between the two, in urban institutions, or in still other factors.

In this article, however, I have concentrated on the narrower topic of the scale and speed of urban growth and the indirect measurement of the gain in agricultural productivity per head, which roughly doubled in England between 1600 and 1800. By the standards of the recent past, a doubling in agricultural productivity over a two-century span of time is not a startling achievement, but, in terms of the preindustiral world, it is much more impressive and set England apart from the great bulk of

34 Chandler and Fox, *3,000 years of Urban Growth*, 20. The first nine in order of size were London, Paris, Constantinople, St. Petersburg, Berlin, Vienna, Liverpool, Naples, and Manchester.

continental Europe, conferring on the former advantages denied to economies unable to release more than a tiny fraction of the labor force from work with the flock or plow.[35]

Measuring agricultural advance through urban growth falls well short of *explaining* either phenomenon. It is easy to see how these developments might reinforce one another, but the fact that similar cases were so rare, and that when they did occur they tended to lose momentum after a while, suggests that noting the logical possbility of such a link is only a first step toward undestanding what took place. The most fundamental and intriguing question may concern the circumstances in which Ricardo's law of declining marginal returns to additional unit inpujts of labor and capital may be circumvented in a land long fully settled. This law, if it had universal and invariable application, would prohibit a sustained development of the kind which took place in England.[36]

The population of England increased more than 3.5 times between the early sixteenth and the early nineteenth century without becoming dependent, other than marginally, on food imports. Some new land was taken into farming use, but the bulk of the increased output must have been obtained from land already under cultivation. Moreover, after 1600 there was little increase in the agricultural labor force.

What served to neutralize the operation of declining marginal returns? The usual answer given to this question is innovation. If other things were not equal—if new methods of farming had been introduced which significantly enhanced the productivity of labor and capital—then the operaiton of Ricardo's principle could have been postponed, indefinitely if the flow of suitable innovations were sufficiently sustained. In a sense this must be the right answer but, because it is inescapable, it may be unilluminating, for, although it may be logically necessary, it was historically contingent. Relief from such a source was not always or even commonly forthcoming. Can anything be found in the circum-

35 Grigg provides information about rates of growth in labor productivity in agriculture in Denmark, France, the United States, and the United Kingdom in the late nineteenth and early twentieth centuries, and for all major subdivisions of the world since 1960. David B. Grigg, *The Dynamics of Agricultural Change* (London, 1982), 171—172. Rates as high as 5% per annum have been commonplace in developed countries in recent years.
36 David Ricardo, *On the Principles of Political Economy and Taxation* (London, 1817).

stances of early modern England which provides an answer to this conundrum?

To attempt a full examination of this issue is beyond the scope of this article, but one of its aspects may bear a slight elaboration. There was a remarkably strong and regular positive relationship between the long-term rate of growth of population and the comparable rate of change in food prices in England throughout the early modern period until 1800. At first sight this relationship suggests a uniform and inhibiting tension between population growth and the capacity to sustain rising numbers. But, although the tension was constant, it does not follow that all the demographic and economic variables involved stood in the same relationship to each other throughout. We have seen, for example, that there were major changes in national occupational structure and in agricultural productivity per head. The relationship between population growth and food prices may prove to have been common to England and other countries, but the changing structure of the component variables was not usually found elsewhere. The tension appears to have been beneficial and dynamic in England, when so often it was static and debilitating.[37]

Innovation may be the key to overcoming Ricardian constraints, but inasmuch as what happened in England was so rarely paralleled in other preindustrial economies, it seems doubtful whether innovation could be counted upon to be forthcoming in response to need. When, as in England, events took a more favorable turn, the explanation may be found not in the urgency of human need, nor in the immediate price dynamics of the market place, nor in the accident of individual inventiveness, but in the unusual structural characteristics of the prevailing situation.

If, on the one hand, it were safe to assume that a community always exploited its knowledge of production methods to the full,

37 On the relationship between the rates of change in population and food prices, see Wrigley and Schofield, *Population History*, Fig. 10.2. From the mid-sixteenth to the late eighteenth centuries the relationship was strikingly linear. This in itself may be suggestive. If there had been very tight constraints upon agricultural expansion, a curvilinear relationship might have been expected, with a steeper rise in price than in population as conditions worsened. Its absence may reflect the existence of an unusual capacity to rise to the challenge of increased demand on the part of English agriculture, especially when it is remembered that the absolute rate of population growth was at times quite high, approaching 1% per annum in the late sixteenth century and exceeding it in the late eighteenth. I owe this comment to Schofield's reflections on the point.

it would follow that innovation meant the introduction of some element into the productive system which was previously unknown. If, on the other hand, there were a range of alternatives to existing practice capable of raising productivity, which were held in reserve because current circumstances did not encourage or even allow of their use, the nature of the case would be changed.[38]

It may not be sufficient to look exclusively at changes in technology to explain a rise in productivity—new crops, altered cultivation practices, improved breeding techniques, and so on; nor to turn to the gains in productivity per head arising from increasing specialization of function, the chief proximate cause of enhanced productivity in Smith's analysis of the question. Other developments need to be taken into account if they encourage the exploitation of previously untapped potential and thereby transform living standards and growth prospects. It is in this context that the exceptional scale and speed of urban growth and its particular nature were so important.[39]

38 Some years ago Boserup made effective use of one form of this argument, but her model was one in which the changed methods of agriculture adopted to meet the challenge of rising population resulted in the broad maintenance of a given level of output per head per annum at the cost of a rise in the number of hours worked per annum. I have in mind changes which substantially increase labor productivity whether measured by the hour or by the year, although a part of the latter may be achieved because the workload becomes less markedly seasonal and therefore there are fewer periods during the year when labor is intermittent. Ester Boserup, *The Conditions of Agricultural Growth* (London, 1965), 41–55.

39 Some further aspects of the gain in agricultural output per man and per acre are examined in Wrigley, "Some Reflections on Corn Yields and and Prices in Preindustrial Economies," in John Walter and Schofield (eds.), *Death and the Social Order*, forthcoming.

Acknowledgements

The editors and publishers wish to thank the following for permission to use copyright material:

I. T. Berend and G. Ranki, 'Foreign trade and the industrialization of the European periphery in the nineteenth century', *Journal of European Economic History*, 9 (1980), pp. 539–84, by permission of the *Journal of European Economic History* and I. T. Berend;

R. Cameron, 'Conclusion', in *idem*, ed., *Banking in the early stages of industrialization: a study in comparative economic history* (Oxford, 1967), pp. 290–321. Copyright © 1967 by Oxford University Press, Inc., by permission of Oxford University Press, Inc.;

N. F. R. Crafts, 'Patterns of development in nineteenth-century Europe', *Oxford Economic Papers*, 36 (1984), pp. 438–58. Copyright © 1984 Oxford University Press, by permission of Oxford University Press;

F. Crouzet, 'Wars, blockades and economic change in Europe, 1792–1815', *Journal of Economic History*, XXIV (1964), pp. 567–90, by permission of Cambridge University Press;

A. Gerschenkron, 'Economic backwardness in historical perspective', in *idem*, *Economic backwardness in historical perspective: a book of essays* (Cambridge, Mass., 1962), pp. 5–30, Belknap Press of Harvard University Press. Copyright © 1962 the President and Fellows of Harvard College, by permission of Harvard University Press;

C. P. Kindleberger, 'Commercial expansion and the industrial revolution', *Journal of European Economic History*, 4 (1975), pp. 613–54, by permission of the *Journal of European Economic History* and the author;

D. C. North and R. P. Thomas, 'An economic theory of the growth of the western world', *Economic History Review*, 2nd ser., XXIII (1970), pp. 1–17, by permission of D. C. North;

P. K. O'Brien, 'Transport and economic development in Europe, 1789–1914', in *idem*, ed., *Railways and the economic development of western Europe, 1830–1914* (1983), pp. 1–27, St Antony's College/Macmillan, by permission of the Warden and Fellows of St Antony's College, Oxford;

W. N. Parker, 'The pre-history of the nineteenth century', in *idem*, *Europe, America and the wider world: essays on the economic history of western capitalism* (Cambridge, 1984), I, pp. 15–54, by permission of Cambridge University Press;

S. Pollard, 'Industrialization and the European economy', *Economic History Review*, 2nd ser., XXVI (1973), pp. 636–48, by permission of the author;

W. W. Rostow, 'The take-off into self-sustained growth', in B. Supple, ed., *The experience of economic growth* (New York, 1963), pp. 81–110, by permission of Cambridge University Press;